Drug and Therapy Development for Triple Negative Breast Cancer

Drug and Therapy Development for Triple Negative Breast Cancer

Edited by
Pravin Kendrekar
Vinayak Adimule
Tara Hurst

WILEY-VCH

The Editors

Dr. Pravin Kendrekar
White Collar Food & Beverages
Pune
India

and

T&T Pharmaceuticals
Mumbai
India

Dr. Vinayak Adimule
Angadi Institute of Technology and Management
Karnataka
India

Dr. Tara Hurst
University of Oxford
Oxford
United Kingdom

Cover Image: © CI Photos/Shutterstock

All books published by **WILEY-VCH** are carefully produced. Nevertheless, authors, editors, and publisher do not warrant the information contained in these books, including this book, to be free of errors. Readers are advised to keep in mind that statements, data, illustrations, procedural details or other items may inadvertently be inaccurate.

Library of Congress Card No.: applied for

British Library Cataloguing-in-Publication Data
A catalogue record for this book is available from the British Library.

Bibliographic information published by the Deutsche Nationalbibliothek
The Deutsche Nationalbibliothek lists this publication in the Deutsche Nationalbibliografie; detailed bibliographic data are available on the Internet at <http://dnb.d-nb.de>.

© 2023 WILEY-VCH GmbH, Boschstraße 12, 69469 Weinheim, Germany

Print ISBN: 978-3-527-35175-6
ePDF ISBN: 978-3-527-84118-9
ePub ISBN: 978-3-527-84117-2
oBook ISBN: 978-3-527-84116-5

Typesetting Straive, Chennai, India
Printing and Binding CPI Group (UK) Ltd, Croydon, CR0 4YY

C122717_050623

Contents

13 A Comprehensive Review of Some Heat-Shock Proteins in the Development and Progression of Human Breast Cancer *237*

Xolani H. Makhoba and Ofentse J. Pooe

14 Nanoparticle-Based Therapeutics for Triple Negative Breast Cancer *249*

Isidore A. Egebe and Kamalinder K. Singh

Preface

The book covers unique topics on the novel development of cancer detection, treatment, and its application. The book describes various tools used in the breast cancer therapy such as use of artificial intelligence, deep learning, and machine learning mammography. The book describes use of prosthetic biomarkers and sophisticated techniques in the early detection and treatment of drug discovery and development for triple-negative breast cancer. The book also describes the fundamentals of the cancer, especially breast cancer, MTT assay studies, flow cytometer studies, use of advanced markers, biomarkers, chemotherapy methods, advancement in the treatment of chemotherapy, and use of different and novel potency heterocyclic molecules in the effective *in vivo* and *in vitro* treatment of cancer. The book and the chapters involved describe novel aspects of the triple-negative breast cancer from the initial diagnosis tools used, innovations achieved in the process of diagnosis, treatment procedures, and advantages and drawbacks of the current methods of treatment of the breast cancer. Also, the book contains chapters from the eminent doctors, academicians, researchers, and scientists who are working in the field of cancer for many years. Their experience and the novel aspects of the treatment procedures are mentioned elaborately in their contributing chapter. Especially the use of artificial intelligence, deep learning methods, cause of food toxicity, and different approaches for the detection level of the breast cancer have been emphasized in the present coverage of the book. It becomes very essential for the readers to understand all the parameters relating to the treatment, diagnosis, and applications of detection of breast cancer.

Pravin Kendrekar
University of Central Lancashire,
Preston, United Kingdom

Vinayak Adimule
Angadi Institute of Technology and Management,
Karnataka, India

Tara Hurst
University of Oxford, Oxford, United Kingdom

September 2023

Part I

History of Breast Cancer

1

Early-Stage Diagnosis of Breast Cancer: Amelioration in Approaches

Nidhi Manhas[1], Lalita S. Kumar[1] and Vinayak Adimule[2]

[1]Indira Gandhi National Open University (IGNOU), School of Sciences, Maidan Garhi, New Delhi, 110068, India
[2]Angadi Institute of Technology and Management, Department of Chemistry, Savgaon road, Belagavi, Karnataka, 590009, India

1.1 Introduction

Breast cancer (BC) has become one of the most prevalent malignant tumors in women and is increasing at an alarming rate. Based on the population growth, experts have predicted that by 2050, there will be roughly 3.2 million new cases per year globally [1]. Not only is the number of patients with BC rising all over the world, but also the age of affected patients is tending to be younger [2]. Many factors contribute to these circumstances including age, family history, lifestyle surroundings, and many others [1, 3, 4]. Although the relative risk of BC is inevitable, it is possible to reduce BC mortality rate. Its survival rate largely depends upon the woman's timely access to effective and affordable detection and treatment processes [5]. The World Health Organization (WHO) has also described two different but related approaches to reduce BC, i.e. *early diagnosis*, which is the recognition of symptomatic cancer at an early stage, and *screening*, which is the identification of asymptomatic disease in a target population of apparently healthy individuals [6]. In many developing countries, women are unaware about the BC because of which it is detected at later stages [7]. However, there are various organizations working for generating awareness and promoting self-examination of the breast among women. Such efforts will promote the early detection and will help in reduction of the BC mortality rate.

Even previous research has shown that early BC detection, if combined with appropriate treatment, could greatly reduce BC death rates in the long run. Therefore, detecting BC at an early stage is vital. There are different techniques used for its diagnosis. Presently, mammography (MG), breast ultrasound, and breast magnetic resonance imaging (MRI) examination are the most common diagnostic techniques available for the detection of BC [8, 9]. These procedures necessitate

Drug and Therapy Development for Triple Negative Breast Cancer, First Edition.
Edited by Pravin Kendrekar, Vinayak Adimule, and Tara Hurst.
© 2023 WILEY-VCH GmbH. Published 2023 by WILEY-VCH GmbH.

specialized equipment, skilled practitioners, and expert analysis but the cost of detection is significantly high. In comparison to these methods, biosensor detection is far more reliable and affordable [10]. Weaver and Leung have summarized the various definitions and applications of biomarkers in imaging BC [11]. On the one hand, breast tumor indicators are critical in the early detection of BC, the characterization of molecular subgroups, the selection of treatment options, and the assessment of survival [12–14], whereas biosensors, on the other hand, provide substantial benefits over standard tumor marker detection methods in terms of specificity, sensitivity, speed, and cost of detection [15, 16] such as chemiluminescence immunoassay [17], enzyme-linked immunosorbent assay [18], proteomics [17], molecular biology methods, and liquid biopsy. Several biosensors with improved sensitivity, selectivity, stability, and low cost have been created in the previous decade [19].

In this chapter, many diagnostic methods such as MG, ultrasonography (US), MRI, microwave BC detection techniques, and various biosensors will be discussed. We herein discuss their most recent advances, as well as their benefits and drawbacks. This will aid researchers and those working on BC diagnostic methods in selecting appropriate approaches for properly diagnosing BC in its early stages.

1.2 Imaging Techniques

The use of imaging techniques reveals the anatomy and position of malignant cells and provides clinicians with valuable clinical information. When contrast agents and high-energy rays are used in imaging procedures, unfortunately, patients may be harmed. As a result, we should discuss different imaging modalities and decide which one is best for BC patients. These techniques mainly include MG, US, MRI, positron emission computed tomography (PET), computed tomography (CT), and single-photon emission computed tomography (SPECT). The benefits and drawbacks of these imaging techniques are listed in Table 1.1.

PET, CT, and SPECT are not advised for diagnosing BC patients due to their high cost, limited practicability, and radiation damage [20]. However, in some circumstances, such as screening for metastatic BC and the presence of bone and lymphatic metastases, these techniques can be employed as additional diagnostic procedures for diagnosing BC. As a result, we solely discuss MG, US, and MRI, which are the primary modalities for detecting BC. These common imaging procedures will be summarized and evaluated to assist clinicians to serve their patients in a better way.

1.2.1 Mammography (MG)

MG is the primary method used for screening and diagnosing BC, and it aids clinicians in gathering clinical data on BC patients. This method is especially advantageous to women between the ages of 40 and 74. Early MG screening may reduce the

Table 1.1 Benefits and drawbacks of imaging techniques.

Imaging techniques	Advantages	Disadvantages
XRM	1) Standard for diagnosing BC patients 2) Suitable as a screening method for BC 3) Finding mammary gland calcification	1) Not for people under 40 2) Not for people with high gland density 3) No more than twice a year
US	1) Screening for young women 2) Noninvasive diagnostic method 3) Finding mammary gland inflammation	1) Not for small mass and a typical tissue 2) Affected by the examining doctor 3) Definition and resolution are not high
MRI	1) High sensitivity and specificity to invasive BC 2) Screening of high-risk groups such as family history of BC 3) For patients with breast-conserving surgery	1) Not for everyone such as patients with claustrophobia 2) Not for wide-scale screening 3) Not for BC staging
PET	1) High sensitivity to BC recurrence and metastasis 2) Helpful for staging of the BC 3) High sensitivity to small breast tumor	1) High cost, not recommended as routine screening 2) Not for patients with hypersensitivity to developer
CT	1) Supplementary diagnostic method for BC, such as identifying BC with or without intrapulmonary metastases	1) Not the first choice for diagnosing BC 2) Radiation damage 3) Poor spatial resolution and need experienced doctors
SPECT	1) High resolution, small field of vision 2) Recommended use when suspects metastasis	1) Obtaining little clinic information 2) Not for patients with inflammatory bone lesions and bone proliferative metabolic abnormalities or variations

CT, computed technology, MRI, magnetic resonance imaging, PET, positron emission tomography, SPECT, single-photon emission computed tomography, US, ultrasonography, XRM, X-ray mammography.

death rate of BC patients by 30% to 40%, according to one of the earlier research [21]. However, MG has a significant rate of false-positive and false-negative results, especially in individuals with dense breasts (for subjects under 40 years old) [22, 23]. But with time, MG is progressing continuously and has shown good results in terms of diagnostic accuracy, sensitivity, and resolution. Presently, two key diagnostic methods are under practice for detection, i.e. contrast-enhanced mammography (CEM) and digital breast tomosynthesis (DBT) [24, 25]. CEM has been found to be superior

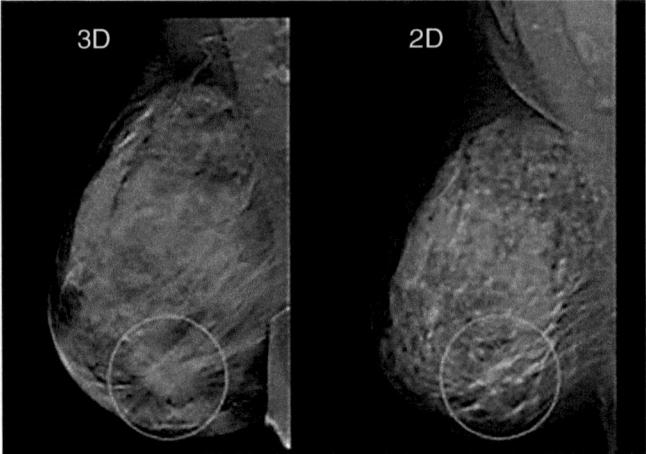

Figure 1.1 3D versus 2D mammography. Source: Andersson et al. [33]/Reproduced with permission from Springer Nature.

to full-field digital mammography (FFDM) in terms of diagnostic accuracy and disease extent assessment, and its efficiency is also comparable to that of MRI as well as US [26–28]. When compared to FFDM, DBT also offers good performance in terms of specificity (96.4%, 57229/59381% versus 97.5%, 23427/24020, $P < 0.001$) [29]. Computer-aided detection (CAD) is an artificial intelligence (AI) technique that has improved the sensitivity of the instrument and decreased human errors as well as false-positive and false-negative results in detection [30]. The combination of CAD with CEM and DBT can significantly improve the performance of these imaging techniques [31, 32]. In individuals who have no indications or symptoms of BC, a 3D MG is utilized to detect the disease as shown in Figure 1.1. The study demonstrated that combining 3D and traditional MG minimizes the need for extra imaging and has increased the accuracy of MVGG up to 94.3% [34]. Various algorithms have been proposed to enhance the MG images. After many experiments and selecting suitable settings, Montaha et al. suggested the BreastNet18 model, which is based on the fine-tuned VGG16. The accuracy of the algorithm grew to 98.02% of the proposed model [35]. Such type of research will help the doctors in efficient and accurate diagnosis of BC.

From the above discussion, we can conclude that MG is an essential component of early diagnosis for BC patients because of its several benefits, including rapid screening, high accuracy, low cost, and suitability for promoted use. Despite these benefits, MG is not suitable for everyone. It requires a hazardous contrast agent and X-ray to perform imaging, cannot be used frequently in a short period of time, and is not suggested for people under the age of 40 [36]. But in the coming years, with significant developments like high resolution, MG will be quite safe. Moreover, advances in AI technology have made it possible to simplify the detection and analysis of BC.

1.2.2 Ultrasonography (US)

US is a technique for evaluating the form and status of tumor tissues, as well as correctly locating lesions. The early grayscale US merely showed whether the tumor existed at the detecting point, and because its resolution was inadequate, it was difficult to discriminate benign and malignant tumors [37, 38]. US images showing normal, benign, and malignant BC are given in Figure 1.2.

The flat photographs of tumors received from the two-dimensional (2D) US might affect the physician's assessment. Therefore, three-dimensional (3D) US technology was introduced so that one can have 3D imaging of tumor anatomy and blood vessel distribution in diagnosed patients [40]. The color **Doppler ultrasound** is one of the 3D techniques of ultrasounds that may vividly display tumor and blood flow information and offer clinicians more useful information, allowing them to discriminate between malignant and benign tumors [41]. Krouskop discovered elastic variations in different tissues laying the theoretical groundwork for constructing elastic US [42]. Furthermore, some studies revealed that employing elastic US to screen suspected diseased tissues considerably enhances the accuracy of diagnosing BC [43, 44]. The elastic US, when paired with 3D US, can diagnose axillary lymphadenopathy and classify the patient's tumor state [45]. Although MG is the best tool for detecting BC calcification, when the calcification is too tiny, it is difficult to identify with MG or normal US; so, **MicroPure**, a new US image-processing method, was developed [46]. By analyzing images of multidimensional array and frequency, this method may decrease random noise and produce high-resolution images and tissue homogeneity [47]. Machado et al. examined *ex vivo* surgical breast specimens with MicroPure and discovered that MicroPure has a high detection rate for BC microcalcifications, whereas conventional US cannot detect [48].

This technique has significant advantages that include the use of less contrast agents, the absence of high-energy rays, and the fact that it is suited for people of all ages. US has been suggested as a supplement to MG for women at high risk of BC, pregnant women, and those who are unable to get MG [49]. Furthermore, it involves the use of skilled radiologists, which has a major impact on sensitivity and specificity. Breast ultrasound has a high false-positive rate while being routinely shown to detect mammographically hidden malignancies. However, using AI, radiologists

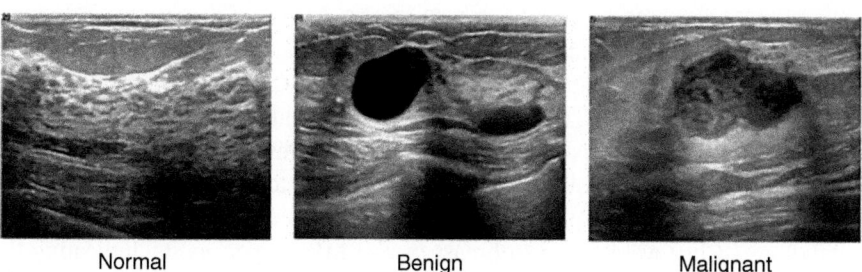

Normal   Benign   Malignant

Figure 1.2 Samples of ultrasonography breast images dataset. Source: Dhabyani et al. [39]/with permission from Elsevier/CC BY 4.0.

were able to cut false positive rates by 37.3% and requested biopsies by 27.8% while retaining the same sensitivity level (Area Under the Receiver Operating Characteristic Curve [AUROC]: 0.962 AI, 0.924±0.02 radiologists) [50]. AI has the potential to improve the accuracy, consistency, and efficiency of breast ultrasound diagnosis in the future, which will help doctors achieve more accurate diagnostic results by reducing errors caused by unprofessional judgments.

1.2.3 Magnetic Resonance Imaging (MRI)

MRI enables the early diagnosis of BC, independent of the patient's age, breast density, or risk status [51]. The common approach for breast imaging is dynamic contrast-enhanced magnetic resonance imaging (DCE-MRI), which focuses on the introduction of contrast agents and displays the malignant vascularity, anatomy, and kinetics of breast tumors [52]. DCE-MRI has been demonstrated as a screening technique for women with a variety of risk levels, with sensitivity ranging from 81% to 100% [53]. The positive prognostic value of DCE-MRI is 98%, which is higher than the positive prognostic value of MRI alone, i.e. 77% and the specificity is 97%, according to the experts [54]. Another recently developed method that allows for excellent spatial and temporal resolution is ultrafast DCE-MRI in which various amplification approaches, such as parallel imaging and compressed sensing, are applied, and its ability to characterize BC aggressiveness and tumor subtypes has been proven [55]. There is another technique called **Magnetic Resonance Diffusion Weighted** (MRDW). It is also used in BC diagnosis that shows clear movement of water molecules in the body, as for different tissues there exist different water dispersion coefficients. Researchers can detect benign and malignant breast tumors by utilizing MRDW to evaluate apparent diffusion coefficient (ADC) values (which reflect diffusion-limited effects) of tumors, i.e. ADC values: normal breast group > benign group > malignant group [56, 57]. DWI has the advantage of being a non-contrast technology with a fast scan time [58]. Moreover, DWI was found to be more accurate than MG in detecting cancer in a sample of asymptomatic women [59].

Magnetic resonance spectroscopy (MRS) is an important method that is used to describe the functional state of malignant, benign, and normal breast tissues in three ways: *in vivo*, *ex vivo*, and *in vitro*. Table 1.2 compares the studies of *in vitro*, *ex vivo*, and *in vivo* MRS and MRI techniques in the diagnosis of BC [60]. MRS is a noninvasive technology that can enhance the rate of BC diagnosis by assessing the risk of BC and leading to BC treatment [61, 62]. Solid-state MR spectroscopic examination of intact biopsied tissues employing the high-resolution magic angle spinning (HRMAS) approach was also employed in studies to monitor metabolite levels for breast tumor diagnosis/prognosis [63–68]. High amounts of choline-containing metabolites (tCho) were found in breast *in vivo* MRS experiments, indicating the rapid proliferation of malignant tumors [69–76]. Hyperpolarized ^{13}C MRI (HP ^{13}C MRI) has recently been used to investigate abnormal tumor metabolism [77].

Table 1.2 Comparison of *in vitro*, *ex vivo*, and *in vivo* magnetic resonance spectroscopy (MRS) and MRI techniques.

Characteristics	Magnetic resonance spectroscopy (MRS)			MRI
	In vitro	*Ex vivo*	*In vivo*	
Information	Biochemical composition (metabolite detection)	Biochemical composition (metabolite detection)	Biochemical composition (metabolite detection)	Anatomy (structure and morphology), functional
Sample/subject	Tissue extract, biofluids, cell lines, aspirates	Excised tissues/biopsies	Living humans/organisms	Living humans/organisms
Equipment	NMR spectrometer	NMR spectrometer with accessories for HRMAS	Human MRI scanner	Human MRI scanner
Field strength	High field strength 9.4–21.1T	High field strength 9.4–18.8T	1.5–7T	1.5–7T
Nuclei of interest	^1H, ^{13}C, ^{31}P, ^{23}Na, ^{19}F	^1H, ^{13}C	^1H, ^{31}P, ^{23}Na, ^{19}F ^{13}C hyperpolarized	^1H from fat and water
Data	1D/2D spectra	1D/2D spectra	SVS 1D, SVS-2D, CSI (MRSI)	Conventional T1, T2-weighted, DCE-MRI, diffusion-weighted, perfusion weighted, MR elastography, fMRI
Advantages	High sensitivity and resolution, detection of a large number of metabolites, easy quantification, easy experimentation	High sensitivity and resolution, detection of a large number of metabolites, quantification not that easy, special experimentation	Organ-specific metabolite composition, and longitudinal studies	Organ-specific structural and functional studies, longitudinal studies possible
Limitations	Tissue excision is invasive	Tissue excision is invasive	Low sensitivity and resolution, detection of a small number of metabolites, claustrophobia of patients	Claustrophobia of patients contrast required in some studies
Reproducibility	Lesser than *in vivo*	Lesser than *in vivo*	High	High

1D, one-dimensional spectrum; 2D, two-dimensional spectrum; HRMAS, high-resolution magic angle spinning; SVS, single voxel spectroscopy; CSI, chemical shift imaging; DCE-MRI, dynamic contrast-enhanced magnetic resonance imaging.
Source: Sharma and Jagannathan [60] ©MDPI/Public Domain CC BY 4.0.

Magnetic resonance elastography (MRE) is a type of magnetic resonance technology that uses the transmission of mechanical waves in tissues to offer information on tissue elasticity. MRE's future tendency is to identify preoperative tumors and predict treatment response and metastatic potential of primary tumors [78]. PET/MRI or PET and MRI can reveal soft tissue structures in the breast and chest wall. PET can offer molecular-level information *in vivo*, and PET/MRI has a high value in evaluating BC metastasis and can improve the positive predictive rate of patients [6, 79, 80]. Moreover, AI-enhanced model of ^{18}F-FDG PET/MRI (^{18}F-fluorodeoxyglucose positron emission tomography magnetic resonance imaging) has accurately shown the difference between benign and malignant breast lesions [81].

MRI is a supplementary method for diagnosing BC that seems to have a number of advantages. Unfortunately, numerous factors influence the widespread use of MRI, i.e. prolonged imaging time, high price, and the fact that it cannot be performed if the patient's body contains metal material. As a result, future research should focus on lowering the cost of MR procedures and reducing the need for contrast agents, so that they can be used at early stages of the BC. Radiomics is a fast-developing area that uses AI algorithms to analyze medical scans digitally, allowing for thorough tumor characterization [82–84]. So, radiomics applications should be thoroughly investigated, and there is a need to improve radiologists' grasp of basic principles and the development of standardized and reproducible methods and data exchange for clinical applications.

1.3 Microwave Breast Imaging Methods

The microwave region ranging between 300 MHz and 300 GHz has received the least attention but has sparked a lot of interest in medical imaging in the last two decades. In this, the interaction of electromagnetic signals with the matter is determined by the dielectric properties of the matter, i.e. electric permittivity and conductivity [85]. Microwave imaging (MWI) has evolved as a technology for creating dielectric maps of various body sections. Basically, a MWI system includes an antenna array, a microwave signal transmitter and receiver, and a radio-frequency switch to switch between the arrays' multiple parts [86]. There are some applications of MWI in brain stroke detection [87], extremity imaging [88, 89], and lung cancer detection [90]. However, in this chapter, we have mainly discussed its application in BC detection. Tumors have high water content when compared to normal cells. This is due to the biology of tumor cells, which retain more fluid than normal cells. The dielectric characteristics of breast tissues are altered by this additional fluid, which is in the form of bound water. Tumors are diagnosed by MWI using scattered or reflected waves emerged from variations in dielectric characteristics between normal and malignant breast tissues [91–93]. Table 1.3 shows the difference in the dielectric properties of the female breast tissue at 3.2 GHz [94].

Table 1.3 Dielectric properties of female breast tissue at 3.2 GHz.

Tissue type	Relative permittivity	Conductivity (mS cm^{-1})	Water content (%)
Fatty tissue	2.8–7.6	0.5–2.9	11–31
Normal tissue	9.8–46	3.7–34	41–76
Benign tissue	15–67	7–49	62–84
Malignant tissue	9–59	2–34	66–79

Source: Adapted from Campbell and Land [94].

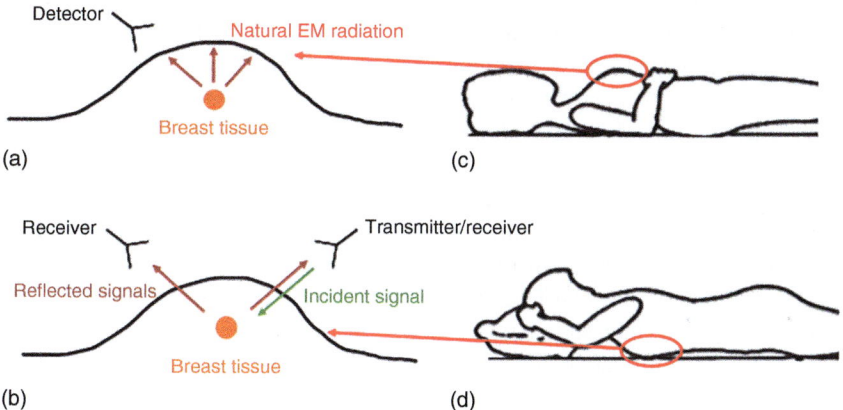

Figure 1.3 Methods of microwave breast imaging. The figures on the left show (a) passive versus (b) active approaches. The figures on the right show patient's orientations for (c) planar systems (supine position) versus (d) cylindrical systems (prone position). Source: AlSawaftah et al. [85] ©MDPI/Public Domain CC BY 4.0.

The MWI techniques can be classified into two groups, active MWI and passive MWI. The active MWI techniques are further subcategorized into microwave tomography (MWT) and radar-based MWI. Active MWI examines the difference in dielectric properties between the healthy and malignant tissues, and passive MWI measures the temperature between the healthy and cancerous tissues using radiometry [95]. The primary distinction between these two approaches is that natural electromagnetic radiation released by living tissues is measured in passive systems, whereas in active systems, an electromagnetic signal from a source is incident on the tissues and the reflected signals are measured (as depicted in Figure 1.3) [85].

1.3.1 Microwave Tomography

MWT is a technique that uses electromagnetic field alterations to obtain 2D slices or pictures of the dielectric characteristics of a sample. Typically, MWT setup includes imaging chamber, which is filled with the matching medium, and to improve the performance of this system matching medium, is selected carefully so that most of the microwave electromagnetic signals can couple with breast tissue. Imaging chamber also has array of antennas surrounding the sample, where each antenna transmits a continuous wave (CW) of single- or multifrequency electromagnetic signals. The electromagnetic signals scattered by matching medium and sample because of differences in their dielectric properties are measured by non-transmitting antennas. After analyzing data using algorithms, 2D images of dielectric properties are created. So, we can say that there are basically three steps to create image, i.e. collecting microwave tomogram, data analyzing, and then image display using MWT [85].

Several theoretical and experimental research studies have been done on the use of this method in BC diagnosis [96–100]. Meaney et al. [99] have done series of experiments to improve the performance of this technique. First multifrequency MWT prototype for breast imaging was set up of a cylindrical array of 16 monopole antennas operating at 300–1000 MHz [96]. This study revealed the relationship between the breast permittivity and radiological breast density. Another experiment with glycerin and water mixture as matching medium was done and found that it helped in the reduction of coupling noises between array elements. The results of these clinical investigations showed that tumors as small as 1 cm in diameter can be diagnosed, indicating that MWT has the ability to detect early-stage BC [101–103].

In recent research, magnetic nanoparticles and compressive sensing (CS) techniques are used as contrast agents to improve the accuracy, sensitivity, and specificity of MWT in BC detection [104, 105]. The results showed that the CS-based MWT with 12 antennas and an MWT with 70 antennas provided equal-quality breast pictures. Thus, we can conclude that CS-based MWT technique lowered operation costs and data-gathering time significantly.

1.3.2 Radio-Based Microwave Imaging

During BC detection, a radio-based MWI technology exploits reflected waves caused by differences in dielectric characteristics between normal and malignant tumor cells and gives valuable information about the location, size, and characteristics of tumor cells. First radar-based MW system was proposed by Bridge [106]. In Bridge's method, ultrawideband (UWB) of microwave frequency ranging from 1 to 10 GHz was used to illuminate BC tissue from array of antennas placed at different positions around the breast [107, 108]. Radar-based MW has more advantages over MWT such as being computationally less expensive, having higher resolution, and having better specificity [108, 109].

As per the developments in radar-based MW, it is classified into five groups [95]:

- **Confocal microwave imaging** (CMI): Hagness et al., first proposed this approach and used pulsed confocal techniques with time gating to improve tumor identification while decreasing tissue asymmetry and absorption effects [110]. Their work involves finite-difference time-domain (FDTD), and its results showed that cancer cells having 2 mm diameter can be detected by the 2D CMI.

By utilizing a resistive bowtie antenna and using 3D-FDTD simulations, Hagness et al. improved the prior design [111]. This technique has the potential to produce high-resolution images, but its limitation is that it can't distinguish between errors and noise.

- **Multi-static adaptive (MSA) system**: The high-impact radar-based MWI system was developed in which real aperture array of UWM antennas was used. It consisted of 16 UWB aperture-coupled stacked-patch antennas located on the section of hemisphere that were arranged in such a way as to improve the conformation to the curve of the breast [112–114]. The results of the clinical trials showed that this system was successful in detecting the 4–6 mm diameter of cancerous cells [113]. The Breast Care Center in Bristol, UK, conducted a substantial initial trial of the group's 31-element prototype radar system in 2010. Despite the fact that this technique produced good results, the clinical trial findings were mixed. The results were shown to be irreproducible when performed by different clinicians. This is due to slight patient movements throughout the 90-second scans, as well as certain ambiguities caused by changes in blood flow and temperature. In order to resolve these flaws, the team developed a 60-antenna array system to improve the system's immunity to clutter and reduce the scan time to 10 seconds. The results showed increase in the accuracy of the images obtained while also delivering a more convenient and acceptable clinical experience for patients [85].

- **Tissue-sensing adaptive radar (TSAR)**: Fear et al. investigated the use of TSAR in BC imaging [115, 116]. This system required the use of two scans of each breast. The first scan specifies the basic location of the breast volume relative to the tank (containing coupling fluid and antennas) obtained at the antenna after the first reflection. The second scan is done in a sagittal direction, from the nipple to the chest wall, providing data for the tumor detection algorithm [116]. The results of these clinical experiments showed that TSAR has the potential to detect and localize tumor with more than 4 mm in diameter [117]. However, the huge reflections created by skin, the construction of adequate antennas, and the desire to develop high-speed electronics for real-time photography all posed hurdles to this system. Development of appropriate sensors, research of practical implementation challenges, enhancement of imaging algorithms, and testing on breast models are all part of the current work on TSAR [95, 118].

- **Microwave imaging via space–time (MIST) beamforming**: This type of technique involves the use of continuous transmission of UWB signals from antenna placed near the breast surface, and the received reflected signals are spatially focused using a space–time beam former. Because of the considerable difference in the dielectric characteristics of normal and malignant tissue, discrete regions of high backscattered energy levels appear in the reconstructed pictures, corresponding to malignant tumors [119]. The first MIST system was introduced by Bond et al. [119], resulting in the detection and localization of very small synthetic tumors embedded in breast phantoms [119]. Bond et al. also developed a MIST system with implementation of a planar array of 16 horn antennas that transmitted UWB microwave signals from each antenna located close to the breast surface [119]. This has resulted in significant improvement in the performance of the UWB-based MI approach. However, further improvements enabled the system to localize, identify, and resolve multiple tumors [118, 119].

- **Holographic microwave imaging** (HMI): HMI has the ability to provide real-time images at a substantially cheaper cost than other radar-based MWI approaches since it does not require expensive ultra-high-speed electronics. In this approach, MWI is performed in two stages: recording of a sampled intensity pattern followed by image reconstruction [120]. Smith and coworkers presented a near-field indirect HMI approach that involves capturing the breast intensity and reconstructing the image from that data [120, 121]. However, before this technique can be used in clinical settings, it needs to be validated further.

Wang et al. proposed far-field HMI in which 3D HMI image was reconstructed using 2D HMI images obtained at different vertical positions with single frequency (12.6 GHz) for early detection of BC [122]. It included one transmitter and an array of 15 receivers placed under the breast phantom. In this system, matching solution medium is not required and air between the antennas and breast phantom. The results of the experiments demonstrated that the suggested 2D HMIA approach could successfully detect tiny tumors with a diameter of less than 5 mm in various places [122]. The other experiment was done by combining the above approach with CS, and results showed that CS-HMIA has the capability of detecting randomly distributed inclusion of various shapes and sizes using smaller number of sensors and lesser scan times [123]. In a recent development, a multifrequency HMI system has been developed by Wang (2019), who checked the feasibility and effectiveness of the proposed algorithm for breast imaging [124]. According to the studies, the multifrequency HMI system has the ability to be used as a microwave diagnostic technology.

A significant amount of research and development is yet to be done in harnessing the full potential of this technology. The future research should focus on improvement of MWI in medical applications, including better designing of hardware, signal processing methods as well as algorithms for image reconstruction.

1.4 Biomarkers and Biosensors for Breast Cancer Detection

The molecular biotechnology studies are done to analyze specific biomarkers such as nucleic acids, proteins, cells, and tissues of patients, and these studies help in early detection of BC than abovementioned imaging techniques or procedures [20]. However, these cannot replace imaging techniques but can be used as auxiliary method to diagnose BC. The examinations help the clinician analyze BC from the level of nucleic acids, proteins, and cells using these biomarkers and biosensors. First, we will discuss biomarkers and then biosensors.

1.4.1 Biomarkers

1.4.1.1 Nucleic Acids

There are different nucleic acid tumor markers viz, BRCA1, BRCA2, microRNA, circulating tumor DNA (ctDNA), circulating cell-free DNA (ccfDNA), circulating RNA (circRNA), long noncoding RNAs (lncRNAs), etc. [125]. MicroRNA is a single-stranded noncoding RNA molecule that has evolved as a significant regulator of BC development, prognosis, and therapeutic response [126]. The study has revealed that MicroRNA has been linked to patients' clinical and biological characteristics and that it can target different genes and alter different pathways. LncRNA is a type of noncoding RNA with a length greater than 200 nt that is produced by RNA polymerase II. The role of lncRNAs in the initiation, development, and metastasis of BC is becoming clearer, and they could represent a new diagnostic marker and therapeutic target for BC [127]. CircRNA is a type of double-stranded closed RNA that is resistant to RNA exonuclease, exhibits steady expression, and is difficult to disintegrate according to the findings, CircRNA expression was linked to tumor cell proliferation, migration, invasion, and treatment resistance [128, 129]. Thus, it can be explored as diagnostic method.

ccfDNA is extracellular DNA found in plasma or serum, and ctDNA is ccfDNA released into the bloodstream by tumor cells [130, 131]. The studies have revealed that primary tumor cells, circulating tumor cells, and hidden and dominant metastatic tumor cells all release more DNA than normal cells, and ctDNA contains mutations that are exclusive to the parent tumor [132]. As a result, ctDNA has the potential to be used as a tumor marker for BC prediction, diagnosis, and prognosis [131, 133]. However, further research is needed to apply it to clinical diagnosis and treatment [133].

1.4.1.2 Proteins

CD24, CD44, and MUC1 are some types of protein tumor biomarkers. CD24 is a glycosylphosphatidylinositol-binding glycoprotein [12, 134], which has been discovered to be an anti-phagocytic signal that protects cancer cells from Siglec-10-expressing macrophage attacks. The expression of CD24 has also been linked to BC grading and staging [135]. As a result, CD24-blocking therapy can significantly improve the therapeutic impact of CD24-positive cancers [134]. On the other hand, CD44 is a complex transmembrane-binding glycoprotein that has been linked to a poor prognosis in patients. It is involved in the regulation of numerous critical signaling pathways, including tumor growth, invasion, metastasis, and treatment resistance [135–137].

MUC1 (CA15-3) is a transmembrane mucin glycoprotein that is found in the majority of epithelial tissues [138]. It was shown to have aberrant profile and glycosylation in 90% of BC cases [139]. MUC1 is also a useful marker for tracking the progression of metastatic BC [140]. Further, serum tumor markers such as CEA, CA19-9, CA125, CA15-3, and TPS play a significant role in the diagnosis and treatment of BC [141].

1.4.1.3 Tumor Cells

The term "circulating tumor cells (CTCs)" refers to BC cells that have broken free from the tumor and entered the bloodstream (CTC). CTCs have the ability to regenerate tumor tissue. So such types of tumor cells themselves are tumor markers [141]. Patients with metastatic BC could be tiered and graded to get tailored treatment by measuring the number of CTC cells [142]. CTC cells can also be used to assess BC patients' prognoses and identify whether they are candidates for further radiation therapy after surgery [132, 143].

Apart from nucleic acid, proteins, and tumor cells, exosomes (membrane-enclosed phospholipid extracellular vesicles) can also be applied in the diagnosis and treatment of cancer due to their high secretion on the surface of cancer cells [144]. From the research findings, it was shown that tumor cells release more exosomes than normal cells, and miRNA-21 and miRNA-1246 in exosomes are upregulated in patients' plasma [9]. Therefore, exosomes have become a research hotspot in recent years because of their great diagnostic potential. In addition to exosomes, estrogen receptor (ER) [145], progesterone receptor (PR) [146], and human epidermal growth receptor 2 (HER2) [147] are the most widely used tumor markers in the early diagnosis and treatment of BC. The diagnosis of BC using tumor markers with high specificity and sensitivity requires more research.

1.4.2 Biosensors

A biosensor is a self-contained, tiny analytical instrument that combines a specific biological system with a physiochemical transducer to detect target molecules by transforming the recognition signal into a detectable output signal [148–152]. When compared to traditional tumor marker detection methods, biosensors offer substantial advantages in terms of specificity, sensitivity, speed, and cost of detection [15, 16]. Biosensors can be divided into electrochemical biosensors, optical biosensors, and other types on the basis of detection principles and signals [19, 151–154].

1.4.2.1 Electrochemical Biosensors

Electrochemical biosensors monitor changes in dielectric characteristics, size, shape, and charge distribution when antibody–antigen complexes form on the electrode surface. By sensing the electrochemical reaction on the electrode's surface, it quantitatively detects the analyte and the signal of the electrochemical reaction, which depends on the concentration of analyte [155–157]. These biosensors have been designed to detect a variety of biomolecules, including proteins, antigens, DNA, antibodies, and heavy metal ions, among others. Electrochemical sensors have previously been shown to have great sensitivity and specificity in buffer and serum samples [158]. Figure 1.4 shows the developed electrochemical biosensor for MCF-7 cells detection [160]. Antibodies against surface proteins of MCF-7 cells were immobilized on nanoparticle-assembled electrode to capture MCF-7 cells at the electrode surface, which increases the interfacial resistance and hence enlarged semicircle in Nyquist plot. Alternatively, cDNA complementary to miRNA can also be immobilized to capture target miRNA released from the cell extracts of MCF-7 cells [95].

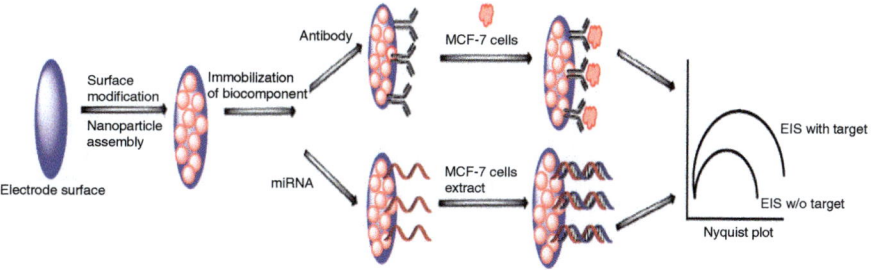

Figure 1.4 Electrochemical biosensor for detection of MCF-7 cells. Source: Mittal et al. [159]/with permission of Elsevier.

There are some methods through which the detection of electrochemical reaction is done: cyclic voltammetry (CV), differential pulse voltammetry (DPV), square wave voltammetry (SWV), linear sweep voltammetry (LSV), electrochemical impedance spectroscopy (EIS), field-effect transistor (FET), and other methods [19, 148, 149] as shown in Table 1.4. Recent development of electrochemical nanobiosensors has reduced the cost, simplified the technique, increased sensitivity and specificity, increased reliability, and provided a quick response in BC detection [178, 179]. Further research in this direction would be helpful in early detection of BC.

1.4.2.2 Optical Sensors

Optical biosensors detect the optical change on the surface of sensing layer of the target. Every optical biosensor detects different optical signals such as refractive index, resonance, wavelength, and intensity [16, 180, 181]. As shown in Figure 1.5, the diagnosis of BC cell (MCF-7) is done using quantum dot optical biosensor. In Figure 1.5, quantum dots are labeled with primary antibodies against MCF-7 cell surface proteins and subjected to sample containing MCF-7 cells. Addition of secondary antibody labeled, magnetic beads enables sensors for their magnetic separation to obtain fluorescence emission spectra [95]. Different types of optical biosensors have been developed that include fiber optics, fluorescence, resonant mirror optical, interferometric, and surface plasmon resonance as given in Table 1.5. Recently, these sensors have been developed using surface chemistry and nanotechnology [194].

In addition to electrochemical and optical biosensors, there are two other types of sensors viz, quartz crystal microbalance (QCM) biosensors (which detect mass change of the target) and photoelectochemical (PEC) biosensors (which detect the effect of the targets on the photoelectric characteristics of materials).

- **QCM biosensor**: The effect of the target on the frequencies of the bulk acoustic waves generated in the piezoelectric quartz crystal is the basis for QCM's sensing mechanism. To achieve the concentration detection of the target, the frequency change of the acoustic wave is connected to the mass change on the chip surface. QCM can detect mass changes on the chip surface at the nanogram level [195].
- **PEC biosensors**: The photoelectric active material in the PEC sensor is stimulated when light is irradiated, resulting in a photocurrent or photovoltage. The target is captured by the recognition sensor on the surface of the photoelectrically

Table 1.4 Developments in electrochemical biosensors.

Electro-chemical biosensor	Target	Detection limit	Linear range	References
CV	CA15	$30.64\,U\,ml^{-1}$	$2.0–240\,U\,ml^{-1}$	[161]
	EGFR	$1\,pg\,ml^{-1}$	$1\,pg\,ml^{-1}$ to $100\,ng\,ml^{-1}$	[162]
	miRNA	$155.2\times10^{-20}\,M$	2×10^{-20} to $2\times10^{-12}\,M$	[163]
DPV	BRCA1 CA15-3	$0.0034\,pM$	$0.01\,pM$ to $1\,nM$	[164]
	BRCA1	$3.34\,mU\,ml^{-1}$	$0.01–1000\,U\,ml^{-1}$	[165]
	miRNA	$3.01\times10^{-16}\,M$	1.0×10^{-15} to $1.0\times10^{-7}\,M$	[166]
		$3.6\,fM$	$0.01–10\,pM$	[167]
		$8.2\,fM$ (miRNA-21)	$0.02–10\,pM$ (miRNA-21)	
SWV	MUC1	$0.33\,pM$	$1.0\,pM$ to $10\,\mu M$	[168]
	miRNA-21	$18.9\,aM$ (miRNA-21)	—	—
	miRNA-155	$39.6\,aM$ (miRNA-155)	$0.1\,fM$ to $10\,nM$	[169]
LCV	HER2-ECD	$4.4\,ng\,ml^{-1}$	$15–100\,ng\,ml^{-1}$	[170]
	HER2	$0.16\,ng\,ml^{-1}$	$7.5–50\,ng\,ml^{-1}$	[171]
	CD44	$2.17\,pg\,ml^{-1}$	$0.01–100\,ng\,ml^{-1}$	—
	CD44 positive cell	$8\,cells\,ml^{-1}$	$10–106\,cells\,ml^{-1}$	[172]
EIS	HER2	$19\,fg\,ml^{-1}$	$0.001–10\,ng\,ml^{-1}$	[173]
	MCF-7 cell MUC1	$23\,cells\,ml^{-1}$	1×10^{2} to	—
	BRCA1	$2.7\,nM$	$1\times10^{5}\,cells\,ml^{-1}$	[174]
		$3\,fM$	$5–115\,nM$	[175]
			$10\,fM$ to $0.1\,\mu M$	
FET	miRNA-155	$0.03\,fM$	$0.1\,fM$ to $10\,nM$	[176]
	CEA	$10\,pg\,ml^{-1}$	$0.1–100\,ng\,ml^{-1}$	[177]

Source: Adapted from Hong et al. [141].

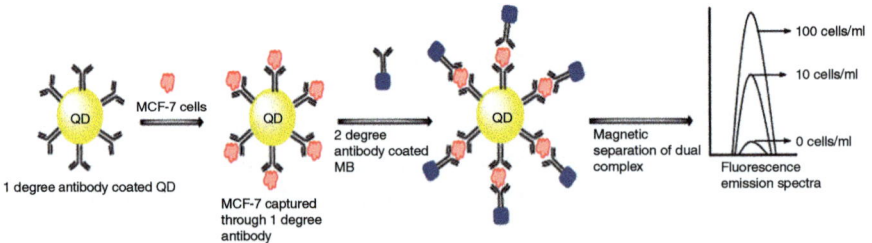

Figure 1.5 Quantum dot based optical biosensor for detection of MCF-7 cells. Source: Mittal et al. [159]/with permission of Elsevier.

active material, causing the photocurrent or photovoltage to change. When the target's concentration changes, so does the photoelectric signal. Therefore, it sets up relationship between the photoelectric signal and concentration. The detection of tumor markers has been reported using several PEC biosensors [141].

Table 1.5 Developments in optical biosensors.

Optical biosensor	Target	Detection limit	Linear range	References
Fluorescence biosensor	CEA	7.9 pg ml^{-1} (Water) 10.7 pg ml^{-1} (human serum samples)	0.03–6 ng ml^{-1} (water) 0.03–6 ng ml^{-1} (human serum samples)	[182]
	miRNA-21	0.03 fM	0.1–125 fM	[183]
Colorimetric biosensor	BRCA1	10^{-18} M	10^{-12} to 10^{-18} M	[184]
	BRCA1	0.34 fM	1 fM to 100 pM	[185]
SPRi	CEA	0.12 ng ml^{-1}	0.40–20 ng ml^{-1}	[186]
	HER2-positive EXO	8280 exosomes µl^{-1}	8280–33 100 exosomes µl^{-1}	[187]
	EXO	5000 exosomes ml^{-1}	—	[188]
SERS	miR-K12-5-5p	884 pM	—	[189]
	MicroRNA	—	—	[190]
ECL	BRCA1	0.71 fM	1.0 fM to 0.1 nM	[191]
	EXO	7.41×104 exosomes	3.4×10^5 to 1.7×10^8 exosomes ml^{-1}	[192]
	miRNA-21	3.2 fM	0.01–10 000 fM	[193]

Source: Adapted from Hong et al. [141].

Both of these are also capable of detecting all types of tumor markers. However, they have trouble identifying multiple targets at the same time, but signal amplification techniques can be used to enhance their detection limits for detecting a single target. The obstacles that biosensors have experienced are mostly due to two factors: detection method and detection equipment. These issues can be overcome by the combination of molecularly imprinted polymers (MIPs) and microfluidic chips with biosensors, and the commercialization of these types of biosensors in the future can change the current trend of diagnosis [196, 197].

1.5 Conclusion

This chapter mainly focused on the most frequent approaches for diagnosing BC. As researchers delve more into imaging technology, they know that a single imaging method is just not enough to meet the requirement of accuracy in BC diagnosis. Thus, combining many imaging modalities will be significant for the emerging approaches [107, 198, 199]. Moreover, developments of AI-based models will help in improving the positive diagnostic rate for BC and reducing the negative diagnosis rate. Additionally, the development of biosensors would lead to the formation of various BC biomarkers. The combination of imaging sensors and biosensors can get unexpected results. Nevertheless, imaging instruments would still be the routine method for screening BC over the next few years as these can be widely applied.

The new BC markers will enable these technologies to achieve more efficiency, speed, sensitivity, and specificity. With the development and application of these approaches in the future, the researchers will be able to not only diagnose BC from multiple perspectives but also monitor the effectiveness of treating BC.

Acknowledgment

All the authors are thankful to the School of Sciences, IGNOU, for their continuous support in using the literature resources.

Conflict of Interest

The authors have no conflict of interest.

Authors Contribution

Dr. Vinayak Adimule conceived the idea; Dr. Lalita S. Kumar and Nidhi Manhas wrote the manuscript. All the authors reviewed the manuscript.

References

1 Tao, Z.Q., Shi, A., Lu, C. et al. (2015). Breast cancer: epidemiology and etiology. *Cell Biochem. Biophys.* 72 (2): 333–338.

2 DeSantis, C.E., Ma, J., and Jemal, A. (2019). Trends in stage at diagnosis for young breast cancer patients in the United States. *Breast Cancer Res. Treat.* 173 (3): 743–747.

3 McPherson, K., Steel, C.M., and Dixon, J.M. (2000). ABC of breast diseases: breast cancer–epidemiology, risk factors, and genetics. *BMJ* 321 (7261): 624–628.

4 Zografos, G.C., Panou, M., and Panou, N. (2004). Common risk factors of breast and ovarian cancer: recent view. *Int. J. Gynecol. Cancer* 14 (5): 721–740.

5 Ginsburg, O., Yip, C.H., Brooks, A. et al. (2020). Breast cancer early detection: a phased approach to implementation. *Cancer* 126: 2379–2393.

6 World Health Organization (WHO) (2017). WHO guide to cancer early diagnosis. WHO. http://www.who.int/iris/handle/10665/254500 (accessed 12 April 2020).

7 Yip, C.H. (2016). Challenges in the early detection of breast cancer in resource-poor settings. *Breast Cancer Manag.* 5: 161–169.

8 Pace, L.E. and Keating, N.L. (2014). A systematic assessment of benefits and risks to guide breast cancer screening decisions. *JAMA* 311: 1327–1335.

9 Jafari, S.H., Saadatpour, Z., Salmaninejad, A. et al. (2018). Breast cancer diagnosis: imaging techniques and biochemical markers. *J. Cell. Physiol.* 233 (7): 5200–5213.

10 Panesar, S. and Neethirajan, S. (2016). Microfluidics: rapid diagnosis for breast cancer. *Nanomicro Lett.* 8: 204–220.

11 Weaver, O. and Leung, J.W.T. (2018). Biomarkers and imaging of breast cancer. *Am. J. Roentgenol.* 210 (2): 271–278.

12 Misek, D.E. and Kim, E.H. (2011). Protein biomarkers for the early detection of breast cancer. *Int. J. Proteomics* 2011: 343582. https://doi .org/10.1155/2011/343582.

13 Harbeck, N. and Gnant, M. (2017). Breast cancer. *Lancet* 389: 1134–1150.

14 Li, G., Hu, J., and Hu, G. (2017). Biomarker studies in early detection and prognosis of breast cancer. *Adv. Exp. Med. Biol.* 1026: 27–39.

15 Jayanthi, V., Das, A.B., and Saxena, U. (2017). Recent advances in biosensor development for the detection of cancer biomarkers. *Biosens. Bioelectron.* 91: 15–23.

16 Tothill, I.E. (2009). Biosensors for cancer markers diagnosis. *Semin. Cell Dev. Biol.* 20: 55–62.

17 Chang, Y., Xu, J., and Zhang, Q. (2017). Microplate magnetic chemiluminescence immunoassay for detecting urinary survivin in bladder cancer. *Oncol. Lett.* 14: 4043–4052.

18 Lakshmipriya, T., Gopinath, S.C.B., Hashim, U., and Murugaiyah, V. (2017). Multi-analyte validation in heterogeneous solution by ELISA. *Int. J. Biol. Macromol.* 105: 796–800.

19 Yang, G., Xiao, Z., Tang, C. et al. (2019). Recent advances in biosensor for detection of lung cancer biomarkers. *Biosens. Bioelectron.* 141: 111416. https://doi .org/10.1016/j.bios.2019.111416.

20 He, Z., Chen, Z., Tan, M. et al. (2020). A review on methods for diagnosis of breast cancer cells and tissues. *Cell Proliferation* 53: e12822. https://doi.org/10.1111/ cpr.12822.

21 Ayer, T. (2015). Inverse optimization for assessing emerging technologies in breast cancer screening. *Ann. Oper. Res.* 230 (1): 57–85.

22 Nelson, H.D., Tyne, K., Naik, A. et al., U.S. Preventive Services Task Force(2009). Screening for breast cancer: an update for the U.S. Preventive Services Task Force. *Ann. Intern. Med.* 151: 727–737.

23 Pisano, E.D., Gatsonis, C., Hendrick, E. et al. (2005). Diagnostic performance of digital versus film mammography for breast-cancer screening. *N. Engl. J. Med.* 353: 1773–1783.

24 Covington, M.F., Pizzitola, V.J., Lorans, R. et al. (2018). The future of contrast-enhanced mammography. *Am. J. Roentgenol.* 210 (2): 292–300.

25 Chong, A., Weinstein, S.P., McDonald, E.S., and Conant, E.F. (2019). Digital breast tomosynthesis: concepts and clinical practice. *Radiology* 292 (1): 1–14.

26 Fallenberg, E.M., Schmitzberger, F.F., Amer, H. et al. (2017). Contrast enhanced spectral mammography vs. mammography and MRI – clinical performance in a multi-reader evaluation. *Eur. Radiol.* 27 (7): 2752–2764.

27 Sorin, V., Yagil, Y., Yosepovich, A. et al. (2018). Contrast-enhanced spectral mammography in women with intermediate breast cancer risk and dense breasts. *Am. J. Roentgenol.* 211 (5): 267–274.

28 Lu, Z., Hao, C., Pan, Y. et al. (2020). Contrast-Enhanced Spectral Mammography Versus Ultrasonography: Diagnostic Performance in Symptomatic Patients with Dense Breasts. *Korean J. Radiol.* 21 (4): 442–449. https://doi.org/10.3348/kjr.2019.0393.

29 Skaane, P., Sebuodegard, S., Bandos, A.I. et al. (2018). Performance of breast cancer screening using digital breast tomosynthesis: results from the prospective population-based Oslo Tomosynthesis Screening Trial. *Breast Cancer Res. Treat.* 169 (3): 489–496.

30 Katzen, J. and Dodelzon, K. (2018). A review of computer aided detection in mammography. *Clin Imaging.* 52: 305–309.

31 Danala, G., Patel, B., Aghaei, F. et al. (2018). Classification of breast masses using a computer-aided diagnosis scheme of contrast enhanced digital mammograms. *Ann. Biomed. Eng.* 46 (9): 1419–1431.

32 Benedikt, R.A., Boatsman, J.E., Swann, C.A. et al. (2018). Concurrent computer-aided detection improves reading time of digital breast tomosynthesis and maintains interpretation performance in a multireader multicase study. *Am. J. Roentgenol.* 210 (3): 685–694.

33 Andersson, I., Ikeda, D.M., Zackrisson, S. et al. (2008). Breast tomosynthesis and digital mammography: a comparison of breast cancer visibility and BIRADS classification in a population of cancers with subtle mammographic findings. *Eur. Radiol. 18*: 2817–2825. (copyright taken, License no.- 5493691267918).

34 Khamparia, A., Bharati, S., Podder, P. et al. (2021). Diagnosis of breast cancer based on modern mammography using hybrid transfer learning. *Multidimension. Syst. Signal Process.* 32: 747–765. https://doi.org/10.1007/s11045-02000756-713.

35 Montaha, S., Azam, S., Rafid, A.K.M.R.H. et al. (2021). BreastNet18: a high accuracy fine-tuned VGG16 model evaluated using ablation study for diagnosing breast cancer from enhanced mammography images. *Biology* 10: 1347.

36 Monticciolo, D.L., Newell, M.S., Hendrick, R.E. et al. (2017). Breast cancer screening for average-risk women: recommendations from the ACR commission on breast imaging. *J. Am. Coll. Radiol.* 14 (9): 1137–1143.

37 Ha, R., Kim, H., Mango, V. et al. (2014). Ultrasonographic features and clinical implications of benign palpable breast lesions in young women. *Ultrasonography* 34 (1): 66–70.

38 Zhang, H., Shi, Q., Gu, J. et al. (2014). Combined value of virtual touch tissue quantification and conventional sonographic features for differentiating benign and malignant thyroid nodules smaller than 10 mm. *J. Ultrasound Med.* 33 (2): 257–264.

39 Dhabyani, W.A., Gomaa, M., Khaled, H., and Fahmy, A. (2020). Dataset of breast ultrasound images. *Data Brief* 28: 104863. https://doi.org/10.1016/j.dib.2019.104863.

40 Helal, M.H., Mansour, S.M., Salaleldin, L.A. et al. (2018). The impact of contrast-enhanced spectral mammogram (CESM) and three-dimensional breast ultrasound (3DUS) on the characterization of the disease extend in cancer patients. *Br. J. Radiol.* 91 (1087): 20170977.

41 Yang, G.C.H. and Fried, K.O. (2017). Most thyroid cancers detected by sonography lack intranodular vascularity on color doppler imaging: review of the literature and sonographic-pathologic correlations for 698 thyroid neoplasms. *J. Ultrasound Med.* 36 (1): 89–94.

42 Krouskop, T.A., Wheeler, T.M., Kallel, F. et al. (1998). Elastic moduli of breast and prostate tissues under compression. *Ultrason Imaging* 20 (4): 260–274.

43 Adimule, V., Yallur, B.C., Pai, M.M. et al. (2022). Biogenic synthesis of magnetic palladium nanoparticles decorated over reduced graphene oxide using *Piper betle* Petiole extract (Pd-rGO@Fe$_3$O$_4$ NPs) as heterogeneous hybrid nanocatalyst for applications in Suzuki–Miyaura coupling reactions of biphenyl compounds. *Top. Catal.* https://doi.org/10.1007/s11244-022-01672-9.

44 Signore, G., Nifosì, R., Albertazzi, L. et al. (2010). Polarity sensitive coumarins tailored to live cell imaging. *J. Am. Chem. Soc.* 132 (4): 1276–1288.

45 Mohan Gift, M.D., Pattnaik, B., Nandi, S.S. et al. (2022). Determination of prohibition mechanism of cationic polymer/SiO$_2$ composite as inhibitor in water using drilling fluid. *Mater. Today Proc.* https://doi.org/10.1016/j.matpr.2022.08.171.

46 Park, A.Y., Seo, B.K., Cha, S.H. et al. (2016). An innovative ultrasound technique for evaluation of tumor vascularity in breast cancers: superb micro-vascular imaging. *J. Breast Cancer* 19 (2): 210.

47 Machado, P., Eisenbrey, J.R., Stanczak, M. et al. (2019). Characterization of breast microcalcifications using a new ultrasound image-processing technique. *J. Ultrasound Med.* 38 (7): 1733–1738.

48 Machado, P., Eisenbrey, J.R., Stanczak, M. et al. (2018). Ultrasound detection of microcalcifications in surgical breast specimens. *Ultrasound Med Biol.* 44 (6): 1286–1290.

49 Hooley, R.J., Scoutt, L.M., and Philpotts, L.E. (2013). Breast ultrasonography: State of the art. *Radiology.* 268: 642–659.

50 Shen, Y., Shamout, F.E., Oliver, J.R. et al. (2021). Artificial intelligence system reduces false-positive findings in the interpretation of breast ultrasound exams. *Nat. Commun.* 12: 5645. https://doi.org/10.1038/s41467-021-26023-2.

51 Riedl, C.C., Luft, N., Bernhart, C. et al. (2015). Triple-modality screening trial for familial breast cancer underlines the importance of magnetic resonance imaging and questions the role of mammography and ultrasound regardless of patient mutation status, age, and breast density. *J. Clin. Oncol.* 33 (10): 1128–1135.

52 Kuhl, C. (2007). The current status of breast MR imaging. Part I. Choice of technique, image interpretation, diagnostic accuracy, and transfer to clinical practice. *Radiology* 244: 356–378.

53 Mann, R.M., Kuhl, C.K., and Moy, L. (2019). Contrast-enhanced MRI for breast cancer screening. *J. Magn. Reson. Imaging.* 50: 377–390.

54 Moy, L., Noz, M.E., Maguire, G.Q. Jr. et al. (2010). Role of fusion of prone FDGPET and magnetic resonance imaging of the breasts in the evaluation of breast cancer. *Breast J.* 16 (4): 369–376.

55 Onishi, N., Sadinski, M., Hughes, M.C. et al. (2020). Ultrafast dynamic contrast-enhanced breast MRI may generate prognostic imaging markers of breast cancer. *Breast Cancer Res.* 22: 58.

56 Orguc, S., Basara, I., and Coskun, T. (2012). Diffusion-weighted MR imaging of the breast: comparison of apparent diffusion coefficient values of normal breast tissue with benign and malignant breast lesions. *Singapore Med. J.* 53 (11): 737–743.

57 Bougias, H., Ghiatas, A., Priovolos, D. et al. (2016). Whole-lesion apparent diffusion coefficient (ADC) metrics as a marker of breast tumour characterization – comparison between ADC value and ADC entropy. *Br. J. Radiol.* 89 (1068): 20160304.

58 Rahbar, H. and Partridge, S.C. (2016). Multiparametric MR imaging of breast cancer. *Magn. Reson. Imaging Clin. N. Am.* 24: 223–238.

59 Yabuuchi, H., Matsuo, Y., Sunami, S. et al. (2011). Detection of non-palpable breast cancer in asymptomatic women by using unenhanced diffusion-weighted and T2-weighted MR imaging: comparison with mammography and dynamic contrast-enhanced MR imaging. *Eur. Radiol.* 21: 11–17.

60 Sharma, U. and Jagannathan, N.R. (2022). Magnetic resonance imaging (MRI) and MR spectroscopic methods in understanding breast cancer biology and metabolism. *Metabolites* 12: 295. https://doi.org/10.3390/metabo12040295.

61 Clauser, P., Marcon, M., Dietzel, M. et al. (2017). A new method to reduce false positive results in breast MRI by evaluation of multiple spectral regions in proton MR-spectroscopy. *Eur. J. Radiol.* 92: 51–57.

62 He, D., Mustafi, D., Fan, X. et al. (2018). Magnetic resonance spectroscopy detects differential lipid composition in mammary glands on low fat, high animal fat versus high fructose diets. *PLoS One* 13 (1): e0190929.

63 Cheng, L.L., Chang, I.W., Smith, B.L., and Gonzalez, R.G. (1998). Evaluating human breast ductal carcinomas with high-resolution magic-angle spinning proton magnetic resonance spectroscopy. *J. Magn. Reson.* 135: 194–202.

64 Sitter, B., Lundgren, S., Bathen, T.F. et al. (2006). Comparison of HR-MAS MR spectroscopic profiles of breast cancer tissue with clinical parameters. *NMR Biomed.* 19: 30–40.

65 Sitter, B., Sonnewald, U., Spraul, M. et al. (2002). High-resolution magic angle spinning MRS of breast cancer tissue. *NMR Biomed.* 15: 327–337.

66 Choi, J., Baek, H.M., Kim, S. et al. (2012). HR-MAS MR spectroscopy of breast cancer tissue obtained with core needle biopsy: correlation with prognostic factors. *PLoS One* 7: e51712.

67 Maria, R.M., Altei, W.F., Selistre-de-Araujo, H.S., and Colnago, L.A. (2017). Impact of chemotherapy on metabolic reprogramming: characterization of the metabolic profile of breast cancer MDA-MB-231 cells using ^{1}H HR-MAS NMR spectroscopy. *J. Pharm. Biomed. Anal.* 146: 324–328.

68 Fuss, T.L. and Cheng, L.L. (2016). Evaluation of cancer metabolomics using ex vivo high resolution magic angle spinning (HRMAS) magnetic resonance spectroscopy (MRS). *Metabolites* 6: 11.

69 Jagannathan, N.R., Kumar, M., Seenu, V. et al. (2001). Evaluation of total choline from in vivo volume localized proton MR spectroscopy and its response to neoadjuvant chemotherapy in locally advanced breast cancer. *Br. J. Cancer* 84: 1016–1022.

70 Sah, R.G., Sharma, U., Parshad, R. et al. (2012). Association of estrogen receptor, progesterone receptor, and human epidermal growth factor receptor 2 status with total choline concentration and tumor volume in breast cancer patients: an MRI and in vivo proton MRS study. *Magn. Reson. Med.* 68: 1039–1047.

71 Katz-Brull, R., Lavin, P.T., and Lenkinski, R.E. (2002). Clinical utility of proton magnetic resonance spectroscopy in characterizing breast lesions. *J. Natl. Cancer Inst.* 94: 1197–1203.

72 Keri, R.S., Adimule, V., Kendrekar, P. et al. (2022). The nano-based catalyst for the synthesis of benzimidazoles. *Top. Catal.* 64: 1–21.

73 Baltzer, P.A. and Dietzel, M. (2013). Breast lesions: diagnosis by using proton MR spectroscopy at 1.5 and 3.0 T – systematic review and meta-analysis. *Radiology* 267: 735–746.

74 Adimule, V., Nandi, S.S., Yallur, B.C., and Shaikh, N. (2021). CNT/graphene-assisted flexible thin-film preparation for stretchable electronics and superconductors. In: *Sensors for Stretchable Electronics in Nanotechnology* (ed. K. Pal), 89–103. CRC Press.

75 Wang, X., Wang, X.J., Song, H.S., and Chen, L.H. (2015). ^{1}H-MRS evaluation of breast lesions by using total choline signal-to-noise ratio as an indicator of malignancy: a meta-analysis. *Med. Oncol.* 32: 160.

76 Adimule, V., Yallur, B.C., Batakurki, S., and Nandi, S.S. (2022). Synthesis, morphology and enhanced optical properties of novel $Gd_xCo_3O_4$ nanostructures. *Adv. Mater. Res.* 1173: 71–82. Trans Tech Publications, Ltd.

77 Gallagher, F.A., Woitek, R., McLean, M.A. et al. (2020). Imaging breast cancer using hyperpolarized carbon-13 MRI. *Proc. Natl. Acad. Sci. U.S.A.* 117: 2092–2098.

78 Bohte, A.E., Nelissen, J.L., Runge, J.H. et al. (2018). Breast magnetic resonance elastography: a review of clinical work and future perspectives. *NMR Biomed.* 31 (10): e3932.

79 Adimule, V., Medapa, S., Rao, P.K., and Kumar, L.S. (2014). Synthesis of Schiff bases of 5-[5-(4-fluorophenyl)thiophen-2-yl]-1,3,4-thiadiazol-2-amine and its anticancer activity. *Int. J. Adv. Pharm. Sci.* 5 (1): 1761–1768.

80 Albano, D., Bosio, G., Orlando, E. et al. (2017). Role of fluorine-18-fluorodeoxyglucose positron emission tomography/computed tomography in evaluating breast mucosa-associated lymphoid tissue lymphoma: a case series. *Hematol. Oncol.* 35 (4): 884–889.

81 Romeo, V., Clauser, P., Rasul, S. et al. (2022). AI-enhanced simultaneous multiparametric ^{18}F-FDG PET/MRI for accurate breast cancer diagnosis. *Eur. J. Nucl. Med. Mol. Imaging* 49: 596–608.

82 Tagliafico, A.S., Piana, M., Schenone, D. et al. (2020). Overview of radiomics in breast cancer diagnosis and prognostication. *Breast* 49: 74–80.

83 Adimule, V., Yallur, B., and Gowda, A. (2022). Crystal structure, morphology, optical and super-capacitor properties of Sr_x: α-Sb_2O_4 nanostructures. *Anal. Bioanal. Electrochem.* 14 (1): 1–17.

84 Crivelli, P., Ledda, R.E., Parascandolo, N. et al. (2018). A new challenge for radiologists: radiomics in breast cancer. *BioMed. Res. Int.* 2018: 6120703.

85 AlSawaftah, N., El-Abed, S., Dhou, S., and Zakaria, A. (2022). Microwave imaging for early breast cancer detection: current state, challenges, and future directions. *J. Imaging* 8: 123. https://doi.org/10.3390/jimaging8050123.

86 Batakurki, S.R., Adimule, V., Pai, M.M. et al. (2022). Synthesis of Cs-Ag/Fe_2O_3 nanoparticles using *Vitis labrusca* Rachis extract as green hybrid nanocatalyst for the reduction of arylnitro compounds. *Top. Catal.* https://doi.org/10.1007/s11244-022-01593-7.

87 Hamidipour, A., Henriksson, T., Hopfer, M. et al. (2018). Electromagnetic tomography for brain imaging and stroke diagnostics: progress towards clinical application. In: *Emerging Electromagnetic Technologies for Brain Diseases Diagnostics, Monitoring and Therapy* (ed. L. Crocco, I. Karanasiou, M.L. James, and R.C. Conceição), 59–86. Springer.

88 Semenov, S., Kellam, J., Sizov, Y. et al. (2011). Microwave tomography of extremities: 1. Dedicated 2D system and physiological signatures. *Phys. Med. Biol.* 56: 2005.

89 Meaney, P.M., Goodwin, D., Golnabi, A.H. et al. (2012). Clinical microwave tomographic imaging of the calcaneus: a first-in-human case study of two subjects. *IEEE Trans. Biomed. Eng.* 59: 3304–3313.

90 Zamani, A., Rezaeieh, S.A., Bialkowski, K.S., and Abbosh, A.M. (2017). Boundary estimation of imaged object in microwave medical imaging using antenna resonant frequency shift. *IEEE Trans. Antennas Propag.* 66: 927–936.

91 Adimule, V., Bhat, V.S., Yallur, B.C. et al. (2022). Facile synthesis of novel $SrO_{0.5}$:$MnO_{0.5}$ bimetallic oxide nanostructure as a high-performance electrode material for supercapacitors. *Nanomater. Nanotechnol.* 12: 1–14. https://doi.org/10.1177/18479804211064028.

92 Grzegorczyk, T.M., Meaney, P.M., Kaufman, P.A., and Paulsen, K.D. (2012). Fast 3-D tomographic microwave imaging for breast cancer detection. *IEEE Trans. Med. Imaging* 31: 1584–1592.

93 Modiri, A., Goudreau, S., Rahimi, A., and Kiasaleh, K. (2017). Review of breast screening: toward clinical realization of microwave imaging. *Med. Phys.* 44: e446–e458.

94 Campbell, A. and Land, D. (1992). Dielectric properties of female human breast tissue measured in vitro at 3.2 GHz. *Phys. Med. Biol.* 37: 193.

95 Shaikh, N.M., Bagihalli, G.B., Adimule, V. et al. (2022). A novel silica immobilized acidic ionic liquid [BMIM][$AlCl_4$] as an effective catalyst for biscoumarine synthesis. *Top. Catal.* https://doi.org/10.1007/s11244-022-01591-9.

96 Meaney, P.M., Fanning, M.W., di Florio-Alexander, R.M. et al. (2010). Microwave tomography in the context of complex breast cancer imaging. In: *Proceedings of the 2010 Annual International Conference of the IEEE Engineering in Medicine and*

Biology Society (EMBC), Buenos Aires, Argentina (31 August–4 September 2010), 3398–3401. http://doi.org/10.1109/IEMBS.2010.5627932.

97 Adimule, V., Nandi, S.S., and Yallur, B.C. (2022). Devices and sensors based on additively manufactured shape-memory of hybrid nanocomposites. In: *Shape Memory Composites Based on Polymers and Metals for 4D Printing* (ed. M.R. Maurya, K.K. Sadasivuni, J.J. Cabibihan, et al.), 341–359. Cham: Springer. https://doi.org/10.1007/978-3-030-94114-7_15.

98 Li, D., Meaney, P.M., Raynolds, T. et al. (2004). Parallel-detection microwave spectroscopy system for breast imaging. *Rev. Sci. Instrum.* 75: 2305–2313.

99 Meaney, P.M., Paulsen, K.D., Hartov, A., and Crane, R.K. (1995). An active microwave imaging system for reconstruction of 2-D electrical property distributions. *IEEE Trans. Biomed. Eng.* 42: 1017–1026.

100 Bulyshev, A.E., Semenov, S.Y., Souvorov, A.E. et al. (2001). Computational modeling of three-dimensional microwave tomography of breast cancer. *IEEE Trans. Biomed. Eng.* 48: 1053–1056.

101 Meaney, P.M., Fanning, M.W., Li, D. et al. (2000). A clinical prototype for active microwave imaging of the breast. *IEEE Trans. Microwave Theory Tech.* 48: 1841–1853.

102 Meaney, P.M., Fanning, M.W., Raynolds, T. et al. (2007). Initial clinical experience with microwave breast imaging in women with normal mammography. *Acad. Radiol.* 14: 207–218.

103 Adimule, V., Yallur, B.C., and Gowda, A.H.J. (2022). Advanced sensors based on carbon nanomaterials. In: *Carbon Nanomaterials-Based Sensors* (ed. J.G. Manjunatha and C.M. Hussain), 259–268. Elsevier. https://doi.org/10.1016/B978-0-323-91174-0.00004-4.

104 Zhu, X., Zhao, Z., Wang, J. et al. (2013). Microwave-induced thermal acoustic tomography for breast tumor based on compressive sensing. *IEEE Trans. Biomed. Eng.* 60: 1298–1307.

105 Bevacqua, M.T. and Scapaticci, R. (2016). A compressive sensing approach for 3D breast cancer microwave imaging with magnetic nanoparticles as contrast agent. *IEEE Trans. Med. Imaging* 35: 665–673.

106 Bridges, J.E. (1998). Non-invasive system for breast cancer detection. US Patent 5704355A.

107 Nandi, S.S., Suryavanshi, A., Adimule, V., and Yallur, B.C. (2020). Super capacitor characteristics of novel rare earth perovskite nanomaterials of $Sr_{0.5}$, $Cu_{0.4}$, $Y_{0.1}$. *AIP Conf. Proc.* 2274: 020007. https://doi.org/10.1063/5.0022454.

108 Fear, E.C. (2005). Microwave imaging of the breast. *Technol. Cancer Res. Treat.* 4: 69–82.

109 Bidhendi, H.K., Jafari, H.M., and Genov, R. (2014). Ultra-wideband imaging systems for breast cancer detection. In: *Ultra-Wideband and 60 GHz Communications for Biomedical Applications* (ed. M.R. Yuce), 83–103. Boston, MA: Springer.

110 Hagness, S.C., Taflove, A., and Bridges, J.E. (1998). Two-dimensional FDTD analysis of a pulsed microwave confocal system for breast cancer detection: fixed-focus and antenna-array sensors. *IEEE Trans. Biomed. Eng.* 45: 1470–1479.

111 Hagness, S.C., Taflove, A., and Bridges, J.E. (1999). Three-dimensional FDTD analysis of a pulsed microwave confocal system for breast cancer detection: Design of an antenna-array element. *IEEE Trans. Antennas Propag.* 47: 783–791.

112 Klemm, M., Craddock, I., Leendertz, J. et al. (2008). Experimental and clinical results of breast cancer detection using UWB microwave radar. In: *Proceedings of the 2008 IEEE Antennas and Propagation Society International Symposium*, San Diego, CA, USA (5–12 July 2008), 1–4. IEEE.

113 Klemm, M., Craddock, I.J., Leendertz, J.A. et al. (2009). Radar-based breast cancer detection using a hemispherical antenna array – experimental results. *IEEE Trans. Antennas Propag.* 57: 1692–1704.

114 Henriksson, T., Klemm, M., Gibbins, D. et al. (2011). Clinical trials of a multistatic UWB radar for breast imaging. In: *Proceedings of the 2011 Loughborough Antennas & Propagation Conference*, Loughborough, UK (14–15 November 2011) (ed. Z. Shen), 1–4. IEEE.

115 Fear, E. and Sill, J. (2003). Preliminary investigations of tissue sensing adaptive radar for breast tumor detection. In: *Proceedings of the 25th Annual International Conference of the IEEE Engineering in Medicine and Biology Society*, Cancun, Mexico (17–21 September 2003), vol. 4, 3787–3790. IEEE (IEEE Cat. No. 03CH37439).

116 Williams, T.C., Fear, E.C., and Westwick, D.T. (2006). Tissue sensing adaptive radar for breast cancer detection – investigations of an improved skin-sensing method. *IEEE Trans. Microwave Theory Tech.* 54: 1308–1314.

117 Adimule, V., Suryavanshi, A., and Nandi, S. (2020). Synthesis, characterization and impedance studies of novel nanocomposites of gadolinium titanate. *IOP Conf. Ser.: Mater. Sci. Eng.* 872: 012099. https://iopscience.iop.org/article/10.1088/1757-899X/872/1/012099/meta.

118 Guang-dong, L. and Ye-rong, Z. (2010). An overview of active microwave imaging for early breast cancer detection. *J. Nanjing Univ. Posts Telecommun.* 30: 64–70.

119 Bond, E.J., Li, X., Hagness, S.C., and Van Veen, B.D. (2003). Microwave imaging via space-time beamforming for early detection of breast cancer. *IEEE Trans. Antennas Propag.* 51: 1690–1705.

120 Elsdon, M., Leach, M., Skobelev, S., and Smith, D. (2007). Microwave holographic imaging of breast cancer. In: *Proceedings of the 2007 International Symposium on Microwave, Antenna, Propagation and EMC Technologies for Wireless Communications*. Hangzhou, China (16–17 August 2007) (ed. G.E. Ponchak), 966–969. IEEE.

121 Adimule, V., Vageesha, P., Bagihalli, G. et al. (2019). Synthesis, characterization of hybrid nanomaterials of strontium, yttrium, copper doped with indole Schiff base derivatives possessing dielectric and semiconductor properties. In: *Emerging Research in Electronics, Computer Science and Technology*, Lecture Notes in Electrical Engineering, vol. 545 (ed. V. Sridhar, M. Padma, and K. Rao). Singapore: Springer. https://doi.org/10.1007/978-981-13-5802-9_97.

122 Wang, L., Simpkin, R., and Al-Jumaily, A. (2013). Holographic microwave imaging array: experimental investigation of breast tumour detection. In: *Proceedings of the 2013 IEEE International Workshop on Electromagnetics, Applications and Student*

Innovation Competition, Hong Kong, China (1–3 August 2013) (ed. K.K. So and
C.K. Leung), 61–64. IEEE.

123 Wang, L. and Fatemi, M. (2018). Compressive sensing holographic microwave
random array imaging of dielectric inclusion. *IEEE Access* 6: 56477–56487.

124 Adimule, V., Yallur, B.C., Challa, M., and Joshi, R.S. (2021). Synthesis of
hierarchical structured Gd doped α-Sb_2O_4 as an advanced nanomaterial for high
performance energy storage devices. *Heliyon* 7 (12): e08541. https://doi
.org/10.1016/j.heliyon.2021.e08541.

125 Zhang, T., Hu, H., Yan, G. et al. (2019). Long non-coding RNA and breast cancer.
Technol. Cancer Res. Treat. 18: 1533033819843889.

126 Wang, H., Peng, R., Wang, J. et al. (2018). Circulating microRNAs as potential
cancer biomarkers: the advantage and disadvantage. *Clin. Epigenet.* 10: 59.

127 Liang, D., Liu, H., Yang, Q. et al. (2020). Long noncoding RNA RHPN1-AS1,
induced by KDM5B, is involved in breast cancer via sponging miR-6884-5p. *J. Cell.
Biochem.* https://doi.org/10.1002/jcb.29645.

128 Jahani, S., Nazeri, E., Majidzadeh, A.K. et al. (2020). Circular RNA; a new
biomarker for breast cancer: a systematic review. *J. Cell. Physiol.* 235: 5501–5510.

129 Zhao, Q., Yang, Y., Ren, G. et al. (2019). Integrating bipartite network projection
and KATZ measure to identify novel circRNA-disease associations. *IEEE Trans.
Nanobiosci.* 18: 578–584.

130 Ye, Z., Wang, C., Wan, S. et al. (2019). Association of clinical outcomes in
metastatic breast cancer patients with circulating tumour cell and circulating
cell-free DNA. *Eur. J. Cancer* 106: 133–143.

131 Openshaw, M.R., Page, K., Fernandez-Garcia, D. et al. (2016). The role of ctDNA
detection and the potential of the liquid biopsy for breast cancer monitoring.
Expert Rev. Mol. Diagn. 16: 751–755.

132 Tellez-Gabriel, M., Knutsen, E., and Perander, M. (2020). Current status of
circulating tumor cells, circulating tumor DNA, and exosomes in breast cancer
liquid biopsies. *Int. J. Mol. Sci.* 21: 9457.

133 Adimule, V., Bowmik, D., and Adarsha, H.J. (2020). A facile synthesis of Cr doped
WO_3 nanocomposites and its effect in enhanced current–voltage and impedance
characteristics of thin films. *Lett. Mater.* 10 (4): 481–485.

134 Barkal, A.A., Brewer, R.E., Markovic, M. et al. CD24 signalling through
macrophage Siglec-10 is a target for cancer immunotherapy. *Nature* 572:
392–396.

135 Wang, Z., Wang, Q., Wang, Q. et al. (2017). Prognostic significance of CD24 and
CD44 in breast cancer: a meta-analysis. *Int. J. Biol. Markers* 32: e75–e82.

136 Adimule, V., Nandi, S.S., and Jagadeesha Gowda, A.H. (2021). Enhanced power
conversion efficiency of the P3BT (poly-3-butyl thiophene) doped nanocomposites
of Gd-TiO_3 as working electrode. In: *Techno-Societal 2020* (ed. P.M. Pawar,
R. Balasubramaniam, B.P. Ronge, et al.). Cham: Springer. https://doi
.org/10.1007/978-3-030-69925-3_6.

137 Senbanjo, L.T. and Chellaiah, M.A. (2017). CD44: a multifunctional cell surface
adhesion receptor is a regulator of progression and metastasis of cancer cells.
Front. Cell Dev. Biol. 5: 18.

138 Adimule, V., Suryavanshi, A., Yallur, B.C., and Nandi, S.S. (2020). A facile synthesis of poly(3-octyl thiophene):Ni$_{0.4}$Sr$_{0.6}$TiO$_3$ hybrid nanocomposites for solar cell applications. *Macromol. Symp.* 392: 2000001.

139 Zhou, R., Yazdanifar, M., Roy, L.D. et al. (2019). CAR T-cells targeting the tumor MUC1 glycoprotein reduce triple-negative breast cancer growth. *Front. Immunol.* 10: –1149.

140 Diaconu, I., Cristea, C., Harceaga, V. et al. (2013). Electrochemical immunosensors in breast and ovarian cancer. *Clin. Chim. Acta* 425: 128–138.

141 Hong, R., Sun, H., Li, D. et al. (2022). A review of biosensors for detecting tumor markers in breast cancer. *Life* 12: 342. https://doi.org/10.3390/life12030342.

142 Cristofanilli, M., Pierga, J.Y., Reuben, J. et al. (2019). The clinical use of circulating tumor cells (CTCs) enumeration for staging of metastatic breast cancer (MBC): international expert consensus paper. *Crit. Rev. Oncol. Hematol.* 134: 39–45.

143 Adimule, V., Yallur, B.C., and Sharma, K. (2022). Studies on crystal structure, morphology, optical and photoluminescence properties of flake-like Sb doped Y$_2$O$_3$ nanostructures. *J. Opt.* 51: 173–183. https://doi.org/10.1007/s12596-021-00746-3.

144 Zhou, Y.U., Xu, H., Wang, H. et al. (2020). Detection of breast cancer-derived exosomes using the horseradish peroxidase-mimicking DNAzyme as an aptasensor. *Analyst* 145 (1): 107–114.

145 Nicolini, A., Ferrari, P., and Duffy, M.J. (2018). Prognostic and predictive biomarkers in breast cancer: past, present and future. *Semin. Cancer Biol.* 52: 56–73.

146 Suryavanshi, A., Adimule, V., and Nandi, S.S. (2020). Synthesis, impedance, and current–voltage characteristics of strontium-manganese titanate hybrid nanoparticles. *Macromol. Symp.* 392: 2000002.

147 Adimule, V., Yallur, B.C., Bhowmik, D. et al. (2021). Morphology, structural and photoluminescence properties of shaping triple semiconductor Y$_x$CoO:ZrO$_2$ nanostructures. *J. Mater. Sci. – Mater. Electron.* 32: 12164–12181. https://doi.org/10.1007/s10854-021-05845-2.

148 Roointan, A., Ahmad, M.T., Ibrahim, W.S. et al. (2019). Early detection of lung cancer biomarkers through biosensor technology: a review. *J. Pharm. Biomed. Anal.* 164: 93–103.

149 Ranjan, P., Parihar, A., Jain, S. et al. (2020). Biosensor-based diagnostic approaches for various cellular biomarkers of breast cancer: a comprehensive review. *Anal. Biochem.* 610: 113996.

150 Kal-Koshvandi, A.T. (2020). Recent advances in optical biosensors for the detection of cancer biomarker α-fetoprotein (AFP). *TrAC, Trends Anal. Chem.* 128: 115920.

151 Piroozmand, F., Mohammadipanah, F., and Faridbod, F. (2020). Emerging biosensors in detection of natural products. *Synth. Syst. Biotechnol.* 5: 293–303.

152 Adimule, V., Revaigh, M.G., and Adarsha, H.J. (2020). Synthesis and fabrication of Y-doped ZnO nanoparticles and their application as a gas sensor for the detection of ammonia. *J. Mater. Eng. Perform.* 29: 4586–4596. https://doi.org/10.1007/s11665-020-04979-4.

153 Ouyang, M., Tu, D., Tong, L. et al. (2021). A review of biosensor technologies for blood biomarkers toward monitoring cardiovascular diseases at the point-of-care. *Biosens. Bioelectron.* 171: 112621.

154 Xu, L., Shoaie, N., Jahanpeyma, F. et al. (2020). Optical, electrochemical and electrical (nano)biosensors for detection of exosomes: a comprehensive overview. *Biosens. Bioelectron.* 161: 112222.

155 Sadighbayan, D., Sadighbayan, K., Khosroushahi, A.Y., and Hasanzadeh, M. (2019). Recent advances on the DNA-based electrochemical biosensing of cancer biomarkers: analytical approach. *TrAC, Trends Anal. Chem.* 119: 115609.

156 Sharifi, M., Avadi, M.R., Attar, F. et al. (2019). Cancer diagnosis using nanomaterials based electrochemical nanobiosensors. *Biosens. Bioelectron.* 126: 773–784.

157 Adimule, V., Kerur, S.S., Chinnam, S. et al. (2022). Guar gum and its nanocomposites as prospective materials for miscellaneous applications: a short review. *Top. Catal.* https://doi.org/10.1007/s11244-022-01587-5.

158 Nandi, S.S., Suryavanshi, A., Adimule, V., and Maradur, S.R. (2020). Semiconductor current–voltage characteristics of some novel perovskite ionic nanocomposites of $Sr_{0.5}$, $Cu_{0.4}$, $Y_{0.1}$ and $Sr_{0.5}$, $Mn_{0.5}$ and their electronic sensor applications. *AIP Conf. Proc.* 2274: 020006: https://doi.org/10.1063/5.0022453.

159 Mittal, S., Kaur, H., Gautam, N., and Mantha, A.K. (2017). Biosensors for breast cancer diagnosis: a review of bioreceptors, biotransducers and signal amplification strategies. *Biosens. Bioelectron.* 88: 217–231.

160 Arya, S.K., Wang, K.Y., Wong, C.C., and Rahman, A.R. (2013). Anti-EpCAM modified LC-SPDP monolayer on gold microelectrode based electrochemical biosensor for MCF-7 cells detection. *Biosens. Bioelectron.* 41: 446.

161 Hong, C., Yuan, R., Chai, Y., and Zhuo, Y. (2019). Ferrocenyl-doped silica nanoparticles as an immobilized affinity support for electrochemical immunoassay of cancer antigen 15-3. *Anal. Chim. Acta* 633: 244–249.

162 Vasudev, A., Kaushik, A., and Bhansali, S. (2013). Electrochemical immunosensor for label free epidermal growth factor receptor (EGFR) detection. *Biosens. Bioelectron.* 39: 300–305.

163 Hakimian, F. and Ghourchian, H. (2020). Ultrasensitive electrochemical biosensor for detection of microRNA-155 as a breast cancer risk factor. *Anal. Chim. Acta* 1136: 1–8.

164 Wang, J., Wang, D., and Hui, N. (2020). A low fouling electrochemical biosensor based on the zwitterionic polypeptide doped conducting polymer PEDOT for breast cancer marker BRCA1 detection. *Bioelectrochemistry* 136: 107595.

165 Han, R., Wang, G., Xu, Z. et al. (2020). Designed antifouling peptides planted in conducting polymers through controlled partial doping for electrochemical detection of biomarkers in human serum. *Biosens. Bioelectron.* 164: 112317.

166 Xia, Y.M., Li, M.Y., Chen, C.L. et al. (2020). Employing label-free electrochemical biosensor based on 3D-reduced graphene oxide and polyaniline nanofibers for ultrasensitive detection of breast cancer BRCA1 biomarker. *Electroanalysis* 32: 2045–2055.

167 Chang, J., Wang, X., Wang, J. et al. (2019). Nucleic acid-functionalized metal-organic framework-based homogeneous electrochemical biosensor for simultaneous detection of multiple tumor biomarkers. *Anal. Chem.* 91: 3604–3610.

168 Wang, H., Sun, J., Lu, L. et al. (2020). Competitive electrochemical aptasensor based on a cDNA ferrocene/MXene probe for detection of breast cancer marker Mucin1. *Anal. Chim. Acta* 1094: 18–25.

169 Xu, S., Chang, Y., Wu, Z. et al. (2020). One DNA circle capture probe with multiple target recognition domains for simultaneous electrochemical detection of miRNA-21 and miRNA-155. *Biosens. Bioelectron.* 149: 111848.

170 Marques, R.C., Viswanathan, S., Nouws, H.P. et al. (2014). Electrochemical immunosensor for the analysis of the breast cancer biomarker HER2 ECD. *Talanta* 129: 594–599.

171 Freitas, M., Nouws, H.P.A., and Delerue-Matos, C. (2019). Electrochemical sensing platforms for HER2-ECD breast cancer biomarker detection. *Electroanalysis* 31: 121–128.

172 Zhao, J., Tang, Y., Cao, Y. et al. (2018). Amplified electrochemical detection of surface biomarker in breast cancer stem cell using self-assembled supramolecular nanocomposites. *Electrochim. Acta* 283: 1072–1078.

173 Gu, C., Guo, C., Li, Z. et al. (2019). Bimetallic ZrHf-based metal-organic framework embedded with carbon dots: ultra-sensitive platform for early diagnosis of HER2 and HER2-overexpressed living cancer cells. *Biosens. Bioelectron.* 134: 8–15.

174 Paimard, G., Shahlaei, M., Moradipour, P. et al. (2019). Impedimetric aptamer based determination of the tumor marker MUC1 by using electrospun core-shell nanofibers. *Mikrochim. Acta* 187: 5.

175 Shahrokhian, S. and Salimian, R. (2018). Ultrasensitive detection of cancer biomarkers using conducting polymer/electrochemically reduced graphene oxide-based biosensor: application toward BRCA1 sensing. *Sens. Actuators, B* 266: 160–169.

176 Majd, S.M., Salimi, A., and Ghasemi, F. (2018). An ultrasensitive detection of miRNA-155 in breast cancer via direct hybridization assay using two-dimensional molybdenum disulfide field-effect transistor biosensor. *Biosens. Bioelectron.* 105: 6–13.

177 Bao, Z., Sun, J., Zhao, X. et al. (2017). Top-down nanofabrication of silicon nanoribbon field effect transistor (Si-NR FET) for carcinoembryonic antigen detection. *Int. J. Nanomed.* 12: 4623–4631.

178 Mostafa, A., Mahdi, R., Navid, N. et al. (2016). An electrochemical nanobiosensor for plasma miRNA-155, based on graphene oxide and gold nanorod, for early detection of breast cancer. *Biosens. Bioelectron.* 77: 99–106.

179 Wang, K., He, M.Q., Zhai, F.H. et al. (2017). A novel electrochemical biosensor based on polyadenine modified aptamer for label-free and ultrasensitive detection of human breast cancer cells. *Talanta* 166: 87–92.

180 Chen, C. and Wang, J. (2020). Optical biosensors: an exhaustive and comprehensive review. *Analyst* 145: 1605–1628.

181 Maya Pai, M., Yallur, B.C., Batakurki, S.R. et al. (2022). Synthesis and catalytic activity of heterogenous hybrid nanocatalyst of copper/palladium MOF, RIT 62-Cu/Pd for Stille polycondensation of thieno[2,3-*b*]pyrrol-5-one derivatives. *Top. Catal.* https://doi.org/10.1007/s11244-022-01618-1.

182 Wang, Y., Wei, Z., Luo, X. et al. (2019). An ultrasensitive homogeneous aptasensor for carcinoembryonic antigen based on upconversion fluorescence resonance energy transfer. *Talanta* 195: 33–39.

183 Mohammadi, S., Mohammadi, S., and Salimi, A. (2021). A 3D hydrogel based on chitosan and carbon dots for sensitive fluorescence detection of microRNA-21 in breast cancer cells. *Talanta* 224: 121895.

184 Bai, Y., Li, H., Xu, J. et al. (2020). Ultrasensitive colorimetric biosensor for BRCA1 mutation based on multiple signal amplification strategy. *Biosens. Bioelectron.* 166: 112424.

185 Choi, J.H., Lim, J., Shin, M. et al. (2021). CRISPR-Cas12a-based nucleic acid amplification-free DNA biosensor via Au nanoparticle-assisted metal-enhanced fluorescence and colorimetric analysis. *Nano Lett.* 21: 693–699.

186 Szymanska, B., Lukaszewski, Z., Hermanowicz-Szamatowicz, K., and Gorodkiewicz, E. (2020). An immunosensor for the determination of carcinoembryonic antigen by surface plasmon resonance imaging. *Anal. Biochem.* 609: 113964.

187 Sina, A.A., Vaidyanathan, R., Wuethrich, A. et al. (2019). Label-free detection of exosomes using a surface plasmon resonance biosensor. *Anal. Bioanal.Chem.* 411: 1311–1318.

188 Wang, Q., Zou, L., Yang, X. et al. (2019). Direct quantification of cancerous exosomes via surface plasmon resonance with dual gold nanoparticle-assisted signal amplification. *Biosens. Bioelectron.* 135: 129–136.

189 Han, Y., Qiang, L., Gao, Y. et al. (2021). Large-area surface-enhanced Raman spectroscopy substrate by hybrid porous GaN with Au/Ag for breast cancer miRNA detection. *Appl. Surf. Sci.* 541: 148456.

190 Wang, H.N., Crawford, B.M., Norton, S.J., and Vo-Dinh, T. (2019). Direct and label-free detection of microRNA cancer biomarkers using SERS-based plasmonic coupling interference (PCI) nanoprobes. *J. Phys. Chem. B* 123: 10245–10251.

191 Wang, H.-M., Wang, A.J., Yuan, P.X., and Feng, J.J. (2020). Flower-like metal-organic framework microsphere as a novel enhanced ECL luminophore to construct the coreactant-free biosensor for ultrasensitive detection of breast cancer 1 gene. *Sens. Actuators, B* 320: 128395.

192 Vinayak, A., Sudha, M., Jagadeesha, A.H. et al. (2014). Synthesis, characterization of some novel 1,3,4-oxadiazole compounds containing 8-hydroxy quinolone moiety as potential antibacterial and anticancer agents. *Int. J. Pharm. Res. 4* (4): 180–185.

193 Cui, A., Zhang, J., Bai, W. et al. (2019). Signal-on electrogenerated chemiluminescence biosensor for ultrasensitive detection of microRNA-21 based on isothermal strand-displacement polymerase reaction and bridge DNA-gold nanoparticles. *Biosens. Bioelectron.* 144: 111664.

194 Adimule, V., Batakurki, S., Yallur, B.C. et al. (2022). Samarium-decorated $ZrO_2@SnO_2$ nanostructures, their electrical, optical and enhanced photoluminescence properties. *J. Mater. Sci. – Mater. Electron.* 33: 18699–18715. https://doi.org/10.1007/s10854-022-08718-4.

195 Hartz, J.S.R., Emanetoglu, N.W., Howell, C., and Vetelino, J.F. (2020). Lateral field excited quartz crystal microbalances for biosensing applications. *Biointerphases* 15: 030801.

196 Ramanavicius, S., Jagminas, A., and Ramanavicius, A. (2021). Advances in molecularly imprinted polymers based affinity sensors (review). *Polymers* 13: 974.

197 Gao, Y., Huo, W., Zhang, L. et al. (2019). Multiplex measurement of twelve tumor markers using a GMR multi-biomarker immunoassay biosensor. *Biosens. Bioelectron.* 123: 204–210.

198 Borin, T.F., Arbab, A.S., Gelaleti, G.B. et al. (2016). Melatonin decreases breast cancer metastasis by modulating Rho-associated kinase protein-1 expression. *J. Pineal Res.* 60 (1): 3–15.

199 Li, T., Yang, J., Ali, Z. et al. (2017). Synthesis of aptamer-functionalized Ag nanoclusters for MCF-7 breast cancer cells imaging. *Sci. China Chem.* 60 (3): 370–376.

2

DNA Replication Stress and Genome Instability in Breast Cancer

Kirti Sinha[1], Pau B. Sang[2], Priyanka Sharma[1], and Rishi K. Jaiswal[3]

[1] Patna University, Patna Science College, Department of Zoology, Patna, Bihar, 800005, India
[2] Delhi University, Department of Microbiology, South Campus, Delhi, 110021, India
[3] Loyola University Chicago, Stritch School of Medicine, Cardinal Bernardin Cancer Centre, Department of Cancer Biology, Maywood, IL, 60153, USA

2.1 Introduction

DNA replication is the process of making a duplicate copy of the genome in the cells, which is a prerequisite for cell division. It is a tightly regulated process involving multiple proteins and there must be a very faithful duplication of the genomes as any errors during replication can cause mutation, which can lead to cancer. The error in the DNA can be passed down to its subsequent daughter cells, which can accumulate more mutations and cause genome instability. During the process of DNA replication, cells can encounter multiple hurdles due to damages in the DNA, which can stall or slow down the process of replication, a phenomenon termed replication stress [1]. Multiple exogenous and endogenous factors can cause damage to the DNA like UV radiation, or reactive oxygen species (ROS) generated during cellular metabolism, which can cause mutations as well as replication stress [1, 2]. Replication stress is the major cause of genome instability, which is a hallmark of cancer [3].

Genome instability is characterized by the increased rate of genomic alteration, including mutation, insertion and deletion, gain or loss of chromosome fragment, or variation in chromosome copy number, and thus promotes further heterogeneity. The cell also has several mechanisms to prevent the onset of genome instability like the DNA damage checkpoints, DNA repair system, and the mitotic checkpoint. However, mutation or alteration in any of the proteins involved in this defense mechanism for genome maintenance will enhance the genomic instability in the cells. Moreover, defects in proteins involved in replication fork maintenance and oncogenes activation can also induce replication stress [4].

The genome of breast cancer cells often has several genomic alterations like aneuploidy, copy number variation, and chromosome aberration. Triple-negative breast cancer (TNBC) is aggressive breast cancer that accounts for ~15% of all breast cancer cases. It is characterized by the absence of estrogen receptor (ER), progesterone receptor (PR), and epidermal growth factor receptor 2 (HER2), thus the name triple negative. While ER-positive or HER2-positive breast cancer is characterized by the overexpression of the receptors, they are druggable by their respective antagonist, however TNBC lacks these receptors and therefore is associated with poor prognosis. Different types of breast cancer can showcase distinct types of genomic alteration signatures. In TNBC, impaired BRCA1/BRCA2 and other homologous recombination genes which are characterized by a phenotype similar to BRCA1/BRCA2 mutants termed BRCAness signature are common [5, 6].

2.2 Causes of Replication Stress and Genomic Instability

2.2.1 Replication Dysfunction

2.2.1.1 Low and Untimely Initiation of Replication
In S phase, the number of authorized origins on eukaryotic chromosomes is often seen to have increased the usual number of active replication origins. Because of this, a lot of licensed origins remain inactive, forming a backup set to be activated in case of replication failures [7, 8]. Several findings suggest that low-density initiation in mammals can also result in incomplete replication, which then causes genome instability. Therefore, human cells lacking excess origin-bound MCM27 perform replication normally but they become more sensitive to checkpoint inactivation and replication stress and undergo micronucleation in these circumstances, which involves chromosome fragmentation [9]. Therefore, in regions with low origin density, an inadequate amount of licensed origin can be switched on to offset fork stagnation. A break might be seen later in G2/M as a result of insufficiently replicated DNA that is going into the M phase [10].

Premature replication can also cause genomic instability via another mechanism, which is the loading of the MCM27 helicase core from the G1 phase [11]. Overexpression of initiation-factor Sld2 causes premature replication initiation in G1 stage cells, which increases gross-chromosomal rearrangements (GCRs), and this is amplified by MCM27 via overexpression of Cdc6 [12]. As a result, the recurrence of fork breakage appears to be increased by premature initiation. The cMYC oncogene, which causes replication stress and genome instability when overexpressed, interacts physically with MCM27 and initiates further activation of the origin, whose outcome is DNA damage [13]. As a result, restarting the replication fork may result in multi-fork structures, increasing the likelihood of replication fork collisions and potential disconnects, which would encourage GCRs [10]. The cell growth regulator, proto-oncogene MYC, promotes proliferation by increasing the production of cyclin-dependent kinase (CDK) inhibitors [14]. Cell-cycle progression is promoted by MYC, it does so by directly controlling replication initiation, in addition to transcriptional regulation. It interacts with the prereplicative complex and colocalizes with early S-phase replication foci independent of transcription, implying that MYC plays a non-transcriptional role in DNA replication initiation. MYC overexpression affects origin activity timing, resulting in premature activation of origin [15]. MYC causes this early S-phase origin overactivation, contributing to replication stress [16].

As we have seen MYC plays a pivotal role in causing replication stress. Similarly, MYC genes essentially regulate the progression of many cancers too. MYC genes have been found to target ornithine decarboxylase (ODC, an important polyamine synthesizing enzyme). Polyamines (putrescine, spermidine, and spermine) are cations that have proved themselves as an essential molecule for normal cellular functioning. In the case of breast cancer, the level of polyamine and the enzymes associated with its synthesis are upregulated to 3- to 6-folds. Often high concentration of acetylated polyamines, along with estrogen signaling that is related to purine and polyamine synthesis is found in tissues of breast cancer patients. In most cases of breast cancer, estrogen is directly involved in cancer progression, by activating ERα. When estradiol (E2) binds with ERα, it gets activated and upregulates the genes responsible for purine and polyamine synthesis. This usually follows the "mitochondrial folate pathway" [17]. Polyamines assist the interaction of various transcription factors like ER and NF-κB to help these bind to their response elements (ERE and NF-κB RE respectively). Polyamines have been found to interact with and regulate the mitogenic effect of insulin by affecting insulin receptor tyrosine kinase (IRTK) activity in breast cancer. Studies have often correlated the rise in the level of ODC mRNA and its protein with tumor and metastases stages of breast cancer. An elevated level of Arginase (assists arginine to produce ornithine), ODC, ADC, and Agmatinase (assists putrescine synthesis from arginine) is directly involved with the increased proliferation of cancer cells. This leads to a poor prognosis of breast cancer in its patients [18].

2.2.1.2 Replication Fork Maintenance

Several obstacles can hinder the development of replication forks, both with the aid of changing the interest of the replicative helicase or the capacity of replicative polymerase to take in nucleotides. Problems like DNA damage, inter/intra crosslinks, and DNA being susceptible to 2° structure formation are regularly called RFBs (replication fork barriers). Few other RFBs also called intrinsic RFBs, are the result of everyday cellular events and appear at some point during each S phase. Natural RFBs can be engineered to prevent replication and transcription machinery collisions [19] and their function and impact on genomic balance are influenced by the replication fork pathway. Furthermore, −H2AX, a histone mark and intrinsic-RFB [20], is abundant in fission yeast heterochromatin, implying that hypo-acetylated chromatin and replication stress are connected [21]. The DNA replication working in mammals is cell-type specific [22], and so is the working and impact of RFBs. Furthermore, the absence of DNA lesions necessitates the participation of supplementary DNA polymerases, such as PRR (post-replication repair) pathways and TLS (Trans-Lesions Synthesis) polymerases [23]. A new idea suggests that for DNA replication, each genome segment may not be identical at all points required for duplication.

Although genome rearrangement and recombination regions are connected with intrinsic RFBs, they don't contribute to cancer development [19]. Most cancer cells, on the other hand, are specifically sensitive to intrinsic RFBs since these cells attempt to replicate their genetic material in unsuitable metabolic conditions [24].

2.2.1.3 Mitotic Defects

2.2.1.3.1 S-Phase Checkpoint Dysfunction

If the lagging-strand polymerase is inhibited during replication fork progression, DNA synthesis continues via the subsequent Okazaki fragment with no significant consequences. Leading-strand polymerase blockage, on the other hand, causes uncoupling of main-strand and lagging-strand synthesis on the replication fork [25]. Here, the DNA helicase may continue beforehand to the polymerase, allowing replicative forks to be reactivated after the damage. The S-phase checkpoints are activated by ssDNA and double-strand breaks (DSBs) formed after stalling in the fork and this ensures genetic integrity prior to chromosome segregation [26]. Mec1 (ATM-related [ATR]), CHK2/Rad53, and Tel1 (ataxia telangiectasia–mutated [ATM]), the first protein kinases to appear in checkpoint activation, regulate replication choice and timing in a damage-dependent and damage-independent manner [7, 27]. An inactivation of S-phase checkpoint genes results in genomic instability in organisms including mammals [28–31] and various inheritable diseases that are cancer prone. S-phase checkpoint malfunction jeopardizes the integrity of the replisome, which causes lesions [32, 33]. ATM deficiency, in human cells, facilitates the L1 factors retrotransposition without endonuclease activity and depends on spontaneous DSBs [34]. As a result, evidence from unusual structures and organisms suggests that replication and S-phase checkpoints protect replication fork balance and prevent the generation of DSBs [10].

2.2.1.3.2 Spindle Assembly Checkpoint (SAC)

Every sister chromatid assembles a multilayered protein structure called the kinetochore. One of the chromatid's kinetochores is attached to one pole of the spindle, while the kinetochore of the other chromatids is attached to the opposite pole of the spindle and after this, a tension on opposite poles detaches the sister kinetochores. This voltage is necessary to make the transitions of the ends (i.e. binding to more extreme microtubules) and fulfilling the SAC, which allows the progression of the cell into anaphase [35]. Detection and correction of misconnected kinetochores are done by an Aurora B-dependent error correction pathway, and many of these kinases, localized in the centromere and kinetochore regions [36], regulate their activity toward downstream targets such as Ndc80 and Dsn1, which then facilitate the detachment of misconnected microtubules. The kinetochores that are not bound to the appropriate recruit the kinase Mps1, which then links this error correction to the SAC by recruiting MCC (mitotic checkpoint complex), such as Mad2 and BubR1 [37]. The ability of the SAC to recognize and respond to DNA damage/replication stress in mitosis appears to depend on the severity of the stress [38].

Abovementioned checkpoint is highly dependent on microtubule dynamics, as evidenced by the research that impaired microtubule stability and/or possibly turnover leads to kinetochore-microtubule misconnection [39]. The stability of microtubules is regulated posttranscriptionally and depends on the stability of their mRNA to ensure that the microtubules can respond rapidly to various defects [40]. The availability of tubulin monomers is dependent on this mechanism that regulates negatively the mature spliced mRNA [41]. The dynamic instability of microtubules, which may adapt quickly to these changes, is likely connected to the level of DNA damage relayed to mitosis. When this adaptation is disrupted, as it is in tubulin mutants, viability suffers [42]. It is observed that DNA repair factors can be incorporated into mitotic spindles [43–45] or the major organizing centers of microtubules, the centrosomes [43, 46–51]. Centrosomes are the nucleation centers for microtubules in cells, and they recruit factors for DNA repair, which can impact microtubule nucleation rates. Based on protein tracking, Bastian's lab discovered that chromosomally unstable (CIN+) colon cancer (CRC) cells generate more faulty microtubules than non-tumor RPE1 and CIN CRC cells. EB3-Microtubules fail in the spatial and temporal end. For microtubule assembly, the major regulators in CIN+ CRC cells have been identified as Aurora A and CHK2. The Aurora A centrosome pool is limited by CHK2-dependent phosphorylation of BRCA1 at centrosomes, which corresponds to standard rates of centrosome-generated microtubule assembly. Loss of CHK2 or Aurora A overexpression resulted in the high buildup of Aurora A in centrosomes and elevated rates of microtubule assembly [52], implying that the CHK2–BRCA1 axis inhibits Aurora A activity in centrosomes to avoid chromosome missegregation [38, 53].

2.2.1.3.3 Centrosomes and Nucleosomes

Duplication of the Centrosome is a semiconservative process that must be closely synched with DNA replication so as to enter mitosis containing two centrosomes, each with two centrioles [54]. It has recently been shown that this vulnerable process is already abnormal or poorly regulated in precancerous lesions, and many cancerous cells show centrosome overamplification [55]. Because multiple centrosomes can disrupt normal segregation, cells avoid this problem by grouping supernumerary centrosomes before anaphase [56, 57]. Each centrosome in G1 contains one centriole. The emergence of the pro-centriole begins in G1/S, and its elongation and separation from the parent centriole occur in S, G2, and mitosis [58]. BRCA1-deficient cells, which are forced into mitosis first demonstrated appropriate spindle bipolar establishment, accompanied by failed anaphasic segregation, resulting in tetraploid cells with twice as many centrosomes. As a result, in the following mitosis, these tetraploid cells are unable to form a bipolar spindle or assemble their centrosomes in an efficient way. Because failures such as centrosome division or excessive duplication are observed after the depletion of numerous repair/replication proteins, the relation between impaired replication/repair and impeded centrosome integrity may be indirect. Some experiments provide direct evidence that in RAD51- or BRCA2-deficient cells, spindle multipolarity was suppressed by simply suppressing these cells' intrinsic replication rate deficiencies [59, 60]. It was found that low doses of APH cause centriole detachment [38, 61].

DNA is packed into chromatin, the nucleosomes being the most basic component. Many attributes of genome dynamics have been discovered to rely on nucleosome remodeling, which is guided by histone modifications. As a result, Lys56 of newly formed histone H3 is acetylated [62], allowing its integration into nucleosomes through the collaboration of the histone acetyltransferase Rtt109 and the chaperones Asf1, CAFI, and Rtt106. Replication and checkpoint defects result from dysregulation of nucleosome assembly, resulting in recombinogenic DSBs and ssDNA gaps [63] and hyperrecombination [64]. And we know chromatin's additional role in DNA repair, in some cases, genome instability related to disrupted nucleosome assembly is caused by imperfections in DSB repair [10, 65].

2.2.2 Transcription-Induced Stress

2.2.2.1 Failed Post-Replication Repair

DSBs are created during replication, HR is the main pathway for their repair. It is required to start again the broken replication forks. And where there is no post-replicative repair, it acts as an alternate way to bypass damages that prevent DNA synthesis. Once sister chromatid crossover is feasible, HR is active during the S and G2 phases [66]. When HR fails, DSBs can cause various types of genome instability, depending on the system and template being used for repair [10]. As proved in bacteria, yeast, and mammals, the transcription machinery is an inbuilt barrier to replication fork advancement, and collisions between the two systems are the main element for replication stress and genomic instability [67]. Exclusively replicating cells experience transcription-associated recombination, and transcription inhibitors can reduce the high recombination rates seen in cells when fork advancement has stalled [68]. Furthermore, in human cells, inhibition of transcription subdues a major subset of replication stress induced by cyclin E [69]. In addition to contributing to slow replication progression, inverted forks can also be the target of nucleases and other enzymes that might promote genome instability. Furthermore in human cells, during transcription newly formed RNA can hybridize back to the complementary DNA strand and form an RNA–DNA hybrid and a displaced ssDNA called Rloop, the accumulation of which leads to obstruction of progression in replication fork and DNA breaks [30, 70, 71]. Although the correlation between tumorigenesis and replication stress mediated by transcription has not been investigated properly, the recent reports association between BRCA2 mutations and Rloop-associated genome instability is coherent with this possibility [16, 72].

2.2.3 Genomic Aberrations and Instability

2.2.3.1 Site-Specific Hotspots

GI doesn't occur at random in genomes; rather, it appears more often in certain areas known as hotspots. Understanding GI requires unraveling the nature and expressions of these hotspots. DNA repeats, including short interspersed nuclear elements (SINEs), long terminal repeats (LTRs), long interspersed nuclear elements (LINEs), trinucleotide repeats (TNRs), and retrotransposons, are among the most unstable natural elements. Repeat expansion can be induced by replication slippage, MMR, or BER [73, 74], or by DSB or ssDNA gaps caused by defective replication [75–77] induced by the capability of the repeats to make non-B DNA structures [74]. These compressions and/or expansions could happen via template switching (FoSTeS) and replication fork stalling, where the newly formed three-terminal strand uses another repeat like a template to synthesize more DNA [78], or via Unequal SCE Repair if somehow the replication fork is defective. It does not necessarily have specific sequences, but commonly fragile sites consist of flexible DNA helices containing broken rows of AT-rich repeats, while very few fragile sites contain extended regions of the DNA repeat composed of CGG repeats or the AT-rich minisatellite [79, 80]. Fragile sites, which include SCE, translocations, deletions, and inclusion sites of oncogenic viruses, account for up to 80% of GCR breakpoints witnessed in early-stage tumors or pre-cancerous cells [81, 82]. When DSB repair factors, namely RAD51, DNAPKcs, ligase IV, and BRCA1 are depleted, fragility is seen [83, 84]. The stability of the delicate site is additionally compromised once calyculin A untimely induces chromosomal condensation, suggesting that these sites would possibly contain partially duplicated sequences in the late S phase [85], making these sites critical points for chromosome breakage and GCR breakpoints under replicative stress and S-phase checkpoint inactivation [10].

2.2.3.2 Amplifier Genome

The proposed mechanism of DNA amplification is episome formation (genome integrated to chromosome) [86], and breakage–fusion–bridge (BFB) cycles, first defined by McClintock [87] in maize. BFB cycles are initiated as a result of stress, telomere loss, or dicentric chromosome formation. Since cells have damaged chromosome ends, the amplification technique may be repeated in the next cellular divisions. Preliminary damage can arise at chromosome-fragile sites [88, 89]. In early breast cancers, Anaphase bridges (forming dicentric chromosomes), had been discovered [90, 91].

2.2.3.3 Replication in Inappropriate Metabolic Conditions

To avoid slow branching progression, a sufficient amount of resources for DNA replication (i.e. dNTPs) are required.

HPV16 E6/E7 (human papillomavirus) protein expression forces cells to undergo DNA replication through defective stimulation of the RbE2F signaling cascade, which leads to replication stress markers like slower bifurcations, chromosomal instability, as a result of lack of dNTPs [92]. This sort of replication stress is caused by imbalanced DNA replication, which occurs when cells activate replication under abnormal metabolic conditions. Likewise, upregulation of the oncogene cyclin E results in replication fork slowing due to insufficient dNTP levels. Genes for nucleotide biosynthesis, like cmyc are upregulated and proliferated, to avoid unbalanced replication. The stock of newly synthesized histones governs the pace of fork elongation in mammals, supposedly by meddling with proliferating cell nuclear antigen (PCNA) [93]. Surprisingly, the rate of replication is slowed down without activating

DDR or triggering additional replication origins in the transient scarcity of nascent histones, likely since the ATR pathway cannot be activated (ssDNA is too less). However, the extended defect in bifurcation histone deposition results in DNA damage [24].

2.3 Molecular Mechanism of Genomic Instability

Breast cancer cells often incorporate both numerical (nCIN) and structural chromosomal (sCIN) changes (Figure 2.1) [95–97]. The two most important techniques for sCIN are, first, abnormalities impacting DNA repair genes promote incorrect repair of DNA DSBs in breast cancer cells, which can lead to duplications, translocations, and deletions (e.g. BRCA1, BRCA2 mutations). Second, oncogene stimulation or upregulation promotes replication-related DNA DSBs, which are alluded to as "oncogene-induced DNA replication stress" [3, 98–100]. Genes responsible for the normal functioning of a cell sometimes follow other than usual mechanisms and therefore are responsible for causing replication stress in a cell leading to breast cancer (Figure 2.2).

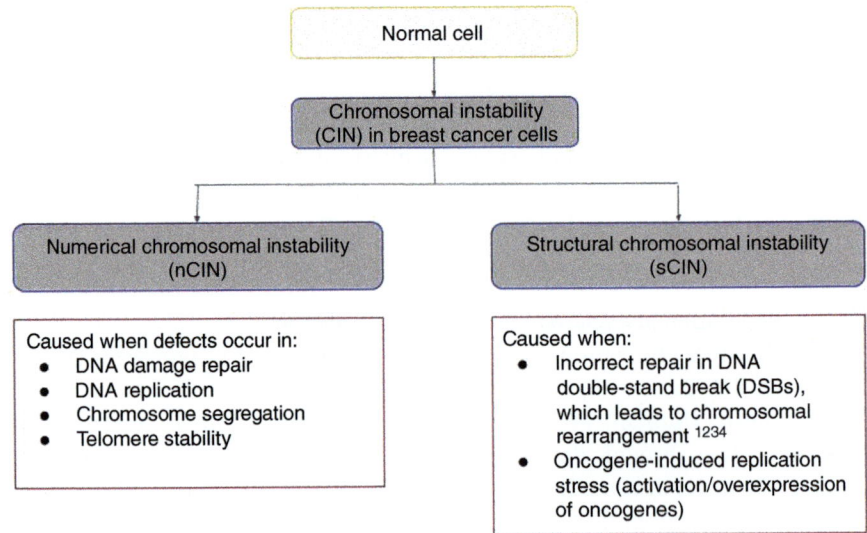

Figure 2.1 A normal cell becomes cancerous when chromosomal instability (CIN) occurs. CIN is of two types nCIN and sCIN. nCIN is seen in a cell when the number of chromosomes increases (Aneuploidy), whereas sCIN is seen when a fragment of a chromosome is removed/changed or multiple fragmentations of a chromosome happen (Chromothripsis). Another reason for CIN is BFB (breakage–fusion–bridge), here, a part of the telomere breaks off and joins to a chromosome forming a dicentric chromosome. Abbreviations: 1. deletion, 2. inversion, 3. translocation, 4. amplification. Source: Adapted from Burrell et al. [94].

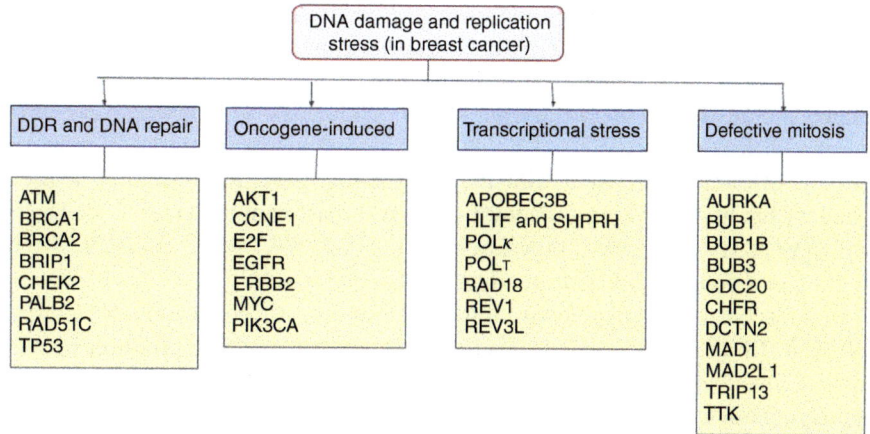

Figure 2.2 When a mutation occurs in genes responsible for normal functioning like DNA repair, transcription, and cell-cycle regulation, in a normal cell, then these genes either get overexpressed or switched off. This causes replication stress and genomic instability. Here is a few genes hallmark for breast cancer that gets mutated during their course of function and does not perform the assigned mechanisms like DNA repair and cell cycle regulation. Some of the common mutations seen in these genes are HR, GCRs, CIN, SCEs, TNR, and Telomere fusion.

2.3.1 Problems Faced During DNA Damage Repair

The DDR pathway quintessentially monitors genome integrity and stability, for cell homeostasis [7]. DDR signals, such as checkpoint activation, cellular cycle arrest, suppression or initiation of transcription and post-transcription elements, and, in rare cases, apoptosis are activated to reply to DNA damage. Cells contain a plethora of DNA repair signaling networks that offer a barely superfluous ability to restore all kinds of DNA damage, including MMR, NHEJ, SSBR, BER, NER, and HR [101]. Failure to restore broken DNA has a significant impact on genomic composition, which is related to terrible cancer diagnosis and therapy [102–104]. In familial breast cancers, germline mutations or lack of key genome caretaker genes like TP53, ATM, and CHEK2 and DNA damage repair genes such as BRCA1, BRCA2, RAD51C, BRIP1, and PALB2 offer significant growth in the normal chance for developing breast cancers [105, 106]. Damage to DNA and DDR faultings in precancerous cells can be caused by initial deformities in tumor suppressor pathways used for DNA damage recognition and repair [3, 100]. However, a study found, that DDR defects are no longer the number one to promote GI in sporadic cancers [107]. Another suggests that, in sporadic cancers, a significant role is of DNA repair abnormalities, in encouraging GI via epigenetic dysregulation of DNA repair genes [108, 109].

Moreover, abnormal transcriptional trickery of DDR gene expression can drastically make a contribution to DNA repair abnormalities in breast cancers. For example, because of cytoplasmic sequestration of CHK1, leading to ubiquitination and degradation, PI3K/AKT flow signaling impedes DNA repair mediated by HR [110]. Furthermore, E2F1, a transcription factor, controls RAD51, and the RB tumor suppressor pathway manages this task and in breast cancer cases instability is observed in this mechanism [95]. It especially happens causing decreased RB expression, through cyclin D1 amplification, and CDK4 aberrations [95]. This affects the misregulation of quite a few E2F main genes, a lot of which can be concerned with DNA repair [111]. HER2 activated, downregulates DNA repair proteins [112], and collectively with the remark that germline mutation in BRCA1 vendors broadly speaking broaden basal-like breast cancers [113, 114]. Lastly, a defective p53 (~23%) pathway (genes consisting of PTEN, BRCA1, and RPA1) affects the cell reaction to DNA damage in more than one way [4].

In breast cancers, hyperactivation of genes like CCNE1, E2F, MYC, AKT1, PIK3CA, ERBB2, and EGFR results in replication stress, which produces breaks by stalling and disintegration of forks [115]. For instance, CCNE1 amplification immediately affects CDK interests and induces stress by obstructing the meeting of prereplication complexes [116]. This additionally promotes centrosome over-duplication, a not-so-unusual purpose of nCIN in most cancer cells [117]. Likewise, stimulation of the RAS pathway reduces replication fork development via growing CDK job, which leads to expanded foundation firing and nucleotide pool depletion [118, 119]. Poorly regulated TSG can also cause stress, for instance, upregulation of the gene p21WAF1/CIP1, discovered in a subset of p53, has been proven to leave DNA replication licensing, allowing stress and DNA lesions, which causes GI, cancer, and chemoresistance [120, 121]. In cells that have oncogenes activated and precancerous cells, when oncogene causes DNA stress, the DDR pathway gets a substantial boom [3, 4].

2.3.2 Transcriptional Stress

In breast cancers, scientists have observed excessive DSBs at the promotor site of estrogen-responsive genes [122] and cytidine deaminase APOBEC3B (AID/APOBEC or A3B) – which is involved in mutating RNA or DNA through deaminating cytidine to uridine [123, 124]. Deoxycytidine to deoxyuridine (C-to-U) deamination catalyzed by A3B is necessary for ER genes, and this process has to be repaired to keep away from the piling of harmful mutations. UNG (Uracil DNA glycosylase), erases dU and hires it to ER binding sites [122, 125]. UNG-structured excision of dU effects in brief DNA DSBs which can be directly repaired through NHEJ. On the other hand, genome instability immediately hinders transcription explicitly by the formation of RNA–DNA hybrids, typically termed R loops [126]. They are everyday transcriptional intermediates in techniques that include transcription activation, class-transfer recombination, and termination [126], and these can act as precursors to DSBs and stress and are related to genome instability in ER+ breast cancers [127]. Several latest research has furnished proof linking genome instability in BRCAless breast cancers to the odd tenacity of R loops [72, 128, 129]. BRCA1 morphed to shape a complicated structure with STEX (senataxin, a helicase associated with R loop) and repress R-loop abundance by taking part in regulating transcription termination [128]. BRCA1 or SETX takes care of and restore ssDNA breaks at transcription-termination site, mainly in lagging strands. Unscheduled R-loop accumulation is caused by BRCA2-deficiency, ensuing in structural chromosomal aberrations [72, 129]. BRCA1 and BRCA2 operate R-loop differently. BRCA2 is observed to have interaction with RNA polymerase II (RNAPII) [129] and a regulated launch of RNAPII from PPP (promoter-proximal pause) sites calls for the recruitment of PAF1, which complicates nucleosome disassembly via ubiquitinating H2B on Lys-120 but it is much needed for transcription elongation downstream of RNAPII [130–134]. BRCA2 depletion effects in impaired recruitment of PAF1 and decreases ubiquitination of histone H2B on lysine-120, for this reason stopping elongation. In BRCA2-less cells fail to hire PAF1 therefore RNAPII is deposited at PPP sites, which leads to the accumulation of site-particular Rloops that therefore cause DNA breaks [4].

2.3.3 CIN: Result of Defective Mitosis

During mitosis, duplicated sister chromatids are connected by cohesins. SAC (spindle assembly checkpoint), guarantees the duplicated chromosomes aren't detached until spindle fibers are attached to the kinetochores at opposite ends, after this is ensured the cleavage of cohesins allows the split of sister chromatids during anaphase [135]. The mitotic checkpoints are very important and are seen to have non-mutated checkpoint genes [136–138]. Instead, mitotic checkpoint genes (like MAD1, MAD2L1, BUB1, BUB1B, TRIP13, TTK, and CDC20), are regularly overexpressed in breast and other different cancers, which leads to over-activation of checkpoints and lengthy mitosis [136, 138–140]. It might also lead to "mitotic slippage" (also referred to as endomitosis), where cells become polyploid (nCIN condition) [53, 138, 141–143]. CHFR, formerly proven to be hyper-methylated in breast cancers [144], is concerned by the mitotic stress caused by the G2–M checkpoint and corresponds to most cancer phenotypes, inclusive of high growth potential, increased invasiveness, better mitotic cells, and multiplied motility, in addition to aneuploidy.

Proteins like BRCA2, CEP55, AURKA, AURKB, and others, guide cells for the cytokinesis, but are often mutated and hyper-expressed in case of breast and other cancers [53, 142, 145, 146]. A failure in cytokinesis makes a cancerous cell tetraploid either by endoreduplication or endomitosis [143, 147]. While the character frequencies of those paths to polyploidy in breast cancers are unknown, tetraploidy (genome doubling), is the number one interim step to aneuploidy in breast cancer advancement [96]. It is likewise properly hooked up that tetraploidy encourages nCIN because of the existence of four centrosomes and the forming of merotelic microtubule attachments [148, 149]. RB pathway (which includes proteins like pRB, p130, and p170) plays a critical role, whose damage causes GI and also breast cancer. These proteins inhibit E2F transcription factors, which were responsible for cell cycle advancement. It is observed that the p53 pathway does not promote CIN, however, additionally holds aneuploid cells to proliferate [4, 126, 150, 151].

In human cancers, the extraordinarily excess degrees of instability, are termed chromothripsis, however, that is probably related to a selected sort of cell stage or differentiation. It is a catastrophic occasion that could arise more regularly in most cancers and elderly cells and wherein chromosomes go through a couple of chromosome fragmentation and rejoining, particularly through NHEJ, which results in a couple of GCRs [152, 153]. Different mechanisms had been stated, along with chromosome pulverization as a result of replication and repair defects going on in micronuclei as a consequence of mistakes in mitosis [154]. However, it's miles too soon to recognize the reasons for this event. Understanding chromothripsis and its molecular foundation will sincerely help us understand GI and its outcomes [10].

2.4 Aftermath of Replication Stress on Cell and Its Fate

An assay of replication stress is the buildup of ssDNA at forks, despite the fact that compromised DNA replication without extra ssDNA has additionally been found [1, 155]. Lengths of ssDNA are uncovered at replication forks attributable to replicative helicases progressing to open the coils of DNA. In addition to the uncoupling between helicases and polymerases, another reason for ssDNA accumulation is the degradation of newly synthesized DNA via the mixed action of nucleases and DNA helicases [17, 156–158]. ATR orchestrates genome instability and the cell reaction to replication stress by phosphorylating numerous targets that manage stabilization of origins, fork, mRNA metabolism, and cell cycle [27]. Two major features of ATR may be distinguished: (i) arrest of cell cycle and repression of late origins (global effects) and (ii) stabilization of stalled replication forks (nearby effect) [24].

2.4.1 Conservation of Stalled Replication Forks

A strong proof of a decrease in stability of forks because of replication stress, is furnished by tumor-susceptible FA and hereditary breast and ovarian cancer syndrome, this is related to BRCA1 and BRCA2 mutations. Cells lacking FA, BRCA1, or BRCA2 genes are extraordinarily sensitive, which means are unable in restoring replication fork development after ICL-induced blockage. The elimination of the ICL damage could be fixed by HR, where BRCA1 recruits ubiquitylated FANCD2 to stalled replication forks [159], here FANCD2 recruits BRCA2 (also known as FANCD1) and RAD51 [160, 161]. Upon UV exposure, BRCA1 represses translesion DNA synthesis (TLS) in inclination toward S-phase checkpoint activation, assisting the re-localization of replication factor C (RFC) at stalled forks [162], and BRCA2 prevents DSBs via means of stabilizing DNA at stalled replication forks [163]. RAD51 additionally appears to start replication fork restart further to its usual feature in HR repair [164]. Similarly, stalled replication forks are protected from endonuclease-dependent degradation by FANCD2, BRCA1, and BRCA2 proteins [158, 165], and the association of FANCD2 with the MCM2–7 helicase permits replication to slow down to stop DNA damage [160]. The recent studies showing that replication forks are guarded in FANCD2-deficient cells by accumulated RAD51 levels or stabilized RAD51 filaments are also behind such an overexpression effect [165] because it might give tumor cells with increased proliferative capability and DNA lesion resistance [16].

TLS is the main route by which mammalian cells replicate through DNA damage [166]. In spite of their mutagenicity, TLS polymerases act as suppressors of tumorigenesis, as suggested by the downregulation of TLS polymerases pol η, pol ι, pol κ, and pol ζ, observed in different types of cancer [167]. Thus, Pol η colocalizes with tumor suppressors BRCA2 and associates with localizers of BRCA2 (PALB2) at blocked replication forks [168]. Individuals with POLH mutations have an increased incidence of mutations caused by UV radiations, due to the incapability of their DNA repair mechanisms to substitute thymine dimers [169] and have an increased incidence of chromosomal breaks and fragile site expression common (CFS) [170]. Pol ζ-deficient cells accumulate replication-dependent DSBs and chromosomal abnormalities [171], and the consequential loss of Rev3l (which encodes Pol ζ) in adult mice increases tumor incidence, which is increased in a p53-deficient environment. A decrease of the human homologs of Rad5, HLTF, and SHPRH (E3 ubiquitin protein ligase), lessens polyubiquitylation of chromatin-bound PCNA after genotoxic agents stop replication [172]. Curiously, in a large number of colon, gastric, and uterine tumors, hypermethylation inactivates HLTF, implying that its silencing acts as a tumor suppressor [16, 173].

2.4.2 Chromosome Segregation Defect Check by HR Repair

Regional chromosome segregation is influenced by replication stress via the creation of anaphase bridges. Aside from this local effect, a low rate of replication stress can result in the production of more centrosomes during mitosis [59]. The mechanisms that link S-phase development and centrosome duplication have started to appear, but it stays doubtful, how persistent is replication stress in promoting centrosome amplification [174]. Because the majority of mitotic added centrosomes function, their presence results in multipolar mitosis and unbalanced chromosome segregation [59] which is systematically linked to chromosome bridges, delayed chromosome condensation, and prolonged metaphase arrest [24, 138].

Persistent stalled replication forks because of replication stress breaks rely on HR for its error-free restoration, consequently, HR disorder can cause GI and tumorigenesis, and BRCA proteins are excellent examples of this. Both the BRCA proteins (tumor suppressors) encourage error-free HR repair and establish that NHEJ is concerned with carcinogenesis [175]. And proven by the fact that LOF mutations in PALB2, which mediates BRCA2 recruitment to broken DNA through BRCA1 binding, purpose an accelerated danger of growing breast cancers much like the predisposition this is visible with BRCA mutations [16, 176, 177].

2.4.3 Aging, Cell Death, and Senescence

Aging is a complicated task that concerns not only DNA damage deposition but also protein damage [178]. Age-triggered LOH is not the outcome of chromosome loss, but rather of mitotic recombination, which could be caused by damage in DNA, implying that older cells have defective replication [10, 179].

Replication stress activates mitotic spindle checkpoint and arrests cell cycle and in a few cases cell death. Endogenous replication stress has been implicated in the very early stages of senescence and/or malignancy initiation [81, 100, 115, 180, 181]. In addition to tumor induction, oncogenes can cause senescence in a process known as oncogene-induced senescence (OIS). It is proposed that oncogene activation results in hyper-replication by stimulating the cellular proliferation program, which increases endogenous replication stress, thus activating the DDR. The DDR's endurance causes cell cycle arrest and senescence [100, 182]. Furthermore, the induction of senescence allows cells capable of bypassing the proliferation restriction [24] (Figure 2.3).

2.5 Therapeutic Approach

Acknowledging the molecular mechanisms that govern GI in cancer development is critical for finding new therapeutic approaches and providing personalized treatment. Extensively reviewed PARP inhibitors, which were discovered in 2005, are a good targeting strategy for genomic instability in breast and gonad cancer [183–185]. Alternative therapeutic methods focus on dividing cells, as loss of this ability, a cell could lead to CIN. Many currently underway clinical trials involve mitotic kinase inhibitors Aurora A/B/C, TTK (MPS1), and PLK1. PLK1 helps promote spindle formation which is found in nearly all tumors, along with breast cancer, therefore considered a great therapeutic target [186], and many PLK1-inhibiting medicines exist, but its medical trials gave mixed results [187]. A current observation shows, that

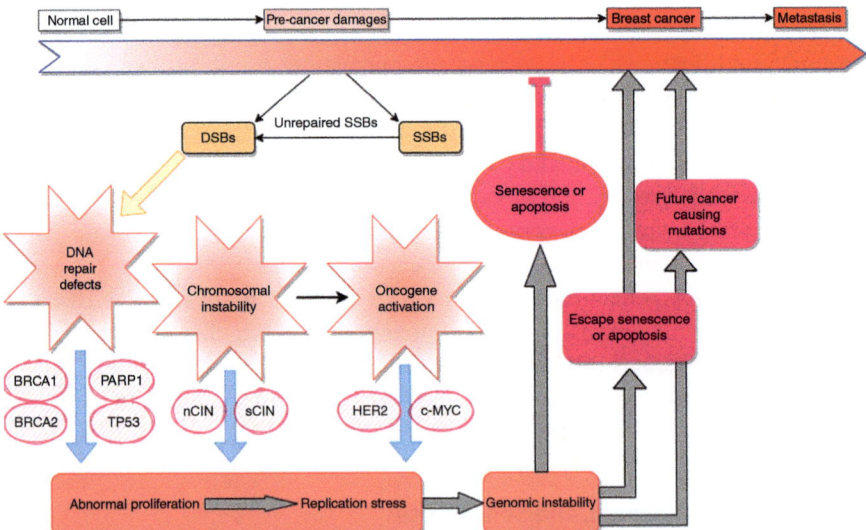

Figure 2.3 A normal cell when suffers from DNA damage or mutation, which gives rise to cancerous cells. A breakage in DNA single or both the strands causes the recruitment of repair proteins by activating DNA repair genes. Defects in DNA repair genes (like BRCA1, BRCA2, ATM, RAD51C, PARP1, TP53) or in genes responsible for cell proliferation (like CDK4, Cyclin D1, Cyclin E1, PTEN) or oncogenes (like HER2, c-MYC, E2F, ATK1) activation results in replication stress. This further leads to genomic instability. After this, the cell can either follow apoptosis or escape cell death and cause cancer.

Plk1 overexpression results in CIN however additionally suppresses tumorigenesis in mice by hindering cytokinesis [188], and its improved expression gave higher results in a few breast cancer subtypes. TTK (MPS1) inhibitors, which act as a crucial mitotic checkpoint, have proven favorable preclinically and are presently in Phase I medical trial [4, 189].

Incorporation of damaged dNTPs is a new and novel approach, in particular to cancer cells. This may be accomplished via way of means focused on NUDT1 (additionally referred to as MTH1), a protein that stops the misincorporation of oxidized dNTPs in the course of replication in cancer cells [150, 190]. As a result, for this reason, in particular, this approach only kills cancer cells. Replication stress can also be prolonged by depleting licensing factors. As normal cells propagate without re-replication, medicines that set off re-replication are useful for anticancer therapy. MLN4924 is however one drug that causes re-replication, CDT1 stabilization, apoptosis, and senescence in checkpoint-lacking cells [16, 191].

For the treatment of many cancers polyamines have been targeted. MYC-ODC-linked polyamine synthesis pathway has been successfully given targets for therapy. DFMO which inhibits ODC working has undergone multiple clinical trials and has proved effective against MYC-driven cancers. Similarly, polyamine like spermidine has entered clinical trials and proved to have induced autophagy by inhibiting acetyltransferase activity of EP300 [192]. Autophagy and apoptosis induced by polyamines are mechanisms that hold potential for treating many types of cancer including breast cancer [193] (Figure 2.4).

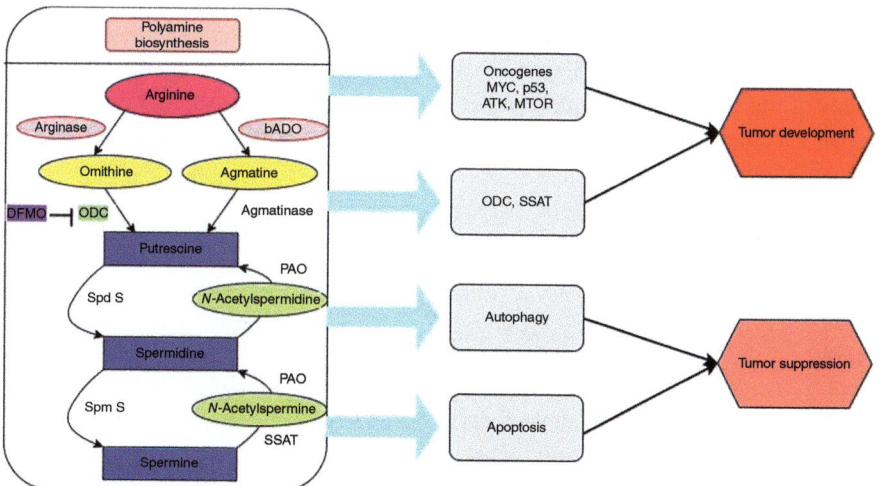

Figure 2.4 Polyamines are synthesized in a cell following this pathway. It is an essential cationic molecule for normal cellular processes. High levels of oncogenes like MYC often leads to high-risk diseases. It upregulates the ODC and polyamine levels and increases the risk of breast cancer. Oncogenes like p53 get destabilized upon polyamine upregulation, causing hyper-proliferation. Studies have found a way of utilizing the impact of polyamine on cancerous cells as an alternate way for focusing on it as a druggable target. Polyamines like spermidine and putrescine can regulate oxidative stress and initiate autophagy in a cancer cell.

The tiny molecule, proTAME inhibits APC, which, like MPS1, helps in controlling mitotic checkpoint function. This roots mitotic arrest *in vitro* but fails *in vivo* [189]. However, different pills that concentrate on mitosis are already normally used to deal with breast cancer patients, such as taxanes, inclusive of paclitaxel (additionally called Taxol) and docetaxel (Taxotere). Because they impede microtubule depolymerization, these chemotherapeutic drugs stabilize the mitotic spindle and suggest high genomic instability can be scientifically beneficial. Another therapeutic technique targets cancer cells that form multipolar spindles with mis-attached kinetochores, thus promoting CIN [194]. These cancerous cells cluster centrosomes to enable bipolar cell division, and impairing centrosome clustering can kill tumor cells, which was proven via an experiment on PDGFR inhibitors and CP-673451. Similarly, different research found that the drugs griseofulvin, EM011, and CCCI-01 set off multipolar spindles and breast cancer and other cells die [195, 196].

2.6 Conclusion

Replication stress and genomic instability in a cell often indicate cancer. It can be well established that breast cancer is a possible outcome of several cascades of mechanisms gone wrong. Mechanisms of genes related to breast cancer, DNA repair, and DNA replication, and their mutations causing pre-mitotic and post-mitotic chaos are the major source of replication stress and GI which is generally observed to lead to breast cancer. There are still a lot of studies related to these genes and their functionalities, to come to the surface. So, by comprehending the molecular mechanisms and different pathways behind the cause of replication fork stress and GI, therapeutic and preventive techniques can be designed against cancer. The study of the complex working of BRCA1 and BRCA2 pathways, and possible DNA repair mechanisms can shed light on the suggestive candidate for drugs in breast cancer patients.

Abbreviations

ATM	ataxia telangiectasia mutated
ATR	ataxia telangiectasia and Rad3-related
BER	base excision repair
BFB	breakage-fusion-bridge
BRCA1	BRest CAncer gene 1
BRCA2	BRest CAncer gene 2
CDK	cyclin-dependent kinase
CIN	chromosomal instability
CRC	colon cancer
DSBs	double-strand breaks

ER	estrogen receptor
GCRs	gross-chromosomal rearrangements
GI	genome instability
HER2	human epidermal growth factor receptor 2
HR	homologous recombination
ICL	interstrand DNA crosslinks
IRTK	insulin receptor tyrosine kinase
LINE	long interspersed nuclear elements
LOH	loss of heterozygosity
LTR	long terminal repeat
MCC	mitotic checkpoint
MMR	DNA mismatch repair
nCIN	numerical chromosomal instability
NER	nucleotide excision repair
NHEJ	non-homologous end joining
ODC	ornithine decarboxylase
OIS	oncogene-induced senescence
PARP	poly ADP-ribose polymerase
PCNA	proliferating cell nuclear antigen
PR	progesterone receptor
RB	retinoblastoma
RFB	replication fork barriers
SAC	spindle assembly checkpoint
SCE	sister chromatid exchange
sCIN	structural chromosomal instability
SINE	short interspersed nuclear elements
SSBR	single strand break repair
TLS	trans-lesions synthesis
TNBC	triple-negative breast cancer
TNRs	trinucleotide repeat expansions and contractions
UNG	uracil DNA glycosylase

References

1 Zeman, M.K. and Cimprich, K.A. (2014). Causes and consequences of replication stress. *Nat. Cell Biol.* 16: 2–9.

2 Mazouzi, A., Velimezi, G., and Loizou, J.I. (2014). DNA replication stress: causes, resolution, and disease. *Exp. Cell. Res.* 329: 85–93.

3 (a)Gift, M.D.M., Pattnaik, B., Nandi, S.S. et al. (2022). Determination of prohibition mechanism of cationic polymer/SiO$_2$ composite as inhibitor in water using drilling fluid. *Mater. Today Proc.* 69: 1080–1086.(b) Duijf, P.H.G., Nanayakkara, D., Nones, K. et al. (2019). Mechanisms of genomic instability in breast cancer. *Mol. Med.* 25 (7): 595–611.

4 Audeh, M.W., Carmichael, J., Penson, R.T. et al. (2010). Oral poly(ADP-ribose) polymerase inhibitor olaparib in patients with BRCA1 or BRCA2 mutations and recurrent ovarian cancer: a proof-of-concept trial. *Lancet* 376: 245–251.

5 Tutt, A., Robson, M., Garber, J.E. et al. (2010). Oral poly(ADP-ribose) polymerase inhibitor olaparib in patients with BRCA1 or BRCA2 mutations and advanced breast cancer: a proof-of-concept trial. *Lancet* 376: 235–244.

6 Batakurki, S.R., Adimule, V., Pai, M.M. et al. (2022). Synthesis of Cs–Ag/Fe$_2$O$_3$ nanoparticles using *Vitis labrusca* Rachis extract as green hybrid nanocatalyst for the reduction of aryl nitro compounds. *Top. Catal.* https://doi.org/10.1007/s11244-022-01593-7.

7 Nandi, S.S., Adimule, V., and Yallur, B.C. (2022). Synthesis, structural and optical properties of Co doped Sm$_2$O$_3$ nanostructures. *Adv. Mater. Res.* 1173: 59–69. Trans Tech Publications, Ltd.

8 Ibarra, A., Schwob, E., and M'endez J. (2008). Excess MCM proteins protect human cells from replicative stress by licensing backup origins of replication. *Proc. Natl. Acad. Sci. U.S.A.* 105 (26): 8956–8961.

9 Adimule, V., Yallur, B.C., Pai, M.M. et al. (2022). Biogenic synthesis of magnetic palladium nanoparticles decorated over reduced graphene oxide using *Piper betle* Petiole extract (Pd-rGO@Fe$_3$O$_4$ NPs) as heterogeneous hybrid nanocatalyst for applications in Suzuki–Miyaura coupling reactions of biphenyl compounds. *Top. Catal.* https://doi.org/10.1007/s11244-022-01672-9.

10 Adimule, V., Jagadeesha Gowda, A.H., Nandi, S.S., and Bowmik, D. (2022). Antimalarial activity of novel class of 1,3-benzoxaborole derivatives containing 1,3,4-oxadiazole moiety. In: *Drug Development for Malaria* (ed. P. Kendrekar), 285–302. Wiley-VCH GmbH.

11 Tanaka, S. and Araki, H. (2011). Multiple regulatory mechanisms to inhibit untimely initiation of DNA replication are important for stable genome maintenance. *PLos Genet.* 7 (6): e1002136.

12 Dominguez-Sola, D., Ying, C.Y., Grandori, C. et al. (2007). Non-transcriptional control of DNA replication by *c-Myc*. *Nature* 448 (7152): 445–451.

13 Adimule, V., Bhat, V.S., Yallur, B.C. et al. (2022). Facile synthesis of novel SrO$_{0.5}$:MnO$_{0.5}$ bimetallic oxide nanostructure as a high-performance electrode material for supercapacitors. *Nanomater. Nanotechnol.* 12: 1–14. https://doi.org/10.1177/18479804211064028.

14 Srinivasan, S.V., Dominguez-Sola, D., Wang, L.C. et al. (2013). Cdc45 is a critical effector of myc-dependent DNA replication stress. *Cell Rep.* 3: 1629–1639.

15 Zhu, D., Zhao, Z., Cui, G. et al. (2018). Single-cell transcriptome analysis reveals estrogen signaling coordinately augments one-carbon, polyamine, and purine synthesis in breast cancer. *Cell Rep.* 25: 2285–2298. https://doi.org/10.1016/j.celrep.2018.10.093, e2284.

16 Singh, R., Avliyakulov, N.K., Braga, M. et al. (2013). Proteomic identification of mitochondrial targets of arginase in human breast cancer. *PLoS One* 8: e79242, https://doi.org/10.1371/journal.pone.0079242.

17 Kadapure, S.A., Kadapure, P., Nandi, S., and Shet, A. (2022). Overview on catalyst and co-solvents for sustainable biodiesel production. *Proc. Inst. Civ. Eng. Energy* https://doi.org/10.1680/jener.21.00092.

18 Adimule, V., Yallur, B.C., Batakurki, S., and Nandi, S.S. (2022). Synthesis, morphology and enhanced optical properties of novel $Gd_xCo_3O_4$ nanostructures. *Adv. Mater. Res.* 1173: 71–82. Trans Tech Publications, Ltd.

19 Adimule, V., Batakurki, S., Yallur, B.C. et al. (2022). Enhanced photoluminescence, optical, structural properties of ZrO_2-incorporated Sm_2O_3:Co_3O_4 nanocomposite and their applications in photocatalytic degradation of methylene blue. *J. Mater. Res.* **37**: 2396–2405.

20 Rozenzhak, S., Mejia-Ramirez, E., Williams, J.S. et al. (2010). Rad3 decorates critical chromosomal domains with gammaH2A to protect genome integrity during S-phase in fission yeast. *PLos Genet.* 6: e1001032.

21 Mechali, M. (2010). Eukaryotic DNA replication origins: many choices for appropriate answers. *Nat. Rev. Mol. Cell Biol.* 11: 728–738.

22 Sale, J.E., Lehmann, A.R., and Woodgate, R. (2012). Y-family DNA polymerases and their role in tolerance of cellular DNA damage. *Nat. Rev. Mol. Cell Biol.* 13: 141–152.

23 Magdalou, I., Lopez, B.S., Pasero, P., and Lambert, S.A.E. (2014). The causes of replication stress and their consequences on genome stability and cell fate. *Semin. Cell Dev. Biol.* 30 (2014): 154–164.

24 Pagès, V. and Fuchs, R.P. (2003). Uncoupling of leading- and lagging-strand DNA replication during lesion bypass in vivo. *Science* 300 (5623): 1300–1303.

25 Bartek, J., Lukas, C., and Lukas, J. (2004). Checking on DNA damage in S phase. *Nat. Rev. Mol. Cell Biol.* 5 (10): 792–804.

26 Ciccia, A. and Elledge, S.J. (2010). The DNA damage response: making it safe to play with knives. *Mol. Cell* 40: 179–204.

27 Alvaro, D., Lisby, M., and Rothstein, R. (2007). Genome-wide analysis of Rad52 foci reveals diverse mechanisms impacting recombination. *PLos Genet.* 3 (12): e228.

28 Myung, K., Datta, A., and Kolodner, R.D. (2001). Suppression of spontaneous chromosomal rearrangements by S phase checkpoint functions in *Saccharomyces cerevisiae*. *Cell* 104 (3): 397–408.

29 Paulsen, R.D., Soni, D.V., Wollman, R. et al. (2009). A genome-wide siRNA screen reveals diverse cellular processes and pathways that mediate genome stability. *Mol. Cell* 35 (2): 228–239.

30 Syljuåsen, R.G., Sørensen, C.S., Hansen, L.T. et al. (2005). Inhibition of human Chk1 causes increased initiation of DNA replication, phosphorylation of ATR targets, and DNA breakage. *Mol. Cell. Biol.* 25 (9): 3553–3562.

31 Casper, A.M., Nghiem, P., Arlt, M.F., and Glover, T.W. (2002). ATR regulates fragile site stability. *Cell* 111 (6): 779–789.

32 Cha, R.S. and Kleckner, N. (2002). ATR homolog Mec1 promotes fork progression, thus averting breaks in replication slow zones. *Science* 297 (5581): 602–606.

33 Coufal, N.G., Garcia-Perez, J.L., Peng, G.E. et al. (2011). Ataxia telangiectasia mutated (ATM) modulates long interspersed element-1 (L1) retrotransposition in human neural stem cells. *Proc. Natl. Acad. Sci. U.S.A.* 108 (51): 20382–20387.

34 Musacchio, A. (2015). The molecular biology of spindle assembly checkpoint signaling dynamics. *Curr. Biol.* 25: R1002–R1018.

35 Broad, A.J., DeLuca, K.F., and DeLuca, J.G. (2020). Aurora B kinase is recruited to multiple discrete kinetochore and centromere regions in human cells. *J. Cell Biol.* *219*: e201905144.

36 Pachis, S.T.; Kops, G.J.P.L. Leader of the SAC: molecular mechanisms of Mps1/TTK regulation in mitosis. *Open Biol.* 2018, *8.* https://doi.org/10.1098/rsob.180109

37 Wilhelm, T., Said, M., and Naim, V. (2020). DNA replication stress and chromosomal instability: dangerous liaisons. *Genes 11*: 642. https://doi.org/10.3390/genes11060642.

38 Shaikh, N.M., Bagihalli, G.B., Adimule, V. et al. (2022). A novel silica immobilised acidic ionic liquid [BMIM][AlCl$_4$] as an effective catalyst for biscoumarine synthesis. *Top. Catal.* https://doi.org/10.1007/s11244-022-01591-9.

39 Gasic, I., Boswell, S.A., and Mitchison, T.J. (2019). Tubulin mRNA stability is sensitive to change in microtubule dynamics caused by multiple physiological and toxic cues. *PLoS Biol. 17*: e3000225.

40 Yen, T.J., Machlin, P.S., and Cleveland, D.W. (1988). Autoregulated instability of β-tubulin mRNAs by recognition of the nascent amino terminus of β-tubulin. *Nature 334*: 580–585.

41 Laflflamme, G., Sim, S., Leary, A. et al. (2019). Interphase microtubules safeguard mitotic progression by suppressing an aurora B-dependent arrest induced by DNA replication stress. *Cell Rep. 26*: 2875–2889.e3.

42 Nalepa, G., Enzor, R., Sun, Z. et al. (2013). Fanconi anemia signaling network regulates the spindle assembly checkpoint. *J. Clin. Investig. 123*: 3839–3847.

43 Zhou, J., Chan, J., Lambelé, M. et al. (2017). NEIL3 repairs telomere damage during S phase to secure chromosome segregation at mitosis. *Cell Rep. 20*: 2044–2056.

44 Xu, R., Xu, Y., Huo, W. et al. (2018). Mitosis-specifific MRN complex promotes a mitotic signaling cascade to regulate spindle dynamics and chromosome segregation. *Proc. Natl. Acad. Sci. U.S.A. 115*: E10079–E10088.

45 Zhang, S., Hemmerich, P., and Grosse, F. (2007). Centrosomal localization of DNA damage checkpoint proteins. *J. Cell. Biochem. 101*: 451–465.

46 Abdul-Sater, Z., Cerabona, D., Potchanant, E.S. et al. (2015). FANCA safeguards interphase and mitosis during hematopoiesis in vivo. *Exp. Hematol. 43*: 1031–1046.e12.

47 Tsvetkov, L., Xu, X., Li, J., and Stern, D.F. (2003). Polo-like kinase 1 and Chk2 interact and co-localize to centrosomes and the midbody. *J. Biol. Chem. 278*: 8468–8475.

48 Wang, C.-Y., Huang, E.Y.-H., Huang, S.-C., and Chung, B.-C. (2015). DNA-PK/Chk2 induces centrosome amplification during prolonged replication stress. *Oncogene 34*: 1263–1269.

49 Krämer, A., Mailand, N., Lukas, C. et al. (2004). Centrosome-associated Chk1 prevents premature activation of cyclin-B-Cdk1 kinase. *Nat. Cell Biol. 6*: 884–891.

50 Adimule, V., Kendrekar, P., and Batakurki, S. (2022). Synthesis, characterization and antimicrobial properties of novel benzimidazole amide derivatives bearing thiophene moiety. In: *Benzimidazole* (ed. P. Kendrekar and V. Adimule). London: IntechOpen https://doi.org/10.5772/intechopen.104908.

51 Jaiswal, R.K., Kumar, P., and Yadava, P.K. (2013). Telomerase and its extracurricular activities. *Cell. Mol. Biol. Lett. 18* (4): 538–554. https://doi.org/10.2478/s11658-013-0105-0.

52 Ertych, N., Stolz, A., Valerius, O. et al. (2016). CHK2-BRCA1 tumor-suppressor axis restrains oncogenic Aurora-A kinase to ensure proper mitotic microtubule assembly. *Proc. Natl. Acad. Sci. U.S.A. 113*: 1817–1822.

53 Adimule, V., Suryavanshi, A., and Nandi, S. (2020). Synthesis, characterization and impedance studies of novel nanocomposites of gadolinium titanate. *IOP Conf. Ser.: Mater. Sci. Eng.* 872: 012099.

54 Lopes, C.A.M., Mesquita, M., Cunha, A.I. et al. (2018). Centrosome amplification arises before neoplasia and increases upon p53 loss in tumorigenesis. *J. Cell Biol. 217*: 2353–2363.

55 Adimule, V., Nandi, S.S., and Yallur, B.C. (2022). Devices and sensors based on additively manufactured shape-memory of hybrid nanocomposites. In: *Shape Memory Composites Based on Polymers and Metals for 4D Printing* (ed. M.R. Maurya, K.K. Sadasivuni, J.J. Cabibihan, et al.). Cham: Springer. https://doi .org/10.1007/978-3-030-94114-7_15.

56 Jaiswal, R.K. and Yadava, P.K. (2019). TGF-β-mediated regulation of plasminogen activators is human telomerase reverse transcriptase dependent in cancer cells. *BioFactors* 45 (5): https://doi.org/10.1002/biof.1543.

57 Adimule, V., Yallur, B.C., and Gowda, A.H.J. (2022). Advanced sensors based on carbon nanomaterials. In: *Carbon Nanomaterials-Based Sensors* (Chapter 14) (ed. J.G. Manjunatha and C.M. Hussain), 259–268. Elsevier https://doi.org/10.1016 /B978-0-323-91174-0.00004-4.

58 Wilhelm, T., Magdalou, I., Barascu, A. et al. (2014). Spontaneous slow replication fork progression elicits mitosis alterations in homologous recombination-deficient mammalian cells. *Proc. Natl. Acad. Sci. U.S.A. 111*: 763.

59 Wilhelm, T., Ragu, S., Magdalou, I. et al. (2016). Slow replication fork velocity of homologous recombination-defective cells results from endogenous oxidative stress. *PLos Genet. 12*: e1006007.

60 Wilhelm, T., Olziersky, A.-M., Harry, D. et al. (2019). Mild replication stress causes chromosome mis-segregation via premature centriole disengagement. *Nat. Commun. 10*: 3585.

61 Suryavanshi, A., Adimule, V., and Nandi, S.S. (2020). Synthesis, impedance, and current–voltage characteristics of strontium-manganese titanate hybrid nanoparticles. *Macromol. Symp.* 392: 2000002. https://doi.org/10.1002/ masy.202000002.

62 Ye, X., Franco, A.A., Santos, H. et al. (2003). Defective S phase chromatin assembly causes DNA damage, activation of the S phase checkpoint, and S phase arrest. *Mol. Cell* 11 (2): 341–351.

63 Adimule, V., Nandi, S.S., and Jagadeesha Gowda, A.H. (2021, 2020). A facile synthesis of gadolinium titanate (GdTiO$_3$) nanomaterial and its effect in enhanced current–voltage characteristics of thin films. In: *Techno-Societal* (ed. P.M. Pawar, R. Balasubramaniam, B.P. Ronge, et al.), 69–78. Springer.

64 Alabert, C. and Groth, A. (2012). Chromatin replication and epigenome maintenance. *Nat. Rev. Mol. Cell Biol.* 13 (3): 153–167.

65 Heyer, W.D., Ehmsen, K.T., and Liu, J. (2010). Regulation of homologous recombination in eukaryotes. *Annu. Rev. Genet.* 44: 113–139.

66 Gaillard, H., Herrera-Moyano, E., and Aguilera, A. (2013). Transcription-associated genome instability. *Chem. Rev.* 113: 8638–8661.

67 Gottipati, P., Cassel, T.N., Savolainen, L., and Helleday, T. (2008). Transcription-associated recombination is dependent on replication in mammalian cells. *Mol. Cell. Biol.* 28: 154–164.

68 Jones, R.M., Mortusewicz, O., Afzal, I. et al. (2013). Increased replication initiation and conflicts with transcription underlie Cyclin E-induced replication stress. *Oncogene* 32: 3744–3753.

69 Tuduri, S., Crabbé, L., Conti, C. et al. (2009). Topoisomerase I suppresses genomic instability by preventing interference between replication and transcription. *Nat. Cell Biol.* 11: 1315–1324.

70 Dominguez-Sanchez, M.S., Barroso, S., Gomez-Gonzalez, B. et al. (2011). Genome instability and transcription elongation impairment in human cells depleted of THO/TREX. *PLos Genet.* 7: e1002386.

71 Bhatia, V., Barroso, S.I., García-Rubio, M.L. et al. (2014). BRCA2 prevents R-loop accumulation and associates with TREX-2 mRNA export factor PCID2. *Nature* 511: 362–365. This work shows that BRCA2-depleted or BRCA2-deficient cancer cells accumulate R-loops and DNA damage, and that genome instability generated in such cells is partially dependent on R-loops. This work provides a novel role for BRCA2 in preventing or helping to remove R-loops and proposes that these R-loops are a major source of replication stress in cancer cells.

72 McMurray, C.T. (2010). Mechanisms of trinucleotide repeat instability during human development. *Nat. Rev. Genet.* 11 (11): 786–799.

73 Pearson, C.E., Nichol Edamura, K., and Cleary, J.D. (2005). Repeat instability: mechanisms of dynamic mutations. *Nat. Rev. Genet.* 6 (10): 729–742.

74 Ireland, M.J., Reinke, S.S., and Livingston, D.M. (2000). The impact of lagging strand replication mutations on the stability of CAG repeat tracts in yeast. *Genetics* 155 (4): 1657–1665.

75 Schweitzer, J.K. and Livingston, D.M. (1998). Expansions of CAG repeat tracts are frequent in a yeast mutant defective in Okazaki fragment maturation. *Hum. Mol. Genet.* 7 (1): 69–74.

76 Shishkin, A.A., Voineagu, I., Matera, R. et al. (2009). Large-scale expansions of Friedreich's ataxia GAA repeats in yeast. *Mol. Cell* 35 (1): 82–92.

77 Adimule, V., Yallur, B., and Gowda, A. (2022). Crystal structure, morphology, optical and super-capacitor properties of Srx: α-Sb_2O_4 nanostructures. *Anal. Bioanal. Electrochem.* 14 (1): 1–17.

78 Gacy, A.M., Goellner, G., Juranić, N. et al. (1995). Trinucleotide repeats that expand in human disease form hairpin structures in vitro. *Cell* 81 (4): 533–540.

79 Adimule, V., Sudha, M., Jagadesha, A.H. et al. (2014). Synthesis, characterization and in vitro anticancer properties of 1-{5-aryl-2-[5-(4-fluoro-phenyl)-thiophen-2-yl]-[1,3,4]oxadiazol-3-yl}-ethanone. *Int. J. Pharm. Res. Rev.* 3 (12): 20–25.

80 Bartkova, J., Rezaei, N., Liontos, M. et al. (2006). Oncogene-induced senescence is part of the tumorigenesis barrier imposed by DNA damage checkpoints. *Nature* 444 (7119): 633–637.

81 Ozeri-Galai, E., Lebofsky, R., Rahat, A. et al. (2011). Failure of origin activation in response to fork stalling leads to chromosomal instability at fragile sites. *Mol. Cell* 43 (1): 122–131.

82 Arlt, M.F., Xu, B., Durkin, S.G. et al. (2004). BRCA1 is required for commonfragile-site stability via its G2/M checkpoint function. *Mol. Cell. Biol.* 24 (15): 6701–6709.

83 Schwartz, M., Zlotorynski, E., Goldberg, M. et al. (2005). Homologous recombination and nonhomologous end-joining repair pathways regulate fragile site stability. *Genes Dev.* 19 (22): 2715–2726.

84 El Achkar, E., Gerbault-Seureau, M., Muleris, M. et al. (2005). Premature condensation induces breaks at the interface of early and late replicating chromosome bands bearing common fragile sites. *Proc. Natl. Acad. Sci. U.S.A.* 102 (50): 18069–18074.

85 Savelyeva, L. and Schwab, M. (2001). Amplification of oncogenes revisited: from expression profiling to clinical application. *Cancer Lett.* 167: 115–123.

86 Coquelle, A., Pipiras, E., Toledo, F. et al. (1997). Expression of fragile sites triggers intrachromosomal mammalian gene amplification and sets boundaries to early amplicons. *Cell* 89: 215–225.

87 McClintock, B. (1941). The stability of broken ends of chromosomes in zea mays. *Genetics* 26 (2): 234–282. https://doi.org/10.1093/genetics/26.2.234.

88 Hellman, A., Zlotorynski, E., Scherer, S.W. et al. (2002). A role for common fragile site induction in amplification of human oncogenes. *Cancer Cell* 1: 89–97.

89 Chin, K., de Solorzano, C.O., Knowles, D. et al. (2004). In situ analyses of genome instability in breast cancer. *Nat. Genet.* 36: 984–988.

90 Kwei, K.A., Kung, Y., Salari, K. et al. (2010). Genomic instability in breast cancer: pathogenesis and clinical implications. https://doi.org/10.1016/j.molonc.2010.04.001. *Mol. Oncol.* 4: 255–266.

91 Bester, A.C., Roniger, M., Oren, Y.S. et al. (2011). Nucleotide deficiency promotes genomic instability in early stages of cancer development. *Cell* 145: 435–446.

92 Mejlvang, J., Feng, Y., Alabert, C. et al. (2014). New histone supply regulates replication fork speed and PCNA unloading. *J. Cell Biol.* 204: 29–43.

93 Cancer Genome Atlas Network (2012). Comprehensive molecular portraits of human breast tumours. *Nature* 490: 61–70.

94 Burrell, R.A., McClelland, S.E., Endesfelder, D. et al. (2013). Replication stress links structural and numerical cancer chromosomal instability. *Nature* 494: 492–496.

95 Zack, T.I., Schumacher, S.E., Carter, S.L. et al. (2013). Pan-cancer patterns of somatic copy number alteration. *Nat. Genet.* 45: 1134–1140.

96 Duijf, P.H., Schultz, N., and Benezra, R. (2013). Cancer cells preferentially lose small chromosomes. *Int. J. Cancer* 132: 2316–2326.

97 Adimule, V., Batakurki, S., Yallur, B.C. et al. (2022). Samarium-decorated ZrO_2@ SnO_2 nanostructures, their electrical, optical and enhanced photoluminescence properties. *J. Mater. Sci. - Mater. Electron.* 33: 18699–18715.

98 Dereli-Öz, A., Versini, G., and Halazonetis, T.D. (2011). Studies of genomic copy number changes in human cancers reveal signatures of DNA replication stress. *Mol. Oncol.* 5: 308–314.

99 Halazonetis, T.D., Gorgoulis, V.G., and Bartek, J. (2008). An oncogene-induced DNA damage model for cancer development. *Science* 319: 1352–1355.

100 Mani, R.S. and Chinnaiyan, A.M. (2010). Triggers for genomic rearrangements: insights into genomic, cellular and environmental influences. *Nat. Rev. Genet.* 11: 819–829.

101 Nik-Zainal, S., Davies, H., Staaf, J. et al. (2016). Landscape of somatic mutations in 560 breast cancer whole-genome sequences. *Nature* 534: 47–54.

102 Adimule, V., Vageesha, P., Bagihalli, G. et al. (2019). Synthesis, characterization of hybrid nanomaterials of strontium, yttrium, copper doped with indole schiff base derivatives possessing dielectric and semiconductor properties. In: *Emerging Research in Electronics, Computer Science and Technology*, Lecture Notes in Electrical Engineering, vol. 545 (ed. V. Sridhar, M. Padma, and K. Rao). Singapore: Springer. https://doi.org/10.1007/978-981-13-5802-9_97.

103 Thomas, A., Routh, E.D., Pullikuth, A. et al. (2018). Tumor mutational burden is a determinant of immune-mediated survival in breast cancer. *Onco immunology* 7: e1490854.

104 Adimule, V., Yallur, B.C., Challa, M., and Joshi, R.S. (2021). Synthesis of hierarchical structured Gd doped α-Sb$_2$O$_4$ as an advanced nanomaterial for high performance energy storage devices. *Heliyon* 7 (12): e08541. https://doi.org/10.1016/j.heliyon.2021.e08541.

105 Lhota, F., Zemankova, P., Kleiblova, P. et al. (2016). Hereditary truncating mutations of DNA repair and other genes in BRCA1/BRCA2/PALB2-negatively tested breast cancer patients. *Clin. Genet.* 90: 324–333.

106 Kumar, M., Jaiswal, R.K., Prasad, R. et al. (2021). PARP-1 induces EMT in non-small cell lung carcinoma cells via modulating the transcription factors Smad4, p65 and ZEB1. *Life Sci.*, Elsevier 269 (15): 118994. https://doi.org/10.1016/j.lfs.2020.118994.

107 Keri, R.S., Adimule, V., Kendrekar, P. et al. (2022). The nano-based catalyst for the synthesis of benzimidazoles. *Top. Catal.* https://doi.org/10.1007/s11244-022-01562-0.

108 Teschendorff, A.E., Menon, U., Gentry-Maharaj, A. et al. (2010). Age-dependent DNA methylation of genes that are suppressed in stem cells is a hallmark of cancer. *Genome Res.* 20: 440–446.

109 Adimule, V., Bowmik, D., and Adarsha, H.J. (2020). A facile synthesis of Cr doped WO$_3$ nanocomposites and its effect in enhanced current–voltage and impedance characteristics of thin films. *Lett. Mater.* 10 (4): 481–485. https://doi.org/10.22226/2410-3535-2020-4-481-485.

110 Jaiswal, R.K., Kumar, P., Kumar, M., and Yadava, P.K. (2018). hTERT promotes tumor progression by enhancing TSPAN13 expression in osteosarcoma cells. *Mol. Carcinog.* 57 (8): 1038–1054. https://doi.org/10.1002/mc.22824.

111 Yaglom, J.A., McFarland, C., Mirny, L., and Sherman, M.Y. (2014). Oncogene-triggered suppression of DNA repair leads to DNA instability in cancer. *Oncotarget* 5: 8367–8378.

112 Larsen, M.J., Kruse, T.A., Tan, Q. et al. (2013). Classifications within molecular subtypes enables identification of BRCA1/BRCA2 mutation carriers by RNA tumor profiling. *PLoS One* 8: e64268.

113 Adimule, V., Nandi, S.S., and Jagadeesha Gowda, A.H. (2021). Enhanced power conversion efficiency of the P3BT (poly-3-butyl thiophene) doped nanocomposites of Gd-TiO$_3$ as working electrode. In: *Techno-Societal 2020* (ed. P.M. Pawar, R. Balasubramaniam, B.P. Ronge, et al.). Cham: Springer https://doi.org/10.1007/978-3-030-69925-3_6.

114 Bartkova, J., Horejsi, Z., Koed, K. et al. (2005). DNA damage response as a candidate anti-cancer barrier in early human tumorigenesis. *Nature* 434: 864–870.

115 Tanaka, S. and Difflfley, J.F. (2002). Interdependent nuclear accumulation of budding yeast Cdt1 and Mcm2-7 during G1 phase. *Nat. Cell Biol.* 4: 198–207.

116 Adimule, V., Suryavanshi, A., BC, Y., and Nandi, S.S. (2020). A facile synthesis of poly(3-octyl thiophene):Ni$_{0.4}$Sr$_{0.6}$TiO$_3$ hybrid nanocomposites for solar cell applications. *Macromol. Symp.* 392: 2000001. https://doi.org/10.1002/masy.202000001.

117 Kotsantis, P., Silva, L.M., Irmscher, S. et al. (2016). Increased global transcription activity as a mechanism of replication stress in cancer. *Nat. Commun.* 7: 13087.

118 Neelsen, K.J., Zanini, I.M.Y., Herrador, R., and Lopes, M. (2013). Oncogenes induce genotoxic stress by mitotic processing of unusual replication intermediates. *J. Cell Biol.* 200: 699–708.

119 Galanos, P., Vougas, K., Walter, D. et al. (2016). Chronic p53-independent p21 expression causes genomic instability by deregulating replication licensing. *Nat. Cell Biol.* 18: 777–789.

120 Galanos, P., Pappas, G., Polyzos, A. et al. (2018). Mutational signatures reveal the role of RAD52 in p53-independent p21-driven genomic instability. *Genome Biol.* 19: 37.

121 Periyasamy, M., Patel, H., Lai, C.-F. et al. (2015). APOBEC3B-mediated cytidine deamination is required for estrogen receptor action in breast cancer. *Cell Rep.* 13: 108–121.

122 Nik-Zainal, S., Alexandrov, L.B., Wedge, D.C. et al. (2012). Mutational processes molding the genomes of 21 breast cancers. *Cell* 149: 979–993.

123 Jaiswal, R.K. and Yadava, P.K. (2020). Assessment of telomerase as drug target in breast cancer. *J. Biosci..* http://link.springer.com/article/10.1007/s12038-020-00045-2.

124 Stavnezer, J. (2011). Complex regulation and function of activation-induced cytidine deaminase. *Trends Immunol.* 32: 194–201.

125 Skourti-Stathaki, K. and Proudfoot, N.J. (2014). A doubleedged sword: R loops as threats to genome integrity and powerful regulators of gene expression. *Genes Dev.* 28: 1384–1396.

126 Nandi, S.S., Suryavanshi, A., Adimule, V., and Yallur, B.C. (2020). Super capacitor characteristics of novel rare earth perovskite nanomaterials of Sr$_{0.5}$, Cu$_{0.4}$, Y$_{0.1}$. *AIP Conf. Proc.* 2274: 020007.

127 Hatchi, E., Skourti-Stathaki, K., Ventz, S. et al. (2015). BRCA1 recruitment to transcriptional pause sites is required for R-loop-driven DNA damage repair. *Mol. Cell* 57: 636–647.

128 Shivji, M.K.K., Renaudin, X., Williams, Ç.H., and Venkitaraman, A.R. (2018). BRCA2 regulates transcription elongation by RNA polymerase II to prevent R-loop accumulation. *Cell Rep.* 22: 1031–1039.

129 Jaiswal, R.K., Varshney, A.K., and Yadava, P.K. (2018). Diversity and functional evolution of the plasminogen activator system. *Biomed. Pharmacother.* 98: 886–898. https://doi.org/10.1016/j.biopha.2018.01.029.

130 Yu, M., Yang, W., Ni, T. et al. (2015). RNA polymerase II–associated factor 1 regulates the release and phosphorylation of paused RNA polymerase II. *Science* 350: 1383–1386.

131 Van Oss, S.B., Shirra, M.K., Bataille, A.R. et al. (2016). The histone modification domain of Paf1 complex subunit Rtf1 directly stimulates H2B ubiquitylation through an interaction with Rad6. *Mol. Cell* 64: 815–825.

132 Wu, L., Li, L., Zhou, B. et al. (2014). H2B ubiquitylation promotes RNA Pol II processivity via PAF1 and pTEFb. *Mol. Cell* 54: 920–931.

133 Xiao, T., Kao, C.-F., Krogan, N.J. et al. (2005). Histone H2B ubiquitylation is associated with elongating RNA polymerase II. *Mol. Cell. Biol.* 25: 637–651.

134 Musacchio, A. and Salmon, E.D. (2007). The spindle-assembly checkpoint in space and time. *Nat. Rev. Mol. Cell Biol.* 8: 379–393.

135 Tighe, A., Johnson, V.L., Albertella, M., and Taylor, S.S. (2001). Aneuploid colon cancer cells have a robust spindle checkpoint. *EMBO Rep.* 2: 609–614.

136 Perez de Castro, I., de Cárcer, G., and Malumbres, M. (2007). A census of mitotic cancer genes: new insights into tumor cell biology and cancer therapy. *Carcinogenesis* 28: 899–912.

137 Gascoigne, K.E. and Taylor, S.S. (2008). Cancer cells display profound intra- and interline variation following prolonged exposure to antimitotic drugs. *Cancer Cell* 14: 111–122.

138 Keri, R., Patil, M., Brahmkhatri, V.P. et al. (2022). Copper (II)-β-cyclodextrin promoted Kabachnik–Fields reaction: an efficient, one-pot synthesis of α-aminophosphonates. *Top. Catal.* https://doi.org/10.1007/s11244-021-01556-4.

139 Sotillo, R., Hernando, E., Díaz-Rodríguez, E. et al. (2007). Mad2 overexpression promotes aneuploidy and tumorigenesis in mice. *Cancer Cell* 11: 9–23.

140 Adimule, V., Yallur, B.C., and Sharma, K. (2022). Studies on crystal structure, morphology, optical and photoluminescence properties of flake-like Sb doped Y_2O_3 nanostructures. *J. Opt.* 51: 173–183. https://doi.org/10.1007/s12596-021-00746-3.

141 Al-Ejeh, F., Simpson, P.T., Sanus, J.M. et al. (2014). Meta-analysis of the global gene expression profile of triple-negative breast cancer identifies genes for the prognostication and treatment of aggressive breast cancer. *Oncogenesis* 3: e124.

142 Carter, S.L., Eklund, A.C., Kohane, I.S. et al. (2006). A signature of chromosomal instability inferred from gene expression profiles predicts clinical outcome in multiple human cancers. *Nat. Genet.* 38: 1043–1048.

143 Privette, L.M., González, M.E., Ding, L. et al. (2007). Altered expression of the early mitotic checkpoint protein, CHFR, in breast cancers: implications for tumor suppression. *Cancer Res.* 67: 6064–6074.

144 Jeffery, J., Sinha, D., Srihari, S. et al. (2015). Beyond cytokinesis: the emerging roles of CEP55 in tumorigenesis. *Oncogene* 35: 683–690.

145 Daniels, M.J., Wang, Y., Lee, M., and Venkitaraman, A.R. (2004). Abnormal cytokinesis in cells deficient in the breast cancer susceptibility protein BRCA2. *Science* 306: 876–879.

146 Adimule, V., Yallur, B.C., Bhowmik, D. et al. (2021). Morphology, structural and photoluminescence properties of shaping triple semiconductor $Y_xCoO:ZrO_2$ nanostructures. *J. Mater. Sci. - Mater. Electron.* 32: 12164–12181. https://doi .org/10.1007/s10854-021-05845-2.

147 Adimule, V., Revaigh, M.G., and Adarsha, H.J. (2020). Synthesis and fabrication of Y-doped ZnO nanoparticles and their application as a gas sensor for the detection of ammonia. *J. Mater. Eng. Perform.* 29: 4586–4596. https://doi.org/10.1007/ s11665-020-04979-4.

148 Davoli, T. and de Lange, T. (2011). The causes and consequences of polyploidy in normal development and cancer. *Annu. Rev. Cell Dev. Biol.* 27: 585–610.

149 Shaikh, N.M., Adimule, V., Bagihalli, G.B. et al. (2022). A novel mixed Ag–Pd nanoparticles supported on SBA silica through [DMAP-TMSP-DABCO]OH basic ionic liquid for Suzuki coupling reaction. *Top. Catal.* https://doi.org/10.1007/ s11244-022-01586-6.

150 Adimule, V., Medapa, S., Rao, P.K., and Kumar, L.S. (2014). Synthesis of Schiff bases of 5-[5-(4-fluorophenyl)thiophen-2-yl]-1,3,4-thiadiazol-2-amine and its anticancer activity. *Int. J. Adv. Pharm. Sci.* 5 (1): 1761–1768.

151 Kloosterman, W.P., Tavakoli-Yaraki, M., van Roosmalen, M.J. et al. (2012). Constitutional chromothripsis rearrangements involve clustered double-stranded DNA breaks and nonhomologous repair mechanisms. *Cell Rep.* 1 (6): 648–655.

152 Liu, P., Erez, A., Nagamani, S.C.S. et al. (2011). Chromosome catastrophes involve replication mechanisms generating complex genomic rearrangements. *Cell* 146 (6): 889–903.

153 Crasta, K., Ganem, N.J., Dagher, R. et al. (2012). DNA breaks and chromosome pulverization from errors in mitosis. *Nature* 482 (7383): 53–58.

154 Groth, A., Corpet, A., Cook, A.J. et al. (2007). Regulation of replication fork progression through histone supply and demand. *Science* 318: 1928–1931.

155 Grabarz, A., Guirouilh-Barbat, J., Barascu, A. et al. (2013). A role for BLM in double-strand break repair pathway choice: prevention of CtIP/Mre11-mediated alternative nonhomologous end-joining. *Cell Rep.* 5: 21–28.

156 Hashimoto, Y., Ray Chaudhuri, A., Lopes, M., and Costanzo, V. (2010). Rad51 protects nascent DNAfromMre11-dependent degradation and promotes continuous DNA synthesis. *Nat. Struct. Mol. Biol.* 17: 1305–1311.

157 Schlacher, K., Christ, N., Siaud, N. et al. (2011). Double-strand break repair independent role for BRCA2 in blocking stalled replication fork degradation by MRE11. *Cell* 145: 529–542.

158 Garcia-Higuera, I., Taniguchi, T., Ganesan, S. et al. (2001). Interaction of the Fanconi anemia proteins and BRCA1 in a common pathway. *Mol. Cell* 7: 249–262.

159 Wang, X., Andreassen, P.R., and D'Andrea, A.D. (2004). Functional interaction of monoubiquitinated FANCD2 and BRCA2/FANCD1 in chromatin. *Mol. Cell. Biol.* 24: 5850–5862.

160 Adimule, V., Kerur, S.S., Chinnam, S. et al. (2022). Guar gum and its nanocomposites as prospective materials for miscellaneous applications: a short review. *Top. Catal.* https://doi.org/10.1007/s11244-022-01587-5.

161 Pathania, S., Nguyen, J., Hill, S.J. et al. (2011). BRCA1 is required for postreplication repair after UV-induced DNA damage. *Mol. Cell* 44: 235–251.

162 Lomonosov, M., Anand, S., Sangrithi, M. et al. (2003). Stabilization of stalled DNA replication forks by the BRCA2 breast cancer susceptibility protein. *Genes Dev.* 17: 3017–3022.

163 Petermann, E., Orta, M.L., Issaeva, N. et al. (2010). Hydroxyurea-stalled replication forks become progressively inactivated and require two different RAD51-mediated pathways for restart and repair. *Mol. Cell* 37: 492–502.

164 Schlacher, K., Wu, H., and Jasin, M. (2012). A distinct replication fork protection pathway connects Fanconi anemia tumor suppressors to RAD51-BRCA1/2. *Cancer Cell* 22: 106–116.

165 Lossaint, G., Larroque, M., Ribeyre, C. et al. (2013). FANCD2 binds MCM proteins and controls replisome function upon activation of S phase checkpoint signaling. *Mol. Cell* 51: 678–690.

166 Pan, Q., Fang, Y., Xu, Y. et al. (2005). Downregulation of DNA polymerases κ, η, ι, and ζ in human lung, stomach, and colorectal cancers. *Cancer Lett.* 217: 139–147.

167 Buisson, R., Niraj, J., Pauty, J. et al. (2014). Breast cancer proteins PALB2 and BRCA2 stimulate polymerase η in recombination associated DNA synthesis at blocked replication forks. *Cell Rep.* 6: 553–564.

168 Nandi, S.S., Suryavanshi, A., Adimule, V., and Maradur, S.R. (2020). Semiconductor current- voltage characteristics of some novel perovskite ionic nanocomposites of $Sr_{0.5}$, $Cu_{0.4}$, $Y_{0.1}$ and $Sr_{0.5}$, $Mn_{0.5}$ and their electronic sensor applications. *AIP Conf. Proc.* 2274: 020006: https://doi.org/10.1063/5.0022453.

169 Rey, L., Sidorova, J.M., Puget, N. et al. (2009). Human DNA polymerase η is required for common fragile site stability during unperturbed DNA replication. *Mol. Cell. Biol.* 29: 3344–3354.

170 Shaikh, N.M., Sawant, A.D., Bagihalli, G.B. et al. (2022). Highly active mixed Au–Pd nanoparticles supported on RHA silica through immobilised ionic liquid for Suzuki coupling reaction. *Top. Catal.* https://doi.org/10.1007/s11244-021-01547-5.

171 Lin, J.R., Zeman, M.K., Chen, J.Y. et al. (2011). SHPRH and HLTF act in a damage specific manner to coordinate different forms of postreplication repair and prevent mutagenesis. *Mol. Cell* 42: 237–249.

172 Moinova, H.R., Chen, W.-D., Shen, L. et al. (2002). *HLTF* gene silencing in human colon cancer. *Proc. Natl. Acad. Sci. U.S.A.* 99: 4562–4567.

173 Hossain, M. and Stillman, B. (2012). Meier–Gorlin syndrome mutations disrupt an Orc1 CDK inhibitory domain and cause centrosome reduplication. *Genes Dev.* 26: 1797–1810.

174 Smith, L., Plug, A., and Thayer, M. (2001). Delayed replication timing leads to delayed mitotic chromosome condensation and chromosomal instability of chromosome translocations. *Proc. Natl. Acad. Sci. U.S.A.* 98: 13300–13305.

175 Tischkowitz, M., Xia, B., Sabbaghian, N., and Foulkes, W.D. (2007). Analysis of *PALB2/FANCN* associated breast cancer families. *Proc. Natl. Acad. Sci. U.S.A.* 104: 6788–6793.

176 Vinayak, A., Sudha, M., Lalita, K.S., and Kumar, R.P. (2015). Synthesis, characterization and cytotoxic evaluation of novel derivatives of 1-[2-(aryl substituted)-5-(4′-fluoro-3-methyl biphenyl-4-yl)-[1,3,4]oxadiazole-3-yl]-ethanone. *Arch. Appl. Sci. Res.* 7 (5): 4–8.

177 Liu, B., Larsson, L., Caballero, A. et al. (2010). The polarisome is required for segregation and retrograde transport of protein aggregates. *Cell* 140 (2): 257–267.

178 McMurray, M.A. and Gottschling, D.E. (2003). An age-induced switch to a hyper-recombinational state. *Science* 301 (5641): 1908–1911.

179 Challa, M., Yallur, B.C., Ambika, M.R., and Adimule, V. (2022). Influence of nano particles on optical properties of Cu-MOFs. *Adv. Mater. Res..* Trans Tech Publications, Ltd. https://doi.org/10.4028/p-vn4hd4.

180 Di Micco, R., Sulli, G., Dobreva, M. et al. (2011). Interplay between oncogene-induced DNA damage response and heterochromatin in senescence and cancer. *Nat. Cell Biol.* 13: 292–302.

181 Sharma, P. and Sharma, R. (2021). Impact of COVID-19 on mental health and aging. *Saudi J. Biol. Sci.* 28: 7046–7053. https://doi.org/10.1016/j.sjbs.2021.07.087.

182 Prasanna, T., Wu, F., Khanna, K.K. et al. (2018). Optimizing poly(ADP-ribose) polymerase inhibition through combined epigenetic and immunotherapy. *Cancer Sci.* 109: 3383–3392.

183 Batakurki, S.R., Adimule, V., Pai, M.M. et al. (2022). Synthesis of Cs–Ag/Fe$_2$O$_3$ nanoparticles using *Vitis labrusca* Rachis extract as green hybrid nanocatalyst for the reduction of arylnitro compounds. *Top. Catal.* https://doi.org/10.1007/s11244-022-01593-7.

184 Maya Pai, M., Yallur, B.C., Batakurki, S.R. et al. (2022). Synthesis and catalytic activity of heterogenous hybrid nanocatalyst of copper/palladium MOF, RIT 62-Cu/Pd for Stille polycondensation of thieno[2,3-*b*]pyrrol-5-one derivatives. *Top. Catal.* https://doi.org/10.1007/s11244-022-01618-1.

185 Liu, Z., Sun, Q., and Wang, X. (2017). PLK1, a potential target for cancer therapy. *Transl. Oncol.* 10: 22–32.

186 Gyawali, B. and Prasad, V. (2016). Negative trials in ovarian cancer: is there such a thing as too much optimism? *ecancermedicalscience* 10: ed58.

187 De Carcer, G., Venkateswaran, S.V., Salgueiro, L. et al. (2018). Plk1 overexpression induces chromosomal instability and suppresses tumor development. *Nat. Commun.* 9: 3012.

188 Zeng, X., Sigoillot, F., Gaur, S. et al. (2010). Pharmacologic inhibition of the anaphasepromoting complex induces a spindle checkpoint-dependent mitotic arrest in the absence of spindle damage. *Cancer Cell* 18: 382–395.

189 Gad, H., Koolmeister, T., Jemth, A.-S. et al. (2014). MTH1 inhibition eradicates cancer by preventing sanitation of the dNTP pool. *Nature* 508: 215–221.

190 Huber, K.V., Salah, E., Radic, B. et al. (2014). Stereospecific targeting of MTH1 by (*S*)-crizotinib as an anticancer strategy. *Nature* 508: 222–227.

191 Bachmann, A.S. and Geerts, D. (2018). Polyamine synthesis as a target of MYC oncogenes. *J. Biol. Chem.* 293 (48): 18757–18769.

192 Kocaturk, N.M., Akkoc, Y., Kig, C. et al. (2019). Autophagy as a molecular target for cancer treatment. *Eur. J. Pharm. Sci.* 134: 116–137. https://doi.org/10.1016/j.ejps.2019.04.011.

193 Konotop, G., Bausch, E., Nagai, T. et al. (2016). Pharmacological inhibition of centrosome clustering by slingshot-mediated cofilin activation and actin cortex destabilization. *Cancer Res.* 76: 6690–6700.

194 Rebacz, B., Larsen, T.O., Clausen, M.H. et al. (2007). Identification of griseofulvin as an inhibitor of centrosomal clustering in a phenotype-based screen. *Cancer Res.* 67: 6342–6350.

195 Karna, P., Rida, P.C.G., Pannu, V. et al. (2011). A novel microtubule-modulating noscapinoid triggers apoptosis by inducing spindle multipolarity via centrosome amplification and declustering. *Cell Death Differ.* 18: 632–644.

196 Kawamura, E., Fielding, A.B., Kannan, N. et al. (2013). Identification of novel small molecule inhibitors of centrosome clustering in cancer cells. *Oncotarget* 4: 1763–1776.

3

Recent Advancement of Nanotherapeutics to Treat Breast Cancer

Devesh U. Kapoor[1], Rajat Goyal[3], Rajiv R. Kukkar[4] and Rupesh K. Gautam[2]

[1] *Dr. Dayaram Patel Pharmacy College, Bardoli, Gujarat, India*
[2] *Indore Institute of Pharmacy, Department of Pharmacology, IIST Campus, Opposite IIM Indore, Rau-Pithampur Road, Indore, 453331, Madhya Pradesh, India*
[3] *Maharishi Markandeshwar (Deemed to be University), MM College of Pharmacy, Mullana-Ambala, 133207, Haryana, India*
[4] *School of Pharmacy, Raffles University, Neemrana, Alwar, Rajasthan, India*

3.1 Introduction

Breast cancer is a public well-being issue and now it is the utmost recurrent tumor worldwide. It is a life-threatening disease that mostly affects women and is the major source of death among them. Awareness about breast tumors, public awareness, and progressions in breast imaging have made an encouraging impact on the detection and screening of breast tumors [1].

Breast cancer is the malignant cell growth that initiates from the breast cells at the innermost linings of the breast ducts or lobules that function in the supply of milk. Breast cancer is diagnosed at different stages, Stage 0 indicates that cancer cells are discovered in the breast ducts or lobules. The size of the tumor and the location, where the tumor cells have been spread, i.e. skin, chest wall, and lymph nodes adjacent to the breast, may be used to classify cancer at different stages including, Stage I, II, and III. The tumor cells have been spread to the other tissues or lymph nodes, which are further removed from the breast at Stage IV (advanced or metastatic stage). The expression of breast cancer markers, including estrogen receptor (ER), progesterone receptor (PR), human epidermal growth factor receptor 2 (EGFR2), and urokinase plasminogen activator (uPA) have been utilized to assess the development and fierceness of the disease [2].

Drug and Therapy Development for Triple Negative Breast Cancer, First Edition.
Edited by Pravin Kendrekar, Vinayak Adimule, and Tara Hurst.
© 2023 WILEY-VCH GmbH. Published 2023 by WILEY-VCH GmbH.

3.2 Pathophysiology of Breast Cancer

Breast tumor is a multifactorial chronic illness. Obesity and metabolic syndrome have long been associated with a higher risk of breast cancer. It was also discovered that estrogens and ER were linked to obesity and breast tumor. Higher blood cholesterol is a typical symptom of obesity and metabolic syndrome, and its role as a breast cancer risk factor is debatable [3]. Two key molecular targets have been recognized in the genesis of breast tumors. Estrogen receptor alpha (ERα), which is expressed in around 70% of the invasive carcinoma, is one of them. ERα is a steroidal hormone and transcription factor that stimulates the pathways of malignancy growth in breast tumor cells when triggered by estrogen. The expression of PR, a similarly related steroidal hormone, is also an indication of ER signaling. The epidermal growth factor 2 (ERBB2, i.e. HER2 or HER2/neu) is another target. It is a transmembrane receptor of tyrosine kinase in the family of EGFR that is augmented or overexpressed in around 20% of breast tumors and is allied to the deprived prognosis in the nonexistence of the systemic therapy [4, 5].

Breast cancer frequently begins from the ductal hyper-proliferation and progresses to the benign or metastatic carcinoma as a consequence of insistent stimulus via several oncogenic mediators. Breast cancer development is influenced via cancer microenvironments, i.e. macrophages and stromal effects. There are two hypothetical theories for the development of breast tumors including, cancer stem cell (CSC) theory and stochastic theory. The CSC theory states that all the tumor subtypes are derivatives of identical stem cells or transit-amplifying cells such as progenitor cells. Whereas, the stochastic theory states that each tumor subtype derives from a single cell type (i.e. stem cell, differentiated cell, or progenitor cell). The arbitrary mutations may progressively accumulate in any breast cells, ultimately leading to their renovation into cancer cells [6].

3.3 Classification of Breast Cancer

Breast tumor is an extremely heterogeneous illness because of their numerous morphological characteristics, unpredictable clinical prognosis, and responses to various therapeutic options. As a result, a clinically evocative classification of the disease must be developed that is scientifically valid, clinically beneficial, and broadly reproducible [7]. The use of gene expression patterns to classify breast cancer has led to attempts to describe clinically significant subgroups with correlations to survival, disease relapse, metastatic spread site preference, and response to chemotherapy [8].

To aid oncologic decision-making, the classification of cancer strives to provide an accurate diagnosis of the illness and prognosis of the tumor behavior. The traditional classification of the breast tumor is based on the clinically pathological criteria and the evaluation of tedious biomarkers. Invasive breast cancers are classified into over 20 different histologic categories. IDC-NST is the most frequent, accounting for about 70–80% of all invasive malignancies, followed by invasive lobular carcinomas (ILC). Mucinous, micropapillary, cribriform, papillary, medullary, tubular, metaplastic, and apocrine carcinomas are among the less common histologic categories [9].

According to the region from which it grows, breast cancer is characterized as noninvasive and invasive. When a carcinoma is curbed within the ducts and lobules, it is called noninvasive carcinoma (*in situ*), whereas, when cancer spreads to the surrounding connective tissues and metastasizes to the other parts of the body, it is called an invasive carcinoma. Furthermore, these carcinomas are categorized as ductal or lobular carcinomas depending upon where the tumor forms. Ductal carcinomas account for about two-thirds of the cases of breast tumors and are caused by epithelial cells in the ducts, while the lobular carcinomas are triggered via epithelial cells in lobules and account for one-third of the breast cancer instances [10]. The classification of breast cancer [11] is depicted in Figure 3.1.

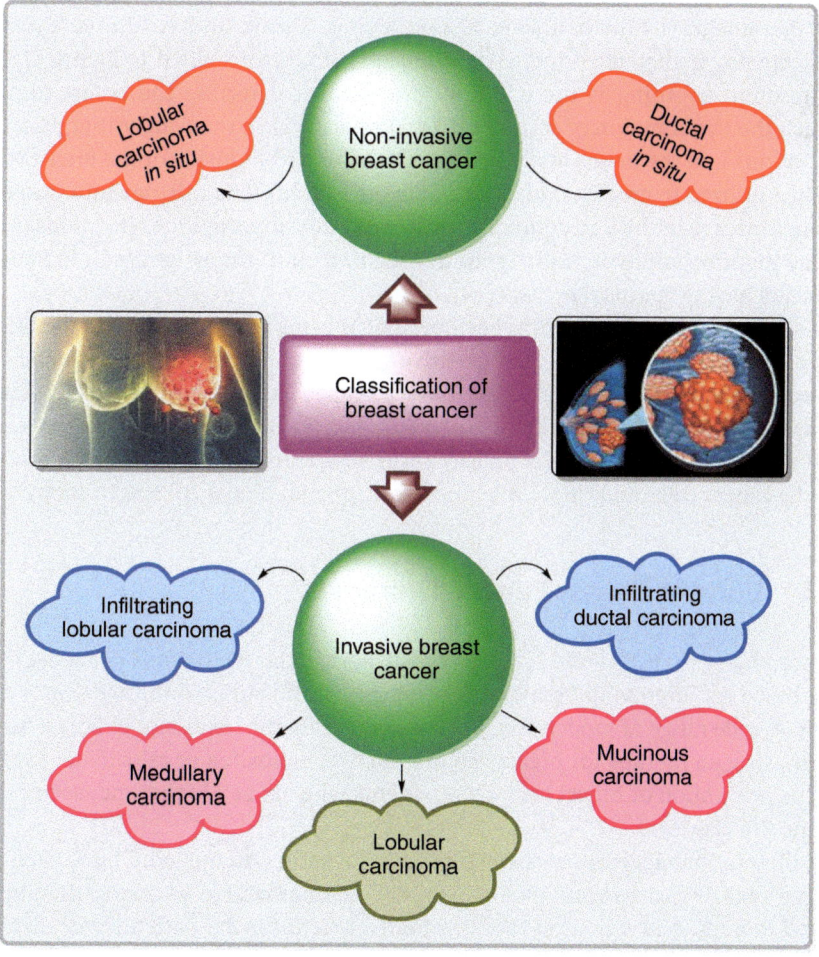

Figure 3.1 Classification of breast cancer.

3.4 Techniques for Breast Cancer Detection

The clinical antiquity of the patients with breast tumors is used to investigate the threat of malignancy and demonstrate the presence or absence of breast illness signs or symptoms [12]. Both personal and family histories should be thoroughly researched. Patients should be scrutinized for specific manifestations like pain in the breast area, frequent loss in weight, pain in bones, tiredness, and discharge from nipples [13].

X-ray mammography, MRI (magnetic resonance imaging), and ultrasound screening are the utmost popular clinical imaging and detection modalities utilized for breast cancer detection. Mammography is one of the USFDA-approved techniques utilized for the screening of breast tumors in women; with no prior symptoms. Mammography, on the other hand, has been imperiled to the enormous analysis because of the relatively higher rate of false-positive and false-negative readings, which can be emotionally draining for patients [14]. Moreover, women who practice the mammography methodology as a screening assessment have an increased risk of acquiring malignancy due to the ionization radiations allied to X-rays [15, 16]. Ultrasound scanning is the additional method of detecting the breast tumor in which the transmission of sound waves takes place via a transducer that directs the pulsations into the breast and perceives the echoes from inside of the breast, which is then utilized to create an ultrasound image [17, 18]. The MRI technique is much more sensitive in the detection of invasive and tiny abnormalities in comparison to X-ray mammography and ultrasound screening, and it may be used efficiently in patients having compact breasts [19].

Each detection technique has been evaluated using four parameters such as accuracy, specificity, sensitivity, and positive predictive value. Electromagnetic (EM) detection techniques, in particular, have attracted a lot of consideration. The EM detection of breast cancer relies on the following: (i) manifestation of the disparity in EM assets among the tissues of breast cancer, (ii) exclusive electrical signals produced by the cancer cells, and (iii) NMR (nuclear magnetic resonance) spectroscopy [20].

3.5 Current Breast Cancer Therapies

According to WHO (World Health Organization), the regulations of breast cancer are based on improving breast cancer outcomes and survival through early detection. A wide range of modern medications is used in the treatment of breast tumors. Anti-estrogen therapy for breast tumors, such as tamoxifen or raloxifene, may help in the prevention of breast tumors in patients who are at a higher menace of developing the disease.

Different management strategies are employed in patients who have been diagnosed with breast cancer, such as surgery, chemotherapy, hormone therapy, targeted therapy, and radiation therapy. Management for the patients with detached metastasis are frequently aimed at enlightening their eminence of life and endurance rate [1].

3.6 Nanotherapeutics for Breast Cancer Treatment and Metastasis

3.6.1 Nanodiamonds (NDs)

These are also called ultra-crystalline diamonds or ultra-dispersed diamonds with noteworthy properties of chemical inertness. These are carbon building blocks of nano-size having diverse chemical, optical, biological, and mechanical properties that make them appropriate carriers for drug delivery against sites of breast cancer. Here we have discussed different delivery systems comprising multifunctional nanodiamonds (NDs) against diverse models of breast cancer.

3.6.2 Intrinsic Toxicity Reduction

Adimule et al., fabricated the NDs delivery system of polyglycerol equipped with doxorubicin (DOX). The researchers evaluated these NDs against 4T1 TNBC (triple-negative breast cancer) murine models which are resistant to DOX. The researchers observed that the fabricated delivery system of NDs exhibited noteworthy anticancer activity. The researchers revealed that NDs exhibited an enhanced pharmacodynamics profile as compared to free DOX [21]. Various ND-based delivery systems against breast cancer models are described in Table 3.1.

Table 3.1 Various ND-based delivery systems against breast cancer models.

Nanodiamonds types	Active pharmaceutical ingredients	Outcomes	References
Pristine NDs	Paclitaxel (PTX) and cetuximab (CET)	PTX-NDs exhibited excellent attenuation of cell viability in MCX-7, MDA-MB-231, and BT474 cell lines as compared to PTX. The PTX-CET-NDs increased the cellular uptake and apoptosis in cell lines of MDA-MB-231 in comparison to PTX but there is no effect against MCF-7 cell lines. The tumor was significantly inhibited in xenograft murine models of MDA-MB-231 when treated with PTX-CET-NDs	[22]
Oxidated gel NDs	Protamine, miR-34a, folic acid	The NDs significantly delivered the miR-34a and considerably suppressed the tumor growth against MDA-MB-231 cell lines and xenograft MDA-MB-231 models. The NDs were found suitable to deliver miR-34a in triple-negative breast cancer (TNBC) therapy	[23]
Water-soluble NDs	PTX	The cytotoxicity efficiency of PTX was enhanced with NDs as compared to free PTX. The 98% PTX was released in a sustained manner at 70 h with NDs	[24]
Pristine NDs	Melittin	The fabricated Melittin-NDs-enhanced cell viability, apoptosis, and necrosis decline as compared to free Melittin against MDA-MB-231 and MCF-7 cell lines	[25]

Li et al., fabricated NDs of PEGylated loaded with DOX using sodium acetate as medium (PEG-DOX-NDs-NaAC). The researchers observed that fabricated NDs showed superior loading efficacy (99%) along with very less leakage of drugs (5%). Moreover, the outcome exhibited that PEG-DOX-NDs-NaAC mainly gathers in cancer cells as compared to normal cells. These NDs demonstrated significant potential to kill the cancer cell. The NDs also inhibit and proliferate the cancer cells and alter the cell cycle as compared to free drugs. The researchers concluded that PEG-DOX-NDs-NaAC could be a good alternative to deliver the drug at the site of action in breast cancer [26].

3.6.3 Diminishing Chemoresistance (CR)

Chemoresistance (CR) is a ubiquitous issue responsible for metastatic breast cancer and is a chief factor in diminishing efficiencies in TNBC chemotherapy. The mechanism of CRs for breast cancer is complicated due to their heterogeneous nature. The nanocarriers are playing a vital role in reducing chemoresistance in breast cancer [27]. Toh et al., prepared a complex of NDs of mitoxantrone (MTHX), which exhibited excellent properties to deliver the drug against TNBC MDA-MB-231 cell line [28].

Yuan et al., fabricated an ND conjugate of polyglycerol loaded with DOX (PG-DOX-NDs) and evaluated its anticancer efficiencies against TNBC cell lines. The researchers observed that key immune cells of tumor-bearing mice exhibited better tolerance with PG-DOX-NDs as compared with DOX severe toxicity. Furthermore, IL-6 and P-gp upregulation were not significantly induced with PG-DOX-NDs, which are the chief mediators of DOX chemoresistance in cells of 4T1. Moreover, the tumor-derived granulocyte colony-stimulating factor (G-CSF) is downregulated with PG-DOX-NDs. The investigators found that myeloid-derived suppressor cells, which are key effectors for cancer-related systemic immunosuppression suppressed by PG-DOX-NDs. The 4T1 cell induction was encouraged by PG-DOX-NDs, which induce the damage-related molecular patterns emission. These patterns stimulate the tumor immune microenvironment by activating immune effector cells, which were associated with an anticancer activity like dendritic cells, macrophages, and lymphocytes in tissues of the tumor. The researchers concluded that PG-DOX-NDs are a significant cytostatic agent with excellent host tolerance and have the potential to evade chemoresistance and reverse cancer-associated immunosuppression in the cancer microenvironment and at the systemic level too in TNBC [29].

3.6.4 Delivery of Combination Therapeutics Through NDs

The fast approval is given by the Food and Drug Administration for different combinational medicines against different kinds of breast cancers. The recent approval is given to the combination of nab-paclitaxel (N-PTX) with atezolizumab (ATZ), which is the very first combination of antineoplastic anticancer immunotherapy against programmed death-ligand (PD-L1) receptor-positive TNBC [30].

3.7 Polymer-Based Nanoparticles (PBNPs)

PLGA is a biodegradable polymer employed in various biomedical products and a very vital carrier to prepare drug-loaded NPs. Mollod et al. fabricated and evaluated PLGA NPs for delivery of methotrexate (MTX) and CUR. The researchers observed that these PLGA NPs exhibited excellent cytotoxicity as compared to MTX and CUR alone. Furthermore, the co-delivery of MTX and CUR made a synergistic effect and diminished the breast cancer progression shown by *in vivo* results. The researchers concluded that fabricated PLGA NPs effectively deliver the MTX and CUR [31].

Yang et al. [32], developed PLGA NPs for co-delivery of widely chemotherapeutic agents PTX and CUR against the breast cancer stem cell (BCSC). The surface of hydrophobic PLGA NPs modified with lipids. The researchers found that bulk tumor cells were eliminated by PTX; however, PLGA NPs hindered the growth of cancer cells by specifically targeting the CSCs [33].

Muntimadugu et al., formulated PLGA NPs loaded with Salinomycin (SLMY) and PTX, with hyaluronic acid (HALC) by employing an emulsion solvent diffusion approach along with a cationic stabilizing agent against MDA-MB-231. The PTX is aimed to eradicate cancer cells however SLMY is used to destroy the cancer stem cells [34].

Amirsaadat et al., formulated PLGA-PEG NPs loaded with Silibinin (SBN) and Metformin (MFM) and evaluated their antitumor efficiency against human breast cancer cells (T47D). The researchers observed identical nanosize distribution in fabricated NPs. In addition, PLGA-PEG NPs-SBN/MFM showed synergistic cytotoxicity against T47D, confirmed by median effect examination and MTT assay. Furthermore, the investigators observed that NPs loaded with dual drugs modify the Bax, Caspase 3,9, Bcl-2, and hTERT expression as compared to NPs loaded with single drugs. The researchers concluded that PLGA-PEG NPs-SBN/MFM might be a good alternative to treat breast cancer in humans [35].

Chen et al., fabricated a synergistic nano-drug delivery system to increase the treatment efficacy for bone metastatic breast cancer. The delivery system consists of PLGA as lipophilic core, conjugated with folic acid which is coated with D-α-tocopheryl polyethylene glycol succinate (TPGS) modified with alendronate (ALT) to deliver the PTX efficiently. The TPGS/PLGA-ALT-PTX NPs exhibited excellent binding affinity toward hydroxyapatite, present in bone tissues. Moreover, fabricated NPs exhibited dual-target delivery of PTX toward bone matrix and folate receptor overexpressing 4T1 tumors. The researchers revealed that TPGS/PLGA-ALT-PTX NPs accumulated considerably in bone metastases and diminished the growth of lung metastasis and 4T1 tumor, resulting in considerable betterment of mice's rate of survival. Additionally, the loss of bone and bone destruction were also retarded in tumor-bearing mice and side effects on normal cells were declined with TPGS/PLGA-ALT-PTX NPs. The investigators concluded that fabricated NPs offer noteworthy results to treat bone metastatic breast cancer effectively [36].

3.8 Inorganic Nanoparticles (IONPs)

The therapeutic and diagnostic use of inorganic nanoparticles (IONPs) has enhanced significantly in the field of breast cancer. The different receptors of breast cancer cell markers for example human epidermal growth factor-2 (HER2), gastrin-releasing peptide (GRP), and CD44 are employed along with magnetic NPs for breast cancer imaging [37].

Lakkakula et al., fabricated gold nanoparticles (AuNPs) of 5-Fluorouracil (5-FU), complexed with cyclodextrin to enhance the solubility and extend the time of circulation of 5-FU. The researchers observed that 5-FU-AuNPs exhibited significant inhibition against MCF-7 and MDA-MB-231 breast cancer cell lines. In addition, this inhibition was the result of increased EPR effect and permeability of specific delivery of 5-FU-AuNPs at the site of the tumor. The investigators concluded that this novel system has the potential to deliver the lipophilic drugs at the tumor site effectively [38].

Gajria et al. [39], fabricated superparamagnetic iron oxide nanoparticles (SPIONPs) conjugated with porphyrin (PP) loaded with Trastuzumab (TRZ). The TRZ was employed as targeting ligand and SPIONPs are used as MRI contrast agents. The fabricated NPs significantly declined the MCF-7 cells. Additionally, *in vitro* PP-SPIONPs-TRZ photothermal ablation showed a noteworthy decline in MCF-7 cells (75%) after 15 minutes of the highest concentration of Fe ($1 \, mg \, ml^{-1}$). The researchers concluded that fabricated NPs could be a favorable bimodal nano-carrier to diagnose and treat breast cancer by employing photothermal and MRI therapy [40].

Samimi et al., fabricated carbon QDs equipped with nitrogen by using a hydro-thermal approach. The researchers have fabricated QDs with a targeting agent (Quininc acid, QA) against the MCF-7 breast cancer cell line. Furthermore, by employing electrostatic interaction these QDs are loaded with Gemcitabine (GCT). The investigators have performed biodistribution studies as well as *in vivo* studies on breast cancer cell line creation in mice. The researchers observed that QA-GCT-CQDs showed excellent luminescent properties and significant antitumor accumulation properties against MCF-7 cancer cell lines. The researchers concluded that fabricated QDs have the potential to treat breast cancer effectively [41].

Bloise et al., fabricated gold nanospheres (AuNSPs) coated with biodegradable and biocompatible polyamidoamine (AGMA1-SH) to deliver the TRZ monoclonal antibody (mAB) against HER2 positive breast cancer. *In vitro* studies revealed significant and specific uptake of fabricated AGMA1-SH/AuNSPs by SKBR-3 and over-expressing cells of HER2. The fabricated AuNSPs of TRZ exhibited excellent efficacy is declining the antiapoptotic components and downregulation of survival proliferation pathways. The researchers concluded that AuNSPs have the potential for selective targeting, which is imperative for breast cancer treatment [42].

DnD is the nanosize structure having three-dimensional nature, effectively dealing with resistance and toxicity of anticancer agents against breast cancer. The DnD are nano-size macromolecules fabricated synthetically by using polymers. The DnD has various properties such as excellent dispersion ability, surface modification, and multivalency play an important for the successful targeted delivery of drugs [43]. The DnD can deliver the chemotherapeutic drugs on the surface and at the site of cavities of selected tumor sites. These nano agents exhibited significant loading efficiency and showed negligible adverse effects with targeted delivery of cytotoxic agents. There are numerous functional groups present at the surface of DnD that can be utilized for conjugation of genetic materials, drugs, and receptor ligands [44] also DnD has the excellent capability to encapsulate these different materials in the cavities of DnD [45].

Kheraldine et al., examined the cytotoxic effect of G4 and G6 polyamidoamine dendrimers (PAMAM DnD) against the different cancer cell lines such as SKBR3, ZR75, and HER2. The researchers found that the cationic type of PAMAM DnD effectively declines the viability of cancer cells. Furthermore, these fabricated DnD support the apoptosis of cells followed by apoptotic markers (Caspases-3, 8, and 9) upregulation and Bcl-2 downregulation. Additionally, the investigators revealed that this cationic DnD diminished the colony formation as compared to other PAMAM. The PAMAM DnD molecular pathway investigation disclosed that PAMAM DnD surges the expression of JNK1/2/2 while hindering the expression of ERL1/2. The researchers concluded that fabricated DnD exhibited noteworthy therapeutic potential against signaling pathways of ERK1/2, HER1/2, and JNK1/2/3 [46].

Torres-Pérez et al. [47], investigated PAMAM DnD equipped with D-glucose and MTX against MDA-MB-231, cell lines of TNBC. The researchers found that PAMAM-DnD-MTX-Glu diminished the cell viability by 25% which was greater than free MTX. The uptake investigation revealed that glycosylation enhanced the PAMAM-DnD-MTX-Glu internalization in tumor cells as compared to non-tumor cells. The investigators recapitulated that PAMAM-DnD-MTX-Glu inhibited TNBC cell lines significantly, and could be a significant approach against breast cancer [48].

Chittasupho et al., investigated the effect of DnD loaded with DOX against T47D and BT-549-Luc breast cancer cells with specific targeting to CXCR4. The researchers functionalized the surface of PAMAM DnD-DOX with LFC-131 peptide for easy recognition of CXCR4. The breast cancer cells bound to PAMAM-DnD-DOX-LFC-131 considerably increased the *in vitro* cellular toxicity. The PAMAM-DnD-DOX-LFC-131 showed a considerable decline in BT-549-Luc migration toward the chemoattractant. The researchers concluded that PAMAM-DnD-DOX-LFC-131 has the potential against breast cancer treatment and metastasis too [49].

LIPs are vesicles having phospholipid bilayers and are natural and synthetic. LIPs can carry both hydrophilic and hydrophobic drugs due to their bilayer structure. Due to their excellent biodegradability, biocompatibility and negligible toxicity are the most promising novel drug delivery system (NDDS) to deliver the drug in breast cancer treatment. There are various LIPs such as long-circulating LIPs, pH-sensitive LIPs, and thermosensitive LIPs that are widely used for breast cancer treatment [50].

Cao et al., investigated the sustained delivery of hydrophilic drug DOX through liposomes (LIPs) equipped with thermogel for breast cancer treatment. The researchers employed a film dispersion approach to fabricate LIPs. The investigators found that there was sustained release of DOX up to 11 days from LIPs-DOX-ThGel without any burst effect as compared to plain DOX-ThGel. In addition, LIPs-DOX-ThGel exhibited excellent anticancer effects against orthotopic breast cancer cell models along with negligible adverse effects [51].

Fu et al., fabricated LIPs for the separate encapsulation of emodin (EMD) and daunorubicin (DANB), modified with arginine$_8$–aspartic acid–glycine (A$_8$–A–G) for the treatment of highly invasive breast cancer cells (MDA-MB-435S). The researchers formulated two targeted LIPs for inhibition of tumor metastasis and destroying vasculogenic mimicry channels (VMC). The amalgamation of these two targeted liposomes exhibited significant toxicity against MDA-MB-435S cells and effectively diminished the development of metastasis tumor cells and VMC channels. The investigators also observed from action mechanism studies that A$_8$–A–G modified DANB-LIPs and A$_8$–A–G-EMD-LIPs downregulated different metastasis-related proteins such as HIF-α, TGF-β1, VE-cad, and MMP-2. These investigations also revealed that chemotherapeutic drugs significantly accumulated at cancer cell lines showed a considerable anticancer effect with these fabricated liposomes. The researchers concluded that targeted LIPs of EMD and DANB have the potential to treat invasive breast cancer effectively [52].

3.9 Hydrogels (HGLs) and Microbubbles (MBs)

HGLs are an emerging drug carrier for breast cancer therapy due to their noteworthy biodegradability, biocompatibility, and lesser toxicity as compared to other nanocarriers. The smart HGs also have the potential to respond to the stimuli present in the environment such as light, heat, pH, and ultrasound. Due to these properties, HGs are widely used for controlled drug release and *in situ* gelation. The HGs as carriers exhibit lesser adverse effects as compared to systemic chemotherapy and also have the potential for sustained drug delivery at tumor sites [53].

Chen et al., developed thermosensitive hydrogels (THGs) for sustained injectable delivery of Herceptin (HCT), which is an antibody targeting HER2. The researchers used THGs for the prevention of local relapse of HER2+ breast cancer, also diminishing systemic adverse effects mainly cardiotoxicity. The intratumoral accumulation of antibodies was enhanced by single hypodermic administration of HCT-THGs at the site of the tumor, revealed by *in vivo* distribution investigation performed on KBR-3 tumor-bearing mice. The investigators found excellent antitumor efficacy with hydrogel formulation as compared to hypodermic injection of HCT solution. Moreover, the sustained release of HCT through hydrogel also prevents cardiotoxicity. The researchers concluded that HCT-THGs offer a good alternative for the prevention of relapse of HER2+ breast cancer [54].

Fathi et al., fabricated hydrogels (HGLs) which are pH-responsive and thermosensitive, loaded with DOX for breast cancer therapy. The HGs were formulated by mixing poly(N-isopropylacrylamide-*co*-itaconic acid) with chitosan by employing ionic crosslinking with glycerophosphate. The researchers found through MTT cytotoxic investigation that HGs are cytocompatible in nature and cytotoxicity was negligible against MCF cells. The diamidino-2-phenylindole (DAPI) staining revealed that tumor cells were significantly lower in numbers when treated with HGs-DOX as compared to drug-free HGs and control [55].

The microbubbles (MB) are very small bubbles filled with gas, having a size range between 0.3 and 5 μm, extensively employed as contrast agents for medical imaging. In addition, they are also used as carriers to deliver the drug to a specific site of action. The core area of MB is a gas, fenced by an outer shell having a combination of proteins, lipids, polymers, and lipoproteins. The researchers are employing MB to target delivery because a small quantity of chemotherapeutics agent is enough to treat. The MB with the ultrasound (ULS) has been widely employed to enhance the chemotherapy treatment in monolayer models and *in vitro* cell suspension. An effective approach known as ULS-MB destruction is very significant in crossing the drugs against the vascular wall and cell membrane.

Hong et al., investigated the growth diminishing and antiangiogenic impact of apatinib (ATN) and ULS-MB destruction (ULS-MB-D) against TNBC. The researchers found that angiogenesis and growth of tumors were considerably diminished with ULS-MB-D-ATN as compared to ATN alone. In addition, expression of VEGF, tumor volume, microvessel density (MvD), and parameters of contrast increased ultrasonography (CIUL) were considerably lesser as compared to ATN alone. However, there was a negligible difference found in body weight, parameters of blood, and rate of heterogeneous vascular positivity with both the treatments. The investigators concluded that ULS-MB-D-ATN has the potential to employ antiangiogenic effects, which are synergistic and also exert growth-diminishing effects against TNBC with negligible adverse effects [56].

Mishra et al. [57], fabricated 3D breast cancer spheroid models (MDA-MB-231) and examined the efficiency of ultrasound (ULS) and MB in escalating drug penetration and drug distribution in the spheroid. The investigators revealed that per cell DOX uptake was enhanced by 25% and there was 1.2 times more DOX uptake found in total DOX fluorescence due to ULS-MB. In addition, there was no considerable difference in the size of the long-term spheroid with ULS-MB-DOX in comparison to DOX alone. However, the dispersion of spheroids was more significant with ULS-MB-DOX as compared to simple DOX. Furthermore, the researchers observed from the tumor model that the effect of chemotherapeutic molecules was significant in 3D tumor spheroids of ULS-MB and penetration was noteworthy along with a considerable accumulation of chemotherapeutics drugs in deeper regions of the spheroid [58].

Al-Radadi et al. [59], formulated AuNPs by using flaxseed aqueous extract. The method is eco-friendly, cheap, and energy efficient. The researchers observed that bioactive material was present in fabricated NPs. In addition, FLX-AuNPs exhibited excellent inhibition against cell lines of breast adenocarcinoma. Moreover, these NPs showed good antioxidant and anticoagulation properties. The researchers recapitulated that FLX-AuNPs enhance the death rate of breast cancer cell lines, but clinical trials have to be performed on humans to justify the outcomes [60].

Khodashenas et al., formulated various sizes and shapes of AuNPs and gold nanorods (AuNRs) loaded with MTX, coated with biopolymer such as gelatin (GTN). The %EE of AuNPs was superior as compared to AuNRs. The researchers observed that the maximum drug release rate was from AuNRs-GTN and it exhibited significant antitumor activity for the MCF-7 cell line. The investigators concluded that AuNPs/AuNRs-GTN could be employed to deliver MTX, which will be very effective against breast carcinoma [61].

Bandyopadhyay et al., studied the effect of AuNPs equipped with chitosan (CH) against hormone-resistant MDA-MB-231 and hormone-responsive MCF-7 breast cancer cell lines. The researchers found that the rate of mortality of these two cancer cell lines was significant with 11 parts per billion (ppb) for 48 hours of treatment without showing any adverse effect on normal human lymphocytes. In addition, apoptotic cell death and nuclear fragmentation were seen by the investigators. There was disturbing homeostasis of certain cellular elements such as Cu, K, Sulfur, Se, and Ca supporting the finding of researchers. The investigators concluded that AuNPs-CH is considerably effective against the above two breast cancer cell lines, and could be suitable candidates against breast cancer in the future [62].

3.10 Recent Patents of Nanotherapeutics for Breast Cancer Treatment

Desai and Patrick, received CA patent no 2672618C for the development of NPs consisting PTX albumin and other chemotherapeutic agents or radiation to treat breast cancer. The therapy was based on estrogen and progesterone hormone receptor status [63].

In 2020, the National Centre of Nanoscience China got approval for CN patent no. 107970226B to treat the TNBC. The investigators employed a nanocarrier loaded with Chlorin E6 and NVP-BEZ235. The carrier surface is coated with lecithin and distearoylphosphatidylethanolamine-PEG [64].

Recently last year another patent was granted to Jinan University where an investigator fabricated gold nanorods and halloysite nanotubes equipped with DOX, bovine serum albumin, and folic acid to treat breast cancer. The researchers employed chemotherapy along with photothermal therapy [65].

3.11 Clinical Trials of Nanotherapeutics for Breast Cancer

The clinical trials described below include the delivery of drugs using different nanocarriers against breast cancer of TNBC. The description of nanocarriers employed for clinical trials of breast cancer is mentioned in Table 3.2.

Table 3.2 Nanocarriers employed for clinical trials of breast cancer.

Nano carriers	Design of study	Aim of study	Side effects	Outcome of trials	References
PTX-albumin NPs	Phase I trial	Investigated the effect of fabricated formulation against patients suffering from breast cancer stage I, II, and III	Fever, neuropathy, vomiting, and diarrhea	The combination exhibited good tolerance with the toxicity of Grade 3 and has shown good efficacy	[66]
NK-105-PTX-NPs	Phase III trial	Examined the impact of NPs of NK-105 (PTX) in the treatment of metastatic or recurrent breast cancer	Myalgia, nausea, fatigue, alopecia, leukopenia, neutropenia, stomatitis, discoloration of nail	The peripheral sensory neuropathy (PSN) was significantly favorable with NK-105-PTXNPs ($P < 0.0002$)	[67]
LIPs of irinotecan (IRT)	Phase I trial	Investigated the impact of IRT-LIPs on metastatic breast cancer (MBC)	Nausea, diarrhea, fatigue, asthenia, hypokalemia	The significant anticancer activity exhibited by the LIPs-IRT against patients suffering heavily from MBC	[68]
Albumin NPs loaded with PTX	Phase III trial	Investigated the albumin NPs equipped with PTX against MBC	Nausea, neutropenia, flushing, hyperglycemia	The NPs exhibited superior efficacy and significant safety profile as compared to PTX against patients suffering from MBC. The enhanced therapeutic index and corticosteroid premedication elimination make albumin-PTX NPs an effective delivery system against MBC	[69]

3.12 Conclusion and Future Perspectives

Nanotherapeutics along with nano-drug delivery systems have been significantly employed for medical diagnosis and therapy of breast cancer due to accurate targeting of tumor cells, diminished toxicity of the drug, and enhanced drug availability. The NDDS become a vital choice to deliver the chemotherapeutic agents against breast cancer cells. In addition, the simultaneous delivery of siRNA along with chemotherapeutic drugs by employing NDDS offers an effective platform for breast cancer therapy. The clinical trials and toxicology investigation for NDDS are crucial for a better understanding of *in vivo* patterns and toxicity profiles. The NDDS application is mainly limited to *in vitro* cell examinations and *in vivo* experiments in breast cancer research, several challenges are present in the NDDS clinical application with humans. The targeted delivery of NDDS has to be improved due to the complexity and different barriers present in the human body. There are some other concerns such as *in vivo* safety, stability; efficacy of NDDS that should be given more attention. The development of new materials and escalation of breast cancer research will surge the importance of NDDS in breast cancer therapy.

References

1 Akram, M., Iqbal, M., Daniyal, M., and Khan, A.U. (2017). Awareness and current knowledge of breast cancer. *Biol. Res.* 50 (1): 1–23. https:/doi.org/10.1186/s40659-017-0140-9.

2 Kamaruzman, N.I., Aziz, N.A., Poh, C.L., and Chowdhury, E.H. (2019). Oncogenic signaling in tumorigenesis and applications of siRNA nanotherapeutics in breast cancer. *Cancers* 11 (5): 632. https://doi.org/10.3390/cancers11050632.

3 Nazih, H. and Bard, J.M. (2020). Cholesterol, oxysterols and LXRs in breast cancer pathophysiology. *Int. J. Mol. Sci.* 21 (4): 1356. https://doi.org/10.3390/ijms21041356.

4 Batakurki, S.R., Adimule, V., Pai, M.M. et al. (2022). Synthesis of Cs–Ag/Fe$_2$O$_3$ nanoparticles using *Vitis labrusca* Rachis extract as green hybrid nanocatalyst for the reduction of arylnitro compounds. *Top. Catal.* https://doi.org/10.1007/s11244-022-01593-7.

5 Mitri, Z., Constantine, T., and O'Regan, R. (2012). The HER2 receptor in breast cancer: pathophysiology, clinical use, and new advances in therapy. *Chemother. Res. Pract.* https://doi.org/10.1155/2012/743193.

6 Adimule, V., Nandi, S.S., Yallur, B.C. et al. (2021). Enhanced photoluminescence properties of Gd$_{(x-1)}$Sr$_x$O:CdO nanocores and their study of optical structural and morphological characteristics. *Mater. Today Chem.* 20: 00438.

7 Adimule, V., Nandi, S.S., Yallur, B.C. et al. (2021). Optical structural and photoluminescence properties of Gd$_x$SrO:CdO nanostructures synthesized by co precipitation method. *J. Fluoresc.* 31 (2): 487–499.

8 Lam, S.W., Jimenez, C.R., and Boven, E. (2014). Breast cancer classification by proteomic technologies: current state of knowledge. *Cancer Treat. Rev.* 40 (1): 129–138. https://doi.org/10.1016/j.ctrv.2013.06.006.

9 Tsang, J. and Tse, G.M. (2020). Molecular classification of breast cancer. *Adv. Anat. Pathol.* 27 (1): 27–35. https://doi.org/10.1097/PAP.0000000000000232.

10 Chowdhury, P., Ghosh, U., Samanta, K. et al. (2021). Bioactive nanotherapeutic trends to combat triple negative breast cancer. *Bioact. Mater.* 6 (10): 3269–3287. https://doi.org/10.1016/j.bioactmat.2021.02.037.

11 Sharma, G.N., Dave, R., Sanadya, J. et al. (2010). Various types and management of breast cancer: an overview. *J. Adv. Pharm. Technol. Res.* 1 (2): 109. PMID: 22247839.

12 Kerlikowske, K., Grady, D., Barclay, J. et al. (1993). Positive predictive value of screening mammography by age and family history of breast cancer. *JAMA* 270 (20): 2444–2450. https://doi.org/10.1001/jama.1993.03510200050031.

13 Adimule, V., Revaiah, R.G., Nandi, S.S., and Jagadeesha, A.H. (2021). Synthesis characterization of Cr doped TeO_2 nanostructures and its application as EGFET pH sensor. *Electroanalysis* 33 (3): 579–590.

14 Adimule, V., Nandi, S.S., Yallur, B.C., and Shaikh, N. (2021). CNT/graphene-assisted flexible thin-film preparation for stretchable electronics and superconductors. In: *Sensors for Stretchable Electronics in Nanotechnology* (ed. K. Pal), 89–103. CRC Press.

15 Gonzalez, A.B. and Darby, S. (2004). Risk of cancer from diagnostic X-rays: estimates for the UK and 14 other countries. *Lancet* 363 (9406): 345–351. https://doi.org/10.1016/S0140-6736(04)15433-0.

16 Ronckers, C.M., Erdmann, C.A., and Land, C.E. (2004). Radiation and breast cancer: a review of current evidence. *Breast Cancer Res.* 7 (1): 1–2. https://doi.org/10.1186/bcr970.

17 Moore, A. (2010). The practice of breast ultrasound: techniques, findings, differential diagnosis. *Ultrasound Med. Biol.* 36 (2): 358. https://doi.org/10.1016/j.ultrasmedbio.2009.01.007.

18 Kwon, S. and Lee, S. (2016). Recent advances in microwave imaging for breast cancer detection. *Int. J. Biomed. Imaging* 2016: https://doi.org/10.1155/2016/5054912.

19 Orel, S.G. and Schnall, M.D. (2001). MR imaging of the breast for the detection, diagnosis, and staging of breast cancer. *Radiology* 220 (1): 13–30. https://doi.org/10.1148/radiology.220.1.r01jl3113.

20 Hassan, A.M. and El-Shenawee, M. (2011). Review of electromagnetic techniques for breast cancer detection. *IEEE Rev. Biomed. Eng.* 29 (4): 103–118. https://doi.org/10.1109/rbme.2011.2169780.

21 Adimule, V., Revaigh, M.G., and Adarsha, H.J. (2020). Synthesis and fabrication of Y-doped ZnO nanoparticles and their application as a gas sensor for the detection of ammonia. *J. Mater. Eng. Perform.* 29 (7): 4586–4596. https://doi.org/10.1007/s11665-020-04979-4.

22 Ligorio, C., Mi, Z., Wychowaniec, J.K. et al. (2019). Graphene oxide containing self-assembling peptide hybrid hydrogels as a potential 3D injectable cell delivery platform for intervertebral disc repair applications. *Acta Biomater.* 92: 92–103.

23 Shaikh, N.M., Bagihalli, G.B., Adimule, V. et al. (2022). A novel silica immobilised acidic ionic liquid [BMIM][AlCl$_4$] as an effective catalyst for biscoumarine synthesis. *Top. Catal.* https://doi.org/10.1007/s11244-022-01591-9.

24 Lim, D.G., Jung, J.H., Ko, H.W. et al. (2016). Paclitaxel–nanodiamond nanocomplexes enhance aqueous dispersibility and drug retention in cells. *ACS*

Appl. Mater. Interfaces 8 (36): 23558–23567. https://doi.org/10.1021/acsami.6b08079.

25 Daniluk, K., Kutwin, M., Grodzik, M. et al. (2019). Use of selected carbon nanoparticles as melittin carriers for MCF-7 and MDA-MB-231 human breast cancer cells. *Materials* 13 (1): 90. https://doi.org/10.3390/ma13010090.

26 Li, L., Tian, L., Zhao, W. et al. (2016). Acetate ions enhance load and stability of doxorubicin onto PEGylated nanodiamond for selective tumor intracellular controlled release and therapy. *Integr. Biol.* 8 (9): 956–967. https://doi.org/10.1039/c6ib00068a.

27 Jones, S.K. and Merkel, O. (2016). Tackling breast cancer chemoresistance with nano-formulated siRNA. *Gene Ther.* 23 (12): 821–828. https://www.nature.com/articles/gt201667.

28 Toh, T.B., Lee, D.K., Hou, W. et al. (2014). Nanodiamond–mitoxantrone complexes enhance drug retention in chemo resistant breast cancer cells. *Mol. Pharmaceutics* 11 (8): 2683–2691. https://dx.doi.org/10.1021%2Fmp5001108.

29 Yuan, S.J., Xu, Y.H., Wang, C. et al. (2019). Doxorubicin–polyglycerol–nanodiamond conjugate is a cytostatic agent that evades chemoresistance and reverses cancer-induced immunosuppression in triple-negative breast cancer. *J. Nanobiotechnol.* 17 (1): 1–25. https://doi.org/10.1186/s12951-019-0541-8.

30 Cyprian, F.S., Akhtar, S., Gatalica, Z., and Vranic, S. (2019). Targeted immunotherapy with a checkpoint inhibitor in combination with chemotherapy: a new clinical paradigm in the treatment of triple-negative breast cancer. *Bosn. J. Basic Med. Sci.* 19 (3): 227. https://doi.org/10.17305/bjbms.2019.4204.

31 Vakilinezhad, M.A., Amini, A., Dara, T., and Alipour, S. (2019). Methotrexate and curcumin co-encapsulated PLGA nanoparticles as a potential breast cancer therapeutic system: in vitro and in vivo evaluation. *Colloids Surf., B* 1 (184): 110515. https://doi.org/10.1016/j.colsurfb.2019.110515.

32 Zhao, Y., Tang, S., Guo, J. et al. (2017). Targeted delivery of doxorubicin by nano-loaded mesenchymal stem cells for lung melanoma metastases therapy. *Sci. Rep.* 7(1): 1–12.

33 Adimule, V., Nandi, S.S., and Yallur, B.C. (2022). Devices and sensors based on additively manufactured shape-memory of hybrid nanocomposites. In: *Shape Memory Composites Based on Polymers and Metals for 4D Printing* (ed. M.R. Maurya, K.K. Sadasivuni, J.J. Cabibihan, et al.), 341–359. Cham: Springer. https://doi.org/10.1007/978-3-030-94114-7_15.

34 Muntimadugu, E., Kumar, R., Saladi, S. et al. (2016). CD44 targeted chemotherapy for co-eradication of breast cancer stem cells and cancer cells using polymeric nanoparticles of salinomycin and paclitaxel. *Colloids Surf., B* 143: 532–546. https://doi.org/10.1016/j.colsurfb.2016.03.075.

35 Amirsaadat, S., Jafari-Gharabaghlou, D., Alijani, S. et al. (2021). Metformin and silibinin co-loaded PLGA-PEG nanoparticles for effective combination therapy against human breast cancer cells. *J. Drug Delivery Sci. Technol.* 61: 102107. https://doi.org/10.1016/j.jddst.2020.102107.

36 Chen, S.H., Liu, T.I., Chuang, C.L. et al. (2020). Alendronate/folic acid-decorated polymeric nanoparticles for hierarchically targetable chemotherapy against bone metastatic breast cancer. *J. Mater. Chem. B* 8 (17): 3789–3800. https://pubs.rsc.org/en/content/articlelanding/2020/tb/d0tb00046a/unauth.

37 Núñez, C., Estévez, S.V., and del Pilar, C.M. (2018). Inorganic nanoparticles in diagnosis and treatment of breast cancer. *J. Biol. Inorg. Chem.* 23 (3): 331–345. https://doi.org/10.1007/s00775-018-1542-z.

38 Lakkakula, J.R., Krause, R.W., Divakaran, D. et al. (2021). 5-Fu inclusion complex capped gold nanoparticles for breast cancer therapy. *J. Mol. Liq.* 341: 117262. https://doi.org/10.1016/j.molliq.2021.117262.

39 Gajria, S., Neumann, T., and Tirrell, M. (2011). Self-assembly and applications of nucleic acid solid-state films. *Wiley Interdiscip. Rev. Nanomed. Nanobiotechnol.* 3(5): 479–500.

40 Kadapure, S.A., Kadapure, P., Nandi, S.S., and Shet, A. (2022). Overview on catalyst and co-solvents for sustainable biodiesel production. *Proc. Inst. Civ. Eng. Energy* 1–9. https://doi.org/10.1680/jener.21.00092.

41 Samimi, S., Ardestani, M.S., and Dorkoosh, F.A. (2021). Preparation of carbon quantum dots-quinic acid for drug delivery of gemcitabine to breast cancer cells. *J. Drug Delivery Sci. Technol.* 61: 102287 https://doi.org/10.1016/j.jddst.2020.102287.

42 Bloise, N., Massironi, A., Della Pina, C. et al. (2020). Extra-small gold nanospheres decorated with a thiol functionalized biodegradable and biocompatible linear polyamidoamine as nanovectors of anticancer molecules. *Front. Bioeng. Biotechnol.* 8: 132. https://doi.org/10.3389/fbioe.2020.00132.

43 Nandi, S.S., Adimule, V., and Yallur, B.C. (2022). Synthesis, structural and optical properties of Co doped Sm_2O_3 nanostructures. *Adv. Mater. Res.* 1173: 59–69. Trans Tech Publications, Ltd.

44 Abu-Qudais, E., Chandrasekaran, B., Samarneh, S., and Kassab, G. (2020). Nanomedicines in cancer therapy. In: *Integrative Nanomedicine for New Therapies* (ed. A. Krishnan and A. Chuturgoon), 321–356. Cham: Springer. https://link.springer.com/chapter/10.1007/978-3-030-36260-7_12.

45 Surekha, B., Kommana, N.S., Dubey, S.K. et al. (2021). PAMAM dendrimer as a talented multifunctional biomimetic nanocarrier for cancer diagnosis and therapy. *Colloids Surf., B* 204: 111837. https://doi.org/10.1016/j.colsurfb.2021.111837.

46 Kheraldine, H., Gupta, I., Alhussain, H. et al. (2021). Substantial cell apoptosis provoked by naked PAMAM dendrimers in HER2-positive human breast cancer via JNK and ERK1/ERK2 signalling pathways. *Comput. Struct. Biotechnol. J.* 19: 2881–2890. https://doi.org/10.1016/j.csbj.2021.05.011.

47 Torres-Perez, S.A., del Pilar Ramos-Godinez, M., and Ramon-Gallegos, E. (2020). Glycosylated one-step PAMAM dendrimers loaded with methotrexate for target therapy in breast cancer cells MDA-MB-231. *J. Drug Deli Sci. Technol.* 58: 101769.

48 Adimule, V.M., Nandi, S.S., Kerur, S.S. et al. (2022). Recent advances in the one-pot synthesis of coumarin derivatives from different starting materials using nanoparticles: a review. *Top. Catal.* https://doi.org/10.1007/s11244-022-01571-z.

49 Chittasupho, C., Anuchapreeda, S., and Sarisuta, N. (2017). CXCR4 targeted dendrimer for anti-cancer drug delivery and breast cancer cell migration inhibition. *Eur. J. Pharm. Biopharm.* 119: 310–321. https://doi.org/10.1016/j.ejpb.2017.07.003.

50 Fang, X., Cao, J., and Shen, A. (2020). Advances in anti-breast cancer drugs and the application of nano-drug delivery systems in breast cancer therapy. *J. Drug Delivery Sci. Technol.* 57: 101662. https://doi.org/10.1016/j.jddst.2020.101662.

51 Cao, D., Zhang, X., Akabar, M.D. et al. (2019). Liposomal doxorubicin loaded PLGA–PEG–PLGA based thermogel for sustained local drug delivery for the treatment of breast cancer. *Artif. Cells Nanomed. Biotechnol.* 47 (1): 181–191. https://doi.org/10.1080/21691401.2018.1548470.

52 Fu, M., Tang, W., Liu, J.J. et al. (2020). Combination of targeted daunorubicin liposomes and targeted emodin liposomes for treatment of invasive breast cancer. *J. Drug Targeting* 28 (3): 245–258. https://doi.org/10.1080/1061186X.2019.1656725.

53 Adimule, V., Batakurki, S., Yallur, B.C. et al. (2022). Enhanced photoluminescence, optical, structural properties of ZrO_2-incorporated Sm_2O_3:Co_3O_4 nanocomposite and their applications in photocatalytic degradation of methylene blue. *J. Mater. Res.* **37**: 2396–2405. https://doi.org/10.1557/s43578-022-00641-y.

54 Chen, X., Wang, M., Yang, X. et al. (2019). Injectable hydrogels for the sustained delivery of a HER2-targeted antibody for preventing local relapse of HER2+ breast cancer after breast-conserving surgery. *Theranostics* 9 (21): 6080. https://dx.doi.org/10.7150%2Fthno.36514.

55 Fathi, M., Alami-Milani, M., Geranmayeh, M.H. et al. (2019). Dual thermo- and pH-sensitive injectable hydrogels of chitosan/(poly(*N*-isopropylacrylamide-*co*-itaconic acid)) for doxorubicin delivery in breast cancer. *Int. J. Biol. Macromol.* 128: 957–964. https://doi.org/10.1016/j.ijbiomac.2019.01.122.

56 Hong, D., Yang, J., Guo, J. et al. (2021). Ultrasound-targeted microbubble destruction enhances inhibitory effect of apatinib on angiogenesis in triple negative breast carcinoma xenografts. *Anal. Cell. Pathol.* 2021: https://doi.org/10.1155/2021/8837950.

57 Mishra, V., Singh, M., and Nayak, P. (2021). Smart functionalised-dendrimeric medicine in cancer therapy. In: *Dendrimers in Nanomedicine*, 233–253. CRC Press.

58 Adimule, V., Yallur, B.C., Kamat, V., and Krishna, P.M. (2021). Characterization studies of novel series of cobalt (II) nickel (II) and copper (II) complexes: DNA binding and antibacterial activity. *J. Pharm. Invest.* 51 (3): 347–359.

59 Al-Radadi, N.S., (2021). Green biosynthesis of flaxseed gold nanoparticles (Au-NPs) as potent anti-cancer agent against breast cancer cells. *J. Saudi.Chem. Soc.* 25 (6): 101243.

60 Shaikh, N.M., Adimule, V., Bagihalli, G.B. et al. (2022). A novel mixed Ag–Pd nanoparticles supported on SBA silica through [DMAP-TMSP-DABCO]OH basic ionic liquid for Suzuki coupling reaction. *Top. Catal.* https://doi.org/10.1007/s11244-022-01586-6.

61 Khodashenas, B., Ardjmand, M., Rad, A.S., and Esfahani, M.R. (2021). Gelatin-coated gold nanoparticles as an effective pH-sensitive methotrexate drug delivery system for breast cancer treatment. *Mater. Today Chem.* 20: 100474. https://doi.org/10.1016/j.mtchem.2021.100474.

62 Bandyopadhyay, A., Roy, B., Shaw, P. et al. (2021). Chitosan-gold nanoparticles trigger apoptosis in human breast cancer cells in vitro. *Nucleus* 64 (1): 79–92. https://link.springer.com/article/10.1007/s13237-020-00328-x.

63 Desai, N.P. and Patrick, S.S. (2008). Breast cancer therapy based on hormone receptor status with nanoparticles comprising taxane. WO2008/076373A1, filed 14 December 2007 and issued 26 June 2008. pp. 1–10. https://patents.google.com/patent/WO2008076373A1/en26.

64 Adimule, V., Yallur, B.C., Pai, M.M. et al. (2022). Biogenic synthesis of magnetic palladium nanoparticles decorated over reduced graphene oxide using *Piper betle* Petiole extract (Pd-rGO@Fe$_3$O$_4$ NPs) as heterogeneous hybrid nanocatalyst for applications in Suzuki–Miyaura coupling reactions of biphenyl compounds. *Top. Catal.* https://doi.org/10.1007/s11244-022-01672-9.

65 Adimule, V., Jagadeesha Gowda, A.H., Nandi, S.S., and Bowmik, D. (2022). Antimalarial activity of novel class of 1,3-benzoxaborole derivatives containing 1,3,4-oxadiazole moiety. In: *Drug Development for Malaria* (ed. P. Kendrekar), 285–302. Wiley-VCH GmbH. https://doi.org/10.1002/9783527830589.ch12.

66 Kaklamani, V.G., Siziopikou, K., Scholtens, D. et al. (2012). Pilot neoadjuvant trial in HER2 positive breast cancer with combination of nab-paclitaxel and lapatinib. *Breast Cancer Res. Treat.* 132 (3): 833–842. https://link.springer.com/article/10.1007/s10549-011-1411-8.

67 Fujiwara, Y., Mukai, H., Saeki, T. et al. (2019). A multi-national, randomised, open-label, parallel, phase III non-inferiority study comparing NK105 and paclitaxel in metastatic or recurrent breast cancer patients. *Br. J. Cancer* 120 (5): 475–480. https://www.nature.com/articles/s41416-019-0391-z.

68 Sachdev, J.C., Munster, P., Northfelt, D.W. et al. (2021). Phase I study of liposomal irinotecan in patients with metastatic breast cancer: findings from the expansion phase. *Breast Cancer Res. Treat.* 185 (3): 759–771. https://doi.org/10.1007/s10549-020-05995-7.

69 Shaikh, N.M., Sawant, A.D., Bagihalli, G.B. et al. (2022). Highly active mixed Au–Pd nanoparticles supported on RHA silica through immobilised ionic liquid for Suzuki coupling reaction. *Top. Catal.* https://doi.org/10.1007/s11244-021-01547-5.

4

HER Receptor in Breast Cancer

Guno S. Chakraborthy

Parul University, Parul Institute of Pharmacy & Research, Pharmacy Department, Limda, 391760, Vadodara, Gujarat, India

4.1 Introduction

The human epidermal growth factor receptors (HERs) are a group of receptors involved in the development of numerous malignancies in humans. These receptors generally assist in the proliferation of cells and differentiation and govern them via numerous signaling pathways like the mitogen-activated protein kinase (MAPK) pathway, phosphoinositide-3-kinase (PI3K)/Akt signaling pathway, and protein kinase C (PKC) activation [1]. The four major family types of these receptors are HER1, HER2, HER3, and HER4, also known as ErbB1, ErbB2, ErbB3, and ErbB4 [2]. The epidermal growth factor (EGF) group of hormones are generally associated with receptors, which are involved in the differentiation and proliferation of various tissue types through a signaling network. Numerous human malignancies, including those of the brain, pulmonary, breast, ovarian, pancreatic, and prostate, are linked to the dysregulation of this system. These findings have sparked research into how this signaling network is controlled and connected to physiological responses, as well as how management and connection are altered in cancer [3, 4] Stanley Cohen of Vanderbilt University was the first person to find the EGF as well as its receptor. For such a discovery of signaling pathways, Cohen and Rita Levi-Montalcini won the Nobel Prize in Medicine in 1986.

The four basic and common members of the ErbB receptor family are as follows: [5]

- EGFR (HER1)
- ErbB2 (HER2)
- ErbB3 (HER3)
- ErbB4 (HER4)

Drug and Therapy Development for Triple Negative Breast Cancer, First Edition.
Edited by Pravin Kendrekar, Vinayak Adimule, and Tara Hurst.
© 2023 WILEY-VCH GmbH. Published 2023 by WILEY-VCH GmbH.

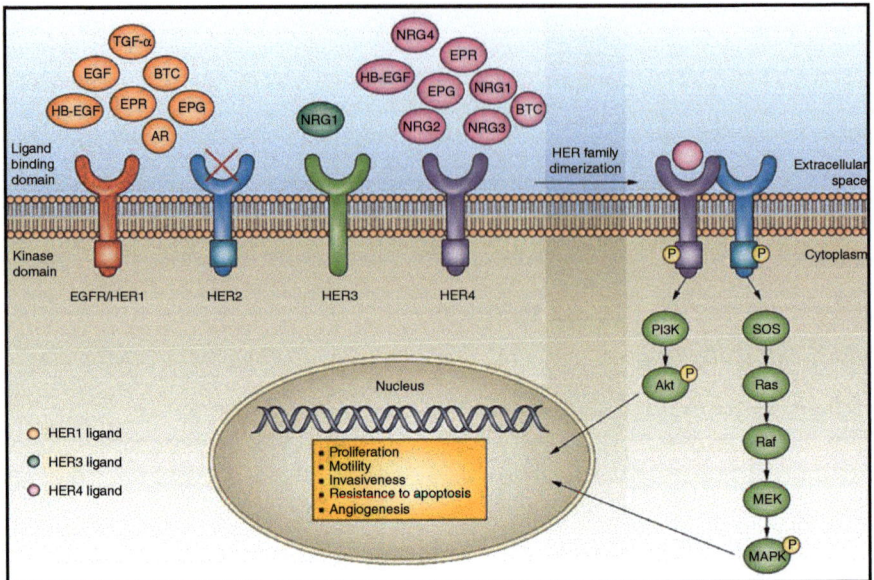

Figure 4.1 Types of HER receptors and their functioning. Source: Arteaga et al. [16]/ Springer Nature.

Such receptors are available in a variety of sizes in humans, ranging from 1210 to 1343 amino acids. Each of these receptors carries a cysteine-rich extracellular ligand, a sole membrane-spanning domain, and a substantial cytoplasmic region with a tyrosine kinase domain and numerous phosphorylated tyrosine residues that are activated by the receptor [6–9]. Ligand interaction causes receptor dimerization and tyrosine phosphorylation at many locations, which allows effector proteins to attach and physiological effects to be coupled [6–9]. The ErbB is derived from the Erb-b gene, which is responsible for the avian erythroblastosis virus. A team of researchers from the Massachusetts Institute of Technology, Rockefeller University, and Harvard University found the new oncogene (referred to as ErbB2, or p185, or HER2) [10, 11]. The HER2 receptor is indeed a 1255-amino-acid, 185-kD transmembrane glycoprotein found on chromosome 17's long arm in humans [12]. HER2 and its role are present in a variety of organs, and its primary function in such tissues is to promote uncontrolled cell proliferation and cancer [13–15] (Figure 4.1).

4.2 Role of HER Receptors in the Human Body

Mostly HER receptors reside on the surface of a cell, as monomers. HER proteins dimerize and trans phosphorylate their intracellular domains when unique ligands adhere to their epitopes. HER2 has no known and documented stimulating ligand but may be activated by itself or through heterodimerization with some other members of the HER family, like HER1 and HER3 [16]. Homo heterodimerization causes autophosphorylation of residues of tyrosine in the receptors' cytoplasmic domains, which activates a number of signaling pathways, including the MAPK, PKC, and PI3K, which lead to cellular proliferation.

Heterodimers produce higher potent impulses compared to homodimers, and all those comprising HER2 have an especially high ligand binding and signaling potency, as HER2 occurs in a free form, enabling it to be the family members' preferred dimerization partner. The HER2–HER3 heterodimer is the most powerful activator of downstream pathways, including PI3K/Akt, which is a key regulator of cellular proliferation and survival. Furthermore, HER2 dimerization enhances cell-cycle inhibitor, protein mislocalization, and fast degradation, resulting in cell-cycle acceleration [17–19]. Other membrane receptors, like insulin-like growth factor receptor 1, also can trigger HER2 [20].

This receptor family's role is basic in *Caenorhabditis elegans*, wherein communication is handled by a solitary ligand and a sole receptor, and marginally more difficult in *Drosophila*, whose four ligands communicate through a given receptor. In mammals, the mechanism is significantly more sophisticated, with at least 12 ligands and 4 receptors performing the tasks of this receptor family [21].

Choosing a partner appears to be a crucial component of HER protein signaling activity, and their signaling functions are organized in a hierarchy structure that favors heterodimers before homodimers. The catalytic kinase action of HER2 is the highest, and heterodimers containing HER2 have the best signaling capabilities [22]. For mammalian systems, the development of the HER receptor family has been linked to functional differentiation that necessitates interdependence instead of fostering separate or redundant roles. HER2 and HER3, which seem to be functionally imperfect receptor molecules, are good examples of this [23].

HER2's domain in the extracellular region, unlike most of the other receptor family members, doesn't really have a pivot among inactive and active conformations and is always in an activated form [24]. HER2 is basically deficient in ligand binding ability, which is associated with its own continuously active conformation, as well as its signaling function is mediated by its own ligand-bound heterodimeric companions. Unlike HER2, HER3 on the other hand is insufficient of ATP coupling inside its catalytic domain, and therefore it is catalytically inactive, unlike most of the other members. The signaling capabilities of HER3 are exclusively mediated by the enzymatic activity of its heterodimeric companions, which is consistent with this [25].

Even though these are incomplete signaling molecules on their own, a substantial body of evidence not just demonstrate HER2 and HER3 as an obligate companion and states that their complex forms are the most active signaling heterodimer of their whole receptor family and therefore is extremely important for several pharmacologic and developmental processes [25–27].

4.3 HER2 Receptor in Breast Cancer Progression

Since we still don't know what exactly causes breast cancer (HER2 positive), many studies have identified and have tried to picture how basically these receptor types lead to oncogenic conditions. The HER2 protein is overexpressed in approximately 25% of breast cancer cells. A mutated gene inside the HER2 gene induces this. Once the HER2 gene is mutated, breast cells multiply rapidly at an uncontrollable rate, resulting in tumor growth [28]. As already discussed, HER2–HER3 heterodimer is thought to be the most powerful activator for downstream pathways, most notably PI3K/Akt, a key regulator of cell-cycle progression. HER2 dimerization promotes altered localization or changes the positioning of the cell cycle and impairment of the p27Kipl protein [1].

The p27Kip1 protein is one of the important enzyme blockers that is encrypted by the human CDKN1B genes. This encrypts a cyclin-dependent kinase (Cdk) inhibitor protein that relates to the Kip/Cip family [13]. The encrypted protein binds and inhibits the stimulation of cyclin E-Cdk2 or cyclin D-Cdk4 clusters, controlling cell growth in the G1 phase. It is frequently referred to as a protein that mainly ceases the cell cycle since its primary function is to halt or slow cell growth and division [29].

For its role as a cell cycle controller, p27 is thought to be a potent inhibitor of the tumor [30]. It is often inhibited in cancers caused by abnormal synthesis, mislocalization, or expedited degradation [31]. Posttranscriptional inhibitory effect of p27 is usually achieved through the oncogenic development of the various routes such as PI3K, receptor tyrosine kinases (RTK), SRC, or Ras-MAPK. Such activities function to speed up the protein degradation of the p27 protein, allowing the cancerous cells to divide quickly and proliferate uncontrollably [31]. When Src phosphorylates p27 at tyrosine 74 or 88, it no longer inhibits cyclinE-Cdk2. Src has also been shown to shorten the life span of p27, implying that it degrades faster [32].

Src stimulation has been linked to low p27 levels in many cases of breast cancer [32]. Breast tumors that were estrogen and progesterone receptor negative were much more likely to have reduced concentrations of p27 but a high tumor grade [33]. Likewise, individuals having breast cancer who had BRCA1/2 mutations were so much more likely to experience lower p27 levels [34].

4.4 Conclusion

HER2 serves as a prognostic and helps to have a predictive biomarker in breast and other cancers like gastric and gastroesophageal. These are directed against and have revolutionized the treatments involved in overexpressing the phenomenon of breast cancer, which will improve the clinical outcomes. Hence, the search for novel HER2 receptors is on, which will help to cure the target with ease.

Conflict of Interest

The authors declare that there is no conflict of interest regarding the publication of this chapter.

References

1 Iqbal, N. and Iqbal, N. (2014). Human epidermal growth factor receptor 2 (HER2) in cancers: overexpression and therapeutic implications. *Mol. Biol. Int.* 2014: 852748. https://doi.org/10.1155/2014/852748.

2 Riese, D.J. and Stern, D.F. (1998). Specificity within the EGF family/ErbB receptor family signaling network. *Bioessays* 20: 41–48. https://doi.org/10.1002/(SICI) 1521-1878(199801)20:1<41::AID-BIES7>3.0.CO;2-V.

3 Adimule, V.M., Nandi, S.S., Kerur, S.S. et al. (2022). Recent advances in the one-pot synthesis of coumarin derivatives from different starting materials using nanoparticles: a review. *Top. Catal.* https://doi.org/10.1007/s11244-022-01571-z.

4 Salomon, D.S., Brandt, R., Ciardiello, F., and Normanno, N. (1995). Epidermal growth factor-related peptides and their receptors in human malignancies. *Crit. Rev. Oncol. Hematol.* 19 (3): 183–232. https://doi.org/10.1016/1040-8428 (94)00144-i.

5 Adimule, V.M. (2014). Design, synthesis and cytotoxic evaluation of novel 2-(4-*N*, *N*-dimethyl)pyridine containing 1,3,4-oxadiazole moiety. *Asian J. Biomed. Pharm. Sci.* 4 (37): 1.

6 Schechter, A.L., Hung, M.C., Vaidyanathan, L. et al. (1985). The neu gene: an erbB-homologous gene distinct from and unlinked to the gene encoding the EGF receptor. *Science* 229 (4717): 976–978. https://doi.org/10.1126/science.2992090.

7 Adimule, V., Kerur, S.S., Chinnam, S. et al. (2022). Guar gum and its nanocomposites as prospective materials for miscellaneous applications: a short review. *Top. Catal.* https://doi.org/10.1007/s11244-022-01587-5.

8 Kraus, M.H., Issing, W., Miki, T. et al. (1989). Isolation and characterization of ERBB3, a third member of the ERBB/epidermal growth factor receptor family: evidence for overexpression in a subset of human mammary tumors. *Proc. Natl. Acad. Sci. U.S.A.* 86 (23): 9193–9197. https://doi.org/10.1073/pnas.86.23.9193.

9 Vinayak, A., Sudha, M., and Lalita, K.S. (2017). Design, synthesis and characterization of novel amine derivatives of 5-[5-(chloromethyl)-1,3, 4-oxadiazol-2-yl]-2-(4-fluorophenyl)-pyridine as a new class of anticancer agents. *Dhaka Univ. J. Pharm. Sci.* 16 (1): 11–19. https://doi.org/10.3329/dujps.v16i1.33377.

10 Adimule, V., Nandi, S.S., Yallur, B.C., and Shaikh, N. (2021). CNT/graphene-assisted flexible thin-film preparation for stretchable electronics and superconductors. In: *Sensors for Stretchable Electronics in Nanotechnology*, 89–103. CRC Press. https://doi.org/10.1201/9781003123781.

11 Schechter, A.L., Stern, D.F., Vaidyanathan, L. et al. (1984). The *neu* oncogene: an *erb-B*-related gene encoding a 185,000-M_r tumour antigen. *Nature* 312 (5994): 513–516. https://doi.org/10.1038/312513a0.

12 Brandt-Rauf, P.W., Pincus, M.R., and Carney, W.P. (1994). The c-erbB2 protein in oncogenesis: molecular structure to molecular epidemiology. *Crit. Rev. Oncog.* 5 (2): 313–329. https://doi.org/10.1155/2014/852748.

13 Adimule, V., Revaigh, M.G., and Adarsha, H.J. (2020). Synthesis and fabrication of Y-doped ZnO nanoparticles and their application as a gas sensor for the detection of ammonia. *J. Mater. Eng. Perform.* 29: 4–5. https://doi.org/10.1007/s11665-020-04979-4.

14 Neve, R.M., Lane, H.A., and Hynes, N.E. (2001). The role of overexpressed HER2 in transformation. *Ann. Oncol.* 12 (Suppl 1): S9–S13. https://doi.org/10.1093/annonc/12.suppl_1.s9.

15 Adimule, V., Nandi, S.S., Yallur, B.C. et al. (2021). Enhanced photoluminescence properties of $Gd_{(x-1)}SrxO:CdO$ nanocores and their study of optical structural and morphological characteristics. *Mater. Today Chem.* 20: 00438. https://doi.org/10.1016/j.mtchem.2021.100438.

16 Arteaga, C., Sliwkowski, M., Osborne, C. et al. (2012). Treatment of HER2-positive breast cancer: current status and future perspectives. *Nat. Rev. Clin. Oncol.* 9: 16–32. https://doi.org/10.1038/nrclinonc.2011.177.

17 Olayioye, M.A. (2001). Update on HER-2 as a target for cancer therapy: intracellular signaling pathways of ErbB2/HER-2 and family members. *Breast Cancer Res.* 3 (6): 385–389. https://doi.org/10.1186/bcr327.

18 Adimule, V., Nandi, S.S., and Adarsha, H.J. (2021). A facile synthesis of Cr doped WO_3 nanostructures study of their current–voltage power dissipation and impedance properties of thin films. *J. Nano Res.* 67: 33–42. https://doi.org/10.4028/www.scientific.net/JNanoR.67.33.

19 Adimule, V., Yallur, B.C., Bhowmik, D., and Gowda, A.H. (2021). Dielectric properties of P3BT doped $ZrY_2O_3/CoZrY_2O_3$ nanostructures for low-cost optoelectronics applications. *Trans. Electr. Electron. Mater.* 1–16. https://doi.org/10.1007/s42341-021-00348-7.

20 Nahta, R., Yuan, L.X., Zhang, B. et al. (2005). Insulin-like growth factor-I receptor/human epidermal growth factor receptor 2 heterodimerization contributes to trastuzumab resistance of breast cancer cells. *Cancer Res.* 65 (23): 11118–11128. https://doi.org/10.1158/0008-5472.CAN-04-3841.

21 Padhy, L.C., Shih, C., Cowing, D. et al. (1982). Identification of a phosphoprotein specifically induced by the transforming DNA of rat neuroblastomas. *Cell* 28 (4): 865–871. https://doi.org/10.1016/0092-8674 (82)90065-4.

22 Vinayak, A., Sudha, M., Jagadeesha, A., and Lalita, K. (2015). Design, synthesis, characterization and cancer cell growth-inhibitory properties of novel derivatives of 2-(4-fluoro-phenyl)-5-(5-aryl substituted-1,3,4-oxadiazol-2-yl)pyridine. *J. Pharm. Res. Int.* 7 (1): 34–43. https://doi.org/10.9734/BJPR/2015/15486.

23 Adimule, V., Vageesha, P., Bagihalli, G. et al. (2019). Synthesis characterization of hybrid nanomaterials of strontium yttrium copper doped with indole schiff base derivatives possessing dielectric and semiconductor properties. In: *Emerging Research in Electronics Computer Science and Technology*, 1131–1140. Singapore: Springer https://doi.org/10.1007/978-981-13-5802-9_97.

24 Garrett, T.P., McKern, N.M., Lou, M. et al. (2003). The crystal structure of a truncated ErbB2 ectodomain reveals an active conformation, poised to interact with other ErbB receptors. *Mol. Cell* 11 (2): 495–505. https://doi.org/10.1016/s1097-2765(03)00048-0.

25 Adimule, V., Revaiah, R.G., Nandi, S.S., and Jagadeesha, A.H. (2021). Synthesis characterization of Cr doped TeO_2 nanostructures and its application as EGFET pH sensor. *Electroanalysis* 33 (3): 579–590. https://doi.org/10.1002/elan.202060329.

26 Horan, T., Wen, J., Arakawa, T. et al. (1995). Binding of neu differentiation factor with the extracellular domain of HER2 and HER3. *J. Biol. Chem.* 270 (41): 24604–24608. https://doi.org/10.1074/jbc.270.41.24604.

27 Britsch, S., Goerich, D.E., Riethmacher, D. et al. (2001). The transcription factor Sox10 is a key regulator of peripheral glial development. *Genes Dev.* 15 (1): 66–78. https://doi.org/10.1101/gad.186601.

28 Adimule, V., Yallur, B.C., Kamat, V., and Krishna, P.M. (2021). Characterization studies of novel series of cobalt (II), nickel (II) and copper (II) complexes: DNA binding and antibacterial activity. *J. Pharm. Invest.* 51 (3): 347–359. https://doi.org/10.1007/s40005-021-00524-0.

29 Polyak, K., Lee, M.H., Erdjument-Bromage, H. et al. (1994). Cloning of p27Kip1, a cyclin-dependent kinase inhibitor and a potential mediator of extracellular antimitogenic signals. *Cell* 78 (1): 59–66. https://doi.org/10.1016/0092-8674 (94)90572-x.

30 Adimule, V., Nandi, S.S., Yallur, B.C. et al. (2021). Optical structural and photoluminescence properties of GdxSrO: CdO nanostructures synthesized by co precipitation method. *J. Fluoresc.* 31 (2): 487–499. https://doi.org/10.1007/s10895-021-02683-7.

31 Adimule, V., Yallur, B.C., and Sharma, K. (2022). Studies on crystal structure, morphology, optical and photoluminescence properties of flake-like Sb doped Y_2O_3 nanostructures. *J. Opt.* **51**: 173–183. https://doi.org/10.1007/s12596-021-00746-3.

32 Chu, I., Sun, J., Arnaout, A. et al. (2007). p27 phosphorylation by Src regulates inhibition of cyclin E-Cdk2. *Cell* 128 (2): 281–294. https://doi.org/10.1016/j.cell.2006.11.049.

33 Adimule, V., Nandi, S.S., and Gowda, A.H.J. (2021). Enhanced power conversion efficiency of the P3BT (poly-3-butyl thiophene) doped nanocomposites of $GdTiO_3$ as working electrode. *Techno-Societal* 2020: 55–68. https://doi.org/10.1007/978-3-030-69925-3_6.

34 Chappuis, P.O., Kapusta, L., Bégin, L.R. et al. (2000). Germline BRCA1/2 mutations and p27 (Kip1) protein levels independently predict outcome after breast cancer. *J. Clin. Oncol.* 18 (24): 4045–4052. https://doi.org/10.1200/JCO.2000.18.24.4045.

5

Human Endogenous Retroviruses in Triple-Negative Breast Cancer

Tara P. Hurst[1], Timokratis Karamitros[2] and Gkikas Magiorkinis[3]

[1]*Pandemic Sciences Institute, University of Oxford, Nuffield Department of Medicine (NDM), Old Road Campus Research Building, Old Road Campus, Roosevelt Drive, Headington, Oxford, OX3 7DQ, UK*
[2]*Hellenic Pasteur Institute, Bioinformatics and Applied Genomics Unit, Department of Microbiology, 127 Vas. Sophias avenue, 11521, Athens, Greece*
[3]*National and Kapodistrian University of Athens, Department of Hygiene, Epidemiology and Medical Statistics, Medical School, Athens, Greece*

5.1 Introduction

Human endogenous retroviruses (HERVs) are an inherent feature of the human genome, found to account for 8% of the total genome [1]. This may be an underestimate of their prevalence since repeat elements, including HERVs, are sensitive to errors in the assembly process. A recent study analyzing an entire reference genome from telomere to telomere (T2T) uncovered missed retrotransposons in previous assemblies, highlighting the possibility of missed complexity in gene regulation and in functions of the transposons [2]. Importantly, this study found that repeat elements comprised >53% of the genome, identified loci that had been missed in previous annotations, and characterized complexity in repeat element biology including variability and composite elements [2]. Thus, while it has been known for two decades that HERVs are prevalent in the genome, this is likely to a greater extent than early predictions from the Human Genome Project. Like other repeat elements, there is evidence for HERV involvement in genome structure and content, as well as in the regulation of gene transcription. There is thus scope for involvement of HERVs in the molecular biology of diseases, including breast cancer.

While HERVs are widespread, they are largely defective proviruses or solo long terminal repeats (LTRs) [3]. Some HERVs have been co-opted or domesticated to act as human genes and HERV LTRs can regulate gene expression [4]. The expression of HERVs is regulated by numerous mechanisms, including epigenetic silencing through histone acetylation, methylation, chromatin packing [5], Krüppel-associated box domain (KRAB)-containing zinc-finger proteins [6], and possibly short RNA molecules derived from tRNAs [7]. In general, HERV expression in

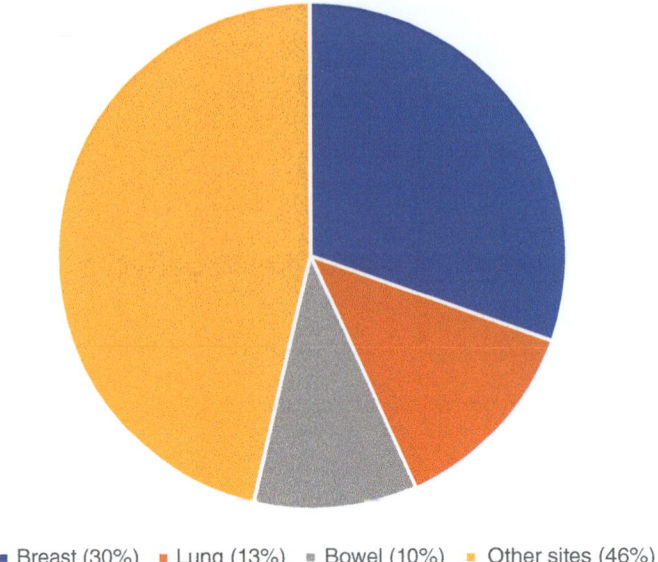

■ Breast (30%)   ■ Lung (13%)   ■ Bowel (10%)   ■ Other sites (46%)

Figure 5.1 Breast cancer is the most common type of cancer in women in the United Kingdom. The percentages are the number of cases out of a total of 179 493. Source: Data obtained from Cancer Research UK [19].

somatic cells is suppressed and this suppression is established in a pattern of epigenetic marks during early development [8]. Despite this, HERV expression is known to occur in normal cells [9, 10] with physiological roles in stem cell identity [11] and reproduction [12]. Further, there is evidence that the recently active HERV-K (human mouse mammary tumor virus-like-2 or HML-2) subgroup that may still be replicating within the genome [13].

The expression of HERVs may also be a marker of or contribute to the pathology of a disease. There is evidence of HERV expression in specific diseases, including autoimmune diseases and cancers [14–16]. Polymorphisms in HERV loci may be pathological through mechanisms such as altered transcription [17, 18]. Alternatively, aberrant epigenetic modifications could allow expression from loci that are normally quiescent. These loci may be expressed as transcripts that encode proteins or are noncoding. The former includes some well-defined HERV proteins, such as HERV-W Env and HERV-K Env, and these have been found to be expressed in pathologies. The latter is a burgeoning area of research, with many noncoding transcripts predicted but not yet studied or with roles that are as yet unclear.

Breast cancer is the most commonly diagnosed cancer in women (Figure 5.1). The disease is complex, with multiple types marked by hormone or growth factor sensitivity, inflammation, inherited genetic mutations, and possibly viral infection [20]. There are 10 types of breast cancer and related diseases [21], with 5 molecular subtypes of breast cancer [22]. It is therefore not simple to identify universal mechanisms behind all forms of cancer, nor to identify single biomarkers that will apply across all cancer types.

Among the types of breast cancer, triple-negative breast cancer (TNBC) is characterized by undetectable estrogen receptor (ER) and progesterone receptor (PR), with human epidermal growth factor receptor 2 (HER2) not being overexpressed [23]. Since therapies for breast cancer typically target HER2, this makes TNBC more difficult to treat. Further, TNBC was found to have a higher mutation load and increased likelihood of producing immunogenic neopeptides compared to non-TNBC samples [23]. The other chapters in the book cover the features of novel treatments for TNBC in more depth. In this chapter, we will examine evidence for HERVs in breast cancer, particularly TNBC.

5.2 HERVs in Breast Cancer and TNBC

There is potential for HERV expression to be linked to breast cancer and thus to act as a biomarker of the disease. HERVs, particularly HERV-K Env, have been hypothesized to be expressed in some types of breast cancer [24]. A recent study found that expression of HERVs was greater in a breast cancer than normal tissues and that expression from the 17p13.1 provirus was associated with a higher risk of breast cancer [25]. They further found that 17p13.1 colocated with TP53 and may regulate TP53 expression in ER+ and HER2+ breast cancers [25]. Locus-specific expression of HERVs was investigated using the TCGA (The Cancer Genome Atlas) RNA-Seq data and this identified 1547 HERVs that were differentially expressed in tumor versus normal samples ($p < 0.05$); of these, 640 were upregulated and 907 were downregulated [26]. In contrast, a small study investigating the fully intact provirus Xq21.33 did not identify a significant association with breast cancer in the study population in West Africa [27]. The study had few participants and so was likely not powered to detect such a difference.

In addition to protein biomarkers, long noncoding RNAs (lncRNAs) have been examined for their association with breast cancer. lncRNAs are defined by being longer than 200 bases but are otherwise diverse in sequence [28–30]. Among the lncRNAs identified to date are some that are expressed from HERV loci; these include elements that may be involved in the regulation of innate immune function as well as stem cell identity. One study identified 215 lncRNAs with altered gene expression in breast cancer [31], while another developed a survival signature based on the expression of 8 lncRNAs [32]. Further, 51 HERVs were found to be associated with TNBC and this includes a HERV-derived lncRNA, TROJAN [33]. In addition to the association with breast cancer generally, some lncRNAs have also been associated with cancer stem cells [34].

Another potential role for lncRNAs to regulate the breast cancer stem cells is the contribution to the immune response. The immune response has been found to be important in tumor progression, with infiltration of the tumor by specific immune cells being associated with disease progression [35]. Breast cancers are predicted to be infiltrated by immune cells (Figure 5.2). For example, tumor-associated macrophages (TAMs) promote tumorigenesis and correlate with worse prognoses in breast cancer [37]. Further, a role for lncRNAs in regulation of the innate immune

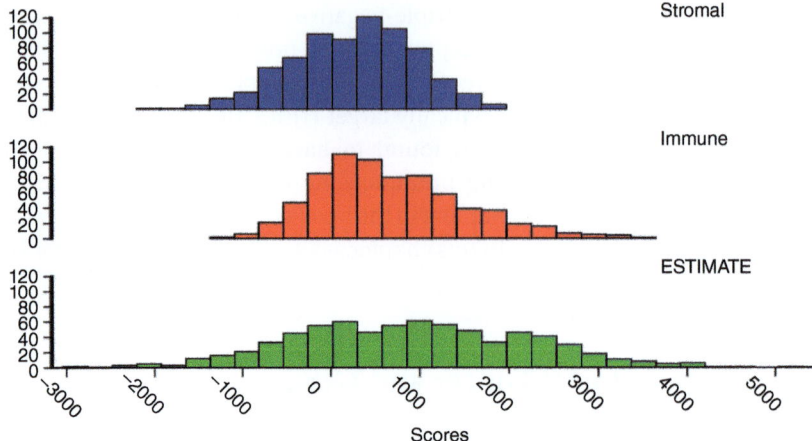

Figure 5.2   Immune infiltration in breast cancer tumors. Source: The data was generated using the ESTIMATE software and RNA-Seq expression data [36]. The stromal score refers to the presence of stroma in the tumor, while the immune score indicates the infiltration of immune cells into the tumor. The overall estimate of tumor purity is given by the estimated score. The website includes data from other platforms that also show similar immune infiltration but only the RNA-Seq data was included here.

response has been identified [38], including at least one that contains an endogenous retrovirus [39].

Interestingly, in a pilot study of people living with HIV who also had breast cancer, a significant difference in expression of human genes and HERVs were detected [40]. The increased HERV expression was hypothesized to drive a pro-inflammatory, tumor-supportive environment, and this was potentially supported by the upregulation of immune genes [40]. Of interest were some human genes associated with oncogenesis, as well as lncRNAs. Of these, the authors identified a significant positive association between the expression of HERVs and the RAD50 gene, as well as the lncRNA AC060780.1 [40].

5.3   TROJAN lncRNA and TNBC

Understanding the molecular biology of TNBC is essential to developing targeted therapeutics. One poorly studied contribution to TNBC is that of endogenous retroviruses. Recently, a study using RNA-Seq identified 51 HERVs that are highly expressed in TNBC [33]. Of those, LTR70 showed the greatest fold change, and this was further found to be part of the 1300 nucleotide lncRNA they named TROJAN [33]. Interestingly, other LTR70-containing transcripts were also identified in TNBC, but TROJAN was predominant. TROJAN was highly expressed in breast cancer cell lines and in TNBC tissues [33]. The authors found its expression to be associated with reduced relapse-free survival in TNBC cohorts [33]. This is evidence for a role for at least one HERV in the TNBC phenotype.

5.4 HERVs and Breast Cancer Treatments

The link between HERV expression and pathology can be shown by the effects of drugs on measures of both. For example, the integrase strand-transfer inhibitor, dolutegravir (DTG) was found to inhibit the proliferation of cancer cell lines and the authors found this correlated with the expression of endogenous retroviruses in the cells [41]. They further demonstrated that overexpression of HERV genes or knockdown of HERV expression were able to render the BT-20 cell line resistant to DTG [41]. The evidence supports the expression of the endogenous retroviruses in the cancer cell lines and that this may be a target for therapeutic intervention by using antiretrovirals.

The expression of HERVs may also lend themselves to targeted treatments, such as the use of HERV-K Env-targeted chimeric antigen receptor T cells (CAR-T) in the treatment of metastatic breast cancer [42]. A single chain fragment variable (SCFV) derived from anti-HERV-K monoclonal antibody to develop CAR-T cells specific for HERV-K Env [42]. K-CAR cells are able to inhibit the proliferation of breast cancer cells in vitro, as well as promote the release of Th1 cytokines, which are associated with improved killing of the target cells. Further, K-CAR cells reduced tumour size, growth, and weight in a xenograft mouse model [42]. Thus, HERV-K Env could be a therapeutic target in breast cancer.

5.5 Conclusion

This chapter explored the biology of HERVs and their association with the pathophysiology of breast cancer. There is evidence for the expression of HERV-K Env and the presence of HERV-K antibodies in breast cancer. Antibodies to HERV-K are being explored as a biomarker for breast cancer, while the use of CAR-T cells that target HERV-K expressing cells could be an effective therapeutic approach. The application of sequencing technologies, such as RNA-Seq, has allowed a more nuanced understanding of the expression of HERVS, including their contribution to lncRNAs. There is more scope for understanding HERVs in lncRNAs and their possible therapeutic applications.

5.6 Search Strategy

Literature searches were conducted in PubMed with several iterations of the search terms "breast cancer HERVs" and "TNBC HERVs" (Table 5.1). The hits were then filtered based on language (English language). Overlapping results from these searches were prevented by copying to the PubMed clipboard, which prevents duplication. This left 106 results from the search and these were then analyzed manually for relevance based on the titles and summaries. Some of the articles (18) were on cancer generally or cancers other than BC or TNBC, with or without a mention of HERVs. Further, some articles (11) were on BC without specifying HERVs, while

Table 5.1 Literature search results.

			Filtered articles	
Term/phrase	Articles (all)	Articles (filtered)	Reviews	Last 10 yr
Breast cancer HERVs	96	94	15	51
Breast cancer HERV	110	107	15	55
Breast cancer endogenous retrovirus	131	128	17	61
Triple-negative breast cancer HERVs	3	3	1	3
Triple-negative breast cancer HERV	3	3	1	3
TNBC HERVs	3	3	1	3
TNBC HERV	3	3	1	3

still others (37) discussed either HERVs or other relevant molecular biology [43–48], without being specifically on breast cancer. This left 40 that discussed HERVs and breast cancer, including studies on mouse models of breast cancer, with 19 of these published since 2010. Analysis of these 40 papers forms the core of this chapter. Of all the articles published on HERVs and breast cancer, only three of these have focused on TNBC, potentially leaving a gap in our knowledge of the mechanisms of TNBC.

References

1 Lander, E.S., Linton, L.M., Birren, B. et al. (2001). Initial sequencing and analysis of the human genome. *Nature* 409: 860–921. https://doi.org/10.1038/35057062.

2 Hoyt, S.J., Storer, J.M., Hartley, G.A. et al. (2022). From telomere to telomere: the transcriptional and epigenetic state of human repeat elements. *Science* 1979: 376.

3 Belshaw, R., Watson, J., Katzourakis, A. et al. (2007). Rate of recombinational deletion among human endogenous retroviruses. *J. Virol.* 81 (17): 9437–9442. https://doi.org/10.1128/JVI.02216-06.

4 Thompson, P.J., Macfarlan, T.S., and Lorincz, M.C. (2016). Long terminal repeats: from parasitic elements to building blocks of the transcriptional regulatory repertoire. *Mol. Cell* 62: 766–776. https://doi.org/10.1016/j.molcel.2016.03.029.

5 Hurst, T.P. and Magiorkinis, G. (2017). Epigenetic control of human endogenous retrovirus expression: focus on regulation of long-terminal repeats (LTRs). *Viruses* 9 (6): 130. https://doi.org/10.3390/v9060130.

6 Ecco, G., Cassano, M., Kauzlaric, A. et al. (2016). Transposable elements and their KRAB-ZFP controllers regulate gene expression in adult tissues. *Dev. Cell* 36: 611–623. https://doi.org/10.1016/j.devcel.2016.02.024.

7 Schorn, A.J., Gutbrod, M.J., LeBlanc, C., and Martienssen, R. (2017). LTR-retrotransposon control by tRNA-derived small RNAs. *Cell* 170 (1): 61–71.e11. https://doi.org/10.1016/j.cell.2017.06.013.

8 Rowe, H.M. and Trono, D. (2011). Dynamic control of endogenous retroviruses during development. *Virology* 411: 273–287. https://doi.org/10.1016/j.virol.2010 .12.007.

9 Becker, J., Pérot, P., Cheynet, V. et al. (2017). A comprehensive hybridization model allows whole HERV transcriptome profiling using high density microarray. *BMC Genomics* 18: https://doi.org/10.1186/s12864-017-3669-7.

10 Pérot, P., Mugnier, N., Montgiraud, C. et al. (2012). Microarray-based sketches of the HERV transcriptome landscape. *PLoS One* 7: e40194. https://doi.org/10.1371/ journal.pone.0040194.

11 Lu, X., Sachs, F., Ramsay, L. et al. (2014). The retrovirus HERVH is a long noncoding RNA required for human embryonic stem cell identity. *Nat. Struct. Mol. Biol.* 21: 423–425. https://doi.org/10.1038/nsmb.2799.

12 Dupressoir, A., Lavialle, C., and Heidmann, T. (2012). From ancestral infectious retroviruses to bona fide cellular genes: role of the captured syncytins in placentation. *Placenta* 33: 663–671.

13 Marchi, E., Kanapin, A., Magiorkinis, G., and Belshaw, R. (2014). Unfixed endogenous retroviral insertions in the human population. *J. Virol.* 88 (17): 9529–9537. https://doi.org/10.1128/JVI.00919-14.

14 de la Hera, B., Varadé, J., García-Montojo, M. et al. (2013). Role of the human endogenous retrovirus HERV-K18 in autoimmune disease susceptibility: study in the Spanish population and meta-analysis. *PLoS One* 8: e62090. https://doi .org/10.1371/journal.pone.0062090.

15 Dolei, A. and Perron, H. (2009). The multiple sclerosis-associated retrovirus and its HERV-W endogenous family: a biological interface between virology, genetics, and immunology in human physiology and disease. *J. Neurovirol.* 15: 4–13. https://doi .org/10.1080/13550280802448451.

16 Jo, J.-O., Kang, Y.-J., Ock, M.S. et al. (2016). Expression profiles of HERV-K Env protein in normal and cancerous tissues. *Genes Genomics* 38: 91–107. https://doi .org/10.1007/s13258-015-0343-9.

17 Karamitros, T., Hurst, T., Marchi, E. et al. (2018). Human endogenous retrovirus-K HML-2 integration within *RASGRF2* is associated with intravenous drug abuse and modulates transcription in a cell-line model. *Proc. Natl. Acad. Sci. U.S.A.* 115: 10434–10439. https://doi.org/10.1073/pnas.1811940115.

18 Li, W., Yang, L., Harris, R.S. et al. (2019). Retrovirus insertion site analysis of LGL leukemia patient genomes. *BMC Med. Genet.* 12: 88. https://doi.org/10.1186/ s12920-019-0549-9.

19 Cancer Research UK (n.d.). Cancer incidence for common cancers. https://www .cancerresearchuk.org/health-professional/cancer-statistics/incidence/common- cancers-compared#heading-Two (accessed 24 August 2021).

20 Joshi, D. and Buehring, G.C. (2012). Are viruses associated with human breast cancer? Scrutinizing the molecular evidence. *Breast Cancer Res. Treat.* 135: 1–15. https://doi.org/10.1007/S10549-011-1921-4.

21 Cancer Research UK (n.d.). Types of breast cancer and related conditions. https:// www.cancerresearchuk.org/about-cancer/breast-cancer/stages-types-grades/types (accessed 25 August 2021).

22 Provenzano, E., Ulaner, G.A., and Chin, S.-F. (2018). Molecular classification of breast cancer. *PET Clin.* 13: 325–338. https://doi.org/10.1016/J.CPET.2018.02.004.

23 Liu, Z., Li, M., Jiang, Z., and Wang, X. (2018). A comprehensive immunologic portrait of triple-negative breast cancer. *Transl. Oncol.* 11: 311–329. https://doi.org/10.1016/J.TRANON.2018.01.011.

24 Zhou, F., Li, M., Wei, Y. et al. (2016). Activation of HERV-K Env protein is essential for tumorigenesis and metastasis of breast cancer cells. *Oncotarget* 7: 84093–84117. https://doi.org/10.18632/oncotarget.11455.

25 Wei, Y., Wei, H., Wei, Y. et al. (2022). Screening and identification of human endogenous retrovirus-K mRNAs for breast cancer through integrative analysis of multiple datasets. *Front. Oncol.* 12: https://doi.org/10.3389/FONC.2022.820883.

26 Steiner, M.C., Marston, J.L., Iñiguez, L.P. et al. (2021). Locus-specific characterization of human endogenous retrovirus expression in prostate, breast, and colon cancers. *Cancer Res.* 81: 3449–3460. https://doi.org/10.1158/0008-5472.CAN-20-3975/670861/AM/LOCUS-SPECIFIC-CHARACTERIZATION-OF-HUMAN.

27 Kaplan, M.H., Contreras-Galindo, R., Jiagge, E. et al. (2020). Is the HERV-K HML-2 Xq21.33, an endogenous retrovirus mutated by gene conversion of chromosome X in a subset of African populations, associated with human breast cancer? *Infect. Agents Cancer* 15: 19. https://doi.org/10.1186/s13027-020-00284-w.

28 Pisignano, G., Pavlaki, I., and Murrell, A. (2019). Being in a loop: how long non-coding RNAs organise genome architecture. *Essays Biochem.* 63: 177–186. https://doi.org/10.1042/EBC20180057.

29 Ransohoff, J.D., Wei, Y., and Khavari, P.A. (2017). The functions and unique features of long intergenic non-coding RNA. *Nat. Rev. Mol. Cell Biol.* 19: 143–157. https://doi.org/10.1038/nrm.2017.104.

30 Young, R.S. and Ponting, C.P. (2013). Identification and function of long non-coding RNAs. *Essays Biochem.* 54: 113–126. https://doi.org/10.1042/bse0540113.

31 Van Grembergen, O., Bizet, M., de Bony, E.J. et al. (2016). Portraying breast cancers with long noncoding RNAs. *Sci. Adv.* 2 (9): e1600220. https://doi.org/10.1126/sciadv.1600220.

32 Sun, M., Wu, D., Zhou, K. et al. (2019). An eight-lncRNA signature predicts survival of breast cancer patients: a comprehensive study based on weighted gene co-expression network analysis and competing endogenous RNA network. *Breast Cancer Res. Treat.* 175: 59–75. https://doi.org/10.1007/S10549-019-05147-6.

33 Jin, X., Xu, X.-E., Jiang, Y.-Z. et al. (2019). The endogenous retrovirus-derived long noncoding RNA TROJAN promotes triple-negative breast cancer progression via ZMYND8 degradation. *Sci. Adv.* 5: eaat9820. https://doi.org/10.1126/sciadv.aat9820.

34 Cruickshank, B.M., Wasson, M.D., Brown, J.M. et al. (2021). lncRNA PART1 promotes proliferation and migration, is associated with cancer stem cells, and alters the miRNA landscape in triple-negative breast cancer. *Cancers (Basel)* 13: https://doi.org/10.3390/CANCERS13112644.

35 Tsujikawa, T., Kumar, S., Borkar, R.N. et al. (2017). Quantitative multiplex immunohistochemistry reveals myeloid-inflamed tumor-immune complexity

associated with poor prognosis. *Cell Rep.* 19: 203–217. https://doi.org/10.1016/j
.celrep.2017.03.037.

36 MD Anderson Cancer Center (n.d.). ESTIMATE: home. https://bioinformatics
.mdanderson.org/estimate/ (accessed 26 August 2021).

37 Qiu, S.Q., Waaijer, S.J.H., Zwager, M.C. et al. (2018). Tumor-associated macrophages
in breast cancer: innocent bystander or important player? *Cancer Treat. Rev.* 70:
178–189. https://doi.org/10.1016/j.ctrv.2018.08.010.

38 Carpenter, S., Aiello, D., Atianand, M.K. et al. (2013). A long noncoding RNA
mediates both activation and repression of immune response genes. *Science* 341:
789–792. https://doi.org/10.1126/science.1240925.

39 Chen, S., Hu, X., Cui, I.H. et al. (2019). An endogenous retroviral element exerts an
antiviral innate immune function via the derived lncRNA lnc-ALVE1-AS1. *Antivir.
Res.* 170: 104571. https://doi.org/10.1016/j.antiviral.2019.104571.

40 Curty, G., Beckerle, G.A., Iñiguez, L.P. et al. (2020). Human endogenous retrovirus
expression is upregulated in the breast cancer microenvironment of HIV infected
women: a pilot study. *Front. Oncol.* 10. https://doi.org/10.3389/FONC.2020.553983.

41 Li, J., Lin, J., Lin, J.R. et al. (2022). Dolutegravir inhibits proliferation and motility
of BT-20 tumor cells through inhibition of human endogenous retrovirus type
K. *Cureus* 14: https://doi.org/10.7759/CUREUS.26525.

42 Zhou, F., Krishnamurthy, J., Wei, Y. et al. (2015). Chimeric antigen receptor T cells
targeting HERV-K inhibit breast cancer and its metastasis through downregulation
of Ras. *Oncoimmunology* 4: e1047582. https://doi.org/10.1080/2162402X.2015
.1047582.

43 Broecker, F., Horton, R., Heinrich, J. et al. (2016). The intron-enriched HERV-K
(HML-10) family suppresses apoptosis, an indicator of malignant transformation.
Mob. DNA 7: 25. https://doi.org/10.1186/s13100-016-0081-9.

44 Peperstraete, E., Lecerf, C., Collette, J. et al. (2020). Enhancement of breast cancer
cell aggressiveness by lncRNA H19 and its Mir-675 derivative: insight into shared
and different actions. *Cancers (Basel)* 12: 1–20. https://doi.org/10.3390/
CANCERS12071730.

45 Fahad Ullah, M. (2019). Breast cancer: current perspectives on the disease status.
Adv. Exp. Med. Biol. 1152: 51–64. https://doi.org/10.1007/978-3-030-20301-6_4.

46 Hohn, O., Hanke, K., and Bannert, N. (2013). HERV-K(HML-2), the best preserved
family of HERVs: endogenization, expression, and implications in health and
disease. *Front. Oncol.* 3: 246. https://doi.org/10.3389/fonc.2013.00246.

47 Krishnamurthy, J., Rabinovich, B.A., Mi, T. et al. (2015). Genetic engineering of T
cells to target HERV-K, an ancient retrovirus on melanoma. *Clin. Cancer Res.* 21:
3241–3251. https://doi.org/10.1158/1078-0432.CCR-14-3197.

48 Zhou, S.-Y., Chen, W., Yang, S.-J. et al. (2019). The emerging role of circular RNAs
in breast cancer. *Biosci. Rep.* 39: BSR20190621. https://doi.org/10.1042/
BSR20190621.

Part II

Novel Drug Discovery and Development

6

Development in Drug Repurposing for the Treatment of Acute Leukemia Complicating Metastatic Breast Cancer

Nilophar M. Shaikh¹, Vinayak Adimule² and Santosh Nandi³

¹*Visvesvaraya Technological University, Recognized Research Centre, Angadi Institute of Technology and Management, Department of Chemistry, Sagaon road, Belagavi, Karnataka, 590018, India*
²*Angadi Institute of Technology and Management, Department of Chemistry, Sagaon road, Belagavi, Karnataka, 590018, India*
³ *KLE Technological University Dr. MSSCET, Chemistry Section, Department of Engineering Science and Humanities, Belagavi, Karnataka, 590008, India*

6.1 Introduction

In the year 2020, a total of 18 989 634 new cases of cancer have been detected, out of which 10 052 507 resulted in death. This number indicates that cancer has become a crucial health problem all over the world [1]. One of the major cases of cancer commonly detected in women is breast cancer. Within the last three years, nearly about two million new cases of breast cancer have been identified, out of which about 627 000 are cases of death [2]. Breast cancer is diagnosed by tissue biopsies, physical examination, and breast scan. It is found that maximum cases of breast cancer were detected in Asia followed by Europe, North America, Latin America, the Caribbean, Africa, and Oceania (Table 6.1) (Figure 6.1) [2].

In very minor cases, it is observed in men [3, 4]. Depending on expressions of ER/PR and HER2, there are four types of breast cancer (i) ER/PR+, HER2+; (ii) ER/PR−, HER2+; (iii) ER/PR+, HER2−; and (iv) ER/PR−, HER2− [5]. Among these, ER/PR−, HER2− is highly aggressive breast cancer, which is also known as triple-negative breast cancer (TNBC). About 70% of breast cancer is positive hormone receptors, which can be cured by respective therapies [5, 6]. There are some biomarkers that are used in different pathological, molecular, and clinical, such as Ki67, CEA, p53, CA153, and BRCA1/BRCA2, for breast cancer characterization. These biomarkers are commonly used in staging, diagnosis, grading, prognosis, clinical management, and therapeutic intervention of recurrent cases and metastatic cases [7].

Drug and Therapy Development for Triple Negative Breast Cancer, First Edition.
Edited by Pravin Kendrekar, Vinayak Adimule, and Tara Hurst.
© 2023 WILEY-VCH GmbH. Published 2023 by WILEY-VCH GmbH.

Table 6.1 Cases of breast cancer detected.

Country	Cases detected
Asia	911 014
Europe	522 512
North America	262 347
Latin America and Caribbean	199 734
Africa	168 690
Oceania	24 551

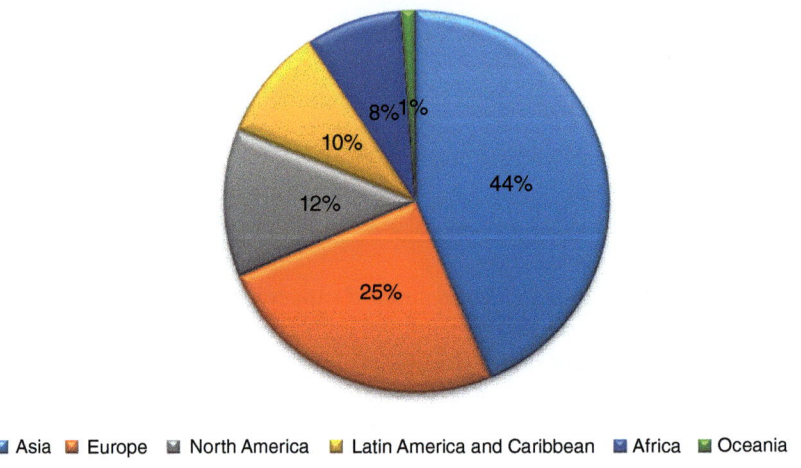

Figure 6.1 Cases of breast cancer detected.

Nowadays, based on the stage, size, grade, aggressiveness, intrinsic molecular subtyping, metastatic behavior of the tumor, overall health, menopausal status, age, preferences of the patient, and comorbidities various choices are available for the treatment of breast cancer [8–10]. This treatment includes immunotherapy, chemotherapy, radiotherapy, hormone therapy, and surgery [8–11]. Surgery is mostly the primary choice to carry out major tumor mass complete resection. To increase the survival chances of the patients and reduce the recurrence chances by destroying the remnant of micrometastatic cancer cells, surgery is commonly followed by radiotherapy or chemotherapy [11, 12]. Based on hormone receptor expression or target protein adjuvant, therapies such as targeted therapy or hormone therapy are also provided to such patients of breast cancer [8]. TNBC patients are treated with standard chemotherapy along with DNA-targeting platinum drug carboplatin or PARP inhibitors as the prognosis is unfavorable in such cases [13, 14]. In some cases, holistic approach is also preferred that involves the recovery of the patient by complementary as well as alternative medicine along with several changes in lifestyle such as meditation, exercise, acupuncture, yoga, and ayurvedic treatments [15]. A number of drugs are approved by the United States FDA for the treatment of breast cancer, which are costly and show various side effects such as headaches, dental issues, pain, numbness, lymphedema, heart problems, menstrual, blood clots, bone loss, memory loss, and infertility [5, 16].

In the last few decades, adjuvant polychemotherapy has been used as an effective treatment for the reduction of cancer recurrence [17]. Consequently, it raises the benefit–risk ratio in which the long-term risk of complications also increases. One of the major complications observed in cancer therapy is leukemia, which is majorly observed in the case of chemotherapy carried out by using topoisomerase II inhibitors like epipodophyllotoxins [18] or by using different alkylating agents like melphalan [19, 20].

According to the report of Chaplain et al carried through the study population based on breast cancer diagnosed between 1 January 1982 to 31 December 1996 in women, from 3450 women 3540 women, the first diagnosed with breast cancer, 195 were detected metastases, 119 were unknown to the surgical treatment, and 133 women were older than 85 years [20]. Chaplain and coworkers carried out a cohort study of over 3093 women who were younger than 85 years. In this study, they reported that till December 1998, 10 cases of acute leukemia were observed. These acute leukemia cases include refractory anemia with excess blasts and nonlymphoid acute leukemia. It is observed that the rate of acute leukemia is significantly more in those women who received chemotherapy and radiotherapy as compared to general females. According to Cox regression analysis, the risk of acute leukemia is high in those patients who were treated with mitoxantrone rather than anthracyclines. The combination of mitoxantrone with adjuvant chemotherapy and radiotherapy increases the high risk of leukemia in patients with breast cancer [20].

6.1.1 Acute Leukemia's

Acute leukemia in terms of morbidity as well as mortality ranks within the first top 10 cancers [21]. Acute leukemia is generally detected by the existence of more than 20% blast cells in the bone marrow or peripheral blood [22]. Acute leukemia have worse prognosis as compared to chronic leukemia because of its inherent characteristics regarding the reduction in differentiation state as well as its ability to propagate rapidly. Acute myelogenous leukemia (AML) is prominently observed in adults more than 65 years of age. About 21 450 cases are annually detected in the United States with 10 920 succumbing cases [21]. Acute leukemia is primarily classified based on translocations with immunophenotyping and the presence of chromosomal abnormalities.

The case study done by Chaplain et al. described that the risk of acute leukemia was observed in those patients who received doses of mitoxantrone in postoperative protocol they are along with other cytostatic drugs. Hence, along with mitoxantrone, other drugs are also responsible partly for their synergistic effect in increasing the risk of acute leukemia. Acute leukemia is initially treated with induction therapy in which the use of chemotherapeutic agents induces the remission of the disease.

The main key role of this treatment is to decrease the concentration of blast cells and induce their differentiation. Drugs that are commonly used in the treatment of cancer and related disease are used; therefore, their mechanism over acute leukemia is not specific. The aim behind the use of these drugs is to kill all cells that interfere with the replication mechanism of the cells as well as proliferates through inducing damage to DNA. In the case of ALL and acute myeloid leukemia (AML) patients, transplant of allogeneic stem cells is also used [22]. Use of cytosine arabinoside along with anthracyclines, such as idarubicin, daunorubicin, and doxorubicin, is also considered in terms of standard care and gives a good response in near about 80% of patients [22, 23]. Another method used is consolidation therapy, which includes a high dose of ArcC [16].

High dose of AraC or FLAG is also used in the treatment of relapsed or refractory AML with a complete remission rate of 47.5% and 32%, respectively. In the case of B-ALL, induction therapy takes four to six weeks of asparaginase, glucocorticoid, vincristine, and anthracycline. This method requires consolidation with some maintenance therapies including treatment with some induction agents like methotrexate, 6-thioguanine, or 6-mercaptopurine [24].

The remission of B-ALL, and AML is determined by minimal residual disease of bone marrow. In the case of B-ALL, the fusion protein ETV6-RUNX1 gives a better treatment response with a good outcome [25], whereas mixed-lineage leukemia (KMT2A) gene rearrangement is comparatively less favorable [25–27]. It has been observed that the response of the different patients is different for various treatments. Patient with similar cytogenetic factors shows the opposite response to the same therapy. In the case treatment of acute leukemia's selectivity for leukemic cells plays a very important role. Lack of selectivity causes systemic damage by the use of leukemia chemotherapeutic agents (LCAs). The use of various chemotherapeutics and anthracyclines causes cardiotoxicity, as cancer cells are metabolically active. ALL survivors increase the high risk of heart disease [28].

Whereas LCA causes damage to the primary organs of metabolism kidneys and liver. Similarly, ALL patients suffer from the risk of osteonecrosis [29, 30]. Leukemia treatment also shows long-term cognitive impairment such as memory, concentration, and attention with executive function deficits [31, 32]. Taking into consideration all these side effects, it becomes very essential to develop drugs that will be highly selective to malignant cells without producing any side effects.

Cyclic adenosine monophosphate (cAMP) is $3',5'$-cyclic adenosine monophosphate, which is the first and second messenger within the cell and is critically important for many cellular processes like survival, regulating growth, differentiation, and transcription of myriad genes. Depending on the type of cell and condition, cAMP signal transduction results either in prosurvival activation or death pathways [33, 34]. Canonical cAMP synthesis includes any one of the nine transmembranes associated with adenylyl cyclases (tmACs) activation stimulated employing Gα-coupled receptors. These receptors are highly active in the stimulation of cAMP within endosomes [35]. These receptors act as inhibitor for the tmAC activity and reduce the intracellular cAMP. The regulation of cAMP levels within the cell membrane vicinity is envisioned by tmACs. In addition to that, cAMP signaling is segregated spatially throughout the compartment, of cytosolic. In this cytosolic compartment, different enzymes such as adenylyl cyclase produce a second messenger. These enzymes also produce cAMP cytosolic microdomains like cytoplasm or in mitochondria or inside the nucleus [36–38]. sAC can be activated by oxidative stress or by carbonates [37, 39–42]. PKA signaling triggers intrinsic apoptosis by elevated icAMP. Where PKA has the ability to trigger upregulation of the proapoptotic protein Bim expression by activating transcription factor CREB [43]. Translocation of proapoptotic Bax is caused due to sAC stimulation of PLA into the mitochondria [44, 45]. The activation of mitochondria related to the intrinsic apoptotic pathway is mainly done by different signaling events.

Two principle mechanisms are to be used for the modulation of icAMP concentrations. Phosphodiesterases (PDEs) is one of the most regularly studied cAMP regulators. Hydrolysis of cyclic nucleotides has been carried out by these enzymes. Active removal of cAMP is implied by the second mechanism from the cytosol. It was mentioned for the first time in 1963 that releasing of cAMP can be done outside the cell [46]. Further, it was reported that these icAMP can be removed actively from cytosol through the ATP-binding cassette transporters. It is also reported that efflux cAMP can be carried out by multidrug resistance family members such as ABCC4(MRP4), ABCC11 (MRP8), and ABCC5 (MRP5) [47]. There are many cAMP-dependent regulatory proteins that can be complexed to facilitate signal transduction, as cAMP activity take place in discrete locations. In the case of A-kinase anchoring proteins, they allow PKA, adenylyl cyclases, and PDEs to localize close to one another [48]. Hence, compartmentalization of icAMP is a critical determinant near mitochondria, plasma membrane, nucleus, and regulatory proteins [49–55].

The increase in icAMP regulation in hematopoietic malignancies reduces the survival of white blood cells. In acute leukemias, many of the cAMP-associated proteins become dysregulated. In the case of AML patients, the overall survival is inversely correlated with the expression of adenylyl cyclase type 7 protein expressed in lymphoid cells [56]. It has been reported that overall PDE activity is 10–20 times higher in lymphoma and leukemia cells as compared to normal blood cells [57]. PDE activity has also been downregulated by the use of glucocorticoids during the treatment of ALL [58]. Adult ALL and primary AML samples on genomic studies show that CREB and its active phosphorylated form are overexpressed [57, 59–63]. On the same contrary physiological antagonist of CREB as well as their downregulation is related to the tumor suppression activity [64]. In the case of the B-ALL sample, it has been observed that an increase in mutation takes place in transcripts of CREB as well as in binding protein for CREB is correlated with deleted or dominant negative activity [41, 65]. Hence, acute leukemia cells reveal increase in CREB signaling that results in an increased activity of prosurvival cAMP as well as proapoptotic factor ICER downregulation [66].

6.1.2 Mitochondrial cAMP–PKA Signaling

cAMP produced by sAC and tmACs within the cytosol can easily reach the inner membrane of mitochondrial space as the outer membrane is permeable to small molecules [21]. cAMP is produced by sAC inside the mitochondrial matrix [36, 37]. Here, in mitochondria, protein import [67–69], oxidative phosphorylation [43], mitochondrial fusion and fission, mitophagy [70], metabolic reprogramming, the anti-Warburg effect [71], and intrinsic apoptotic pathway [43] can be modulated by cAMP–PKA signaling pathway. Most of these methods easily get affected by cAMP that diffuses the outer membrane of mitochondrial. Whereas anti-Warburg effect, metabolic reprogramming, and oxidative reprogramming need changes in cAMP level in the mitochondrial matrix. There is a direct effect of blocking of cAMP efflux over the import of protein, mitophagy as well as mitochondria-induced apoptosis. ICE has tend to trigger four apoptotic endpoints such as activation of effector caspases 3, 7 and mitochondrial membrane depolarization [72] Hence, it is to be considered as cAMP which is produced by tmAC triggers the AML cell apoptosis at the outer mitochondrial membrane. TOMM 7 regulates the assembly as well as stability of the main translocase complex [73]. The aberrant protein import in case of cancer is related to the upregulation of this TOM7 within the AML cell. About 99% of precursor proteins inside the mitochondria provided entry points from TOM complexes hence protein translocation defects in mitochondria devast the organelle function [74, 75]. Protein importing through TOM can be diminished by cAMP–PKA signaling that further triggers switches to glycolysis from the OXPHOS [34].

6.1.3 Nuclear Compartment

Activation of target genes, which takes place through cyclic AMP response elements (CREs), has been carried out by classical cAMP effectors such as nuclear compartment CREB as well as activating transcription factor 1. This directly shows its effect on cAMP-induced apoptosis of leukemia [37]. As mitochondrial proteome originates largely from nuclear DNA as well as CREB with nuclear respiratory factors (NRFs), downstream transcription factors and peroxisome proliferator-activated receptor (PGC-1α) promote the biogenesis of mitochondria by activating numerous mitochondrial genes activation [76, 77].

6.1.4 Cytosolic Compartment and Plasma Membrane

Before the introduction of compartmentalization of cAMP signaling , cytosol and plasma membrane were to be considered as their main site. As the activity of tmAC is to be controlled by Gαi-coupled related to stimulating cAMP production and Gαs-coupled related with inhibition receptor are localized majorly in the plasma membrane. The cAMP that is produced by tmACs gets diffused in the cytosol. In the case of CXCR4 and CXCR7, both are relevant to leukemia as well as other cancers [78]. Gα-coupled H2-histamine receptor signaling is used in the treatment of

AML [79]. The use of ICE in cAMP elevation in the cytosolic compartment is also beneficial on the same [72]. To sort out the problem regarding the elevation of cAMP level resulting from the H2-histamine receptor stimulation, they are combined with ICE. Modulation of cAMP level can be done by cAMP efflux through the MRP transporters [79]. Activation of Gαs-coupled G protein-coupled with estrogen receptor also improves the immune checkpoint therapy [22]. The presence of a definite number of immature phenotype cells shows hematological malignancies. Near about 1/5th of cells shows blast morphology in peripheral blood in case of acute leukemias [22]. From these, it has been clear that leukemia cells have some defects among their adhesion molecules, which causes a premature release in peripheral blood from the bone marrow niche.

6.2 Conclusion

Here, we have described different treatments for the acute leukemia. We highlighted the main role of cAMP signaling in cell proliferation, apoptosis, and survival. The synergistic effect of cAMP efflux transporter along with different therapeutics in combination with ICE produces chemotherapy resistance against acute leukemia. This combination effectively reduces the leukemia cell burden. There is a large scope of development of novel therapeutics through the repurposing pathway as various approved drugs show ICE activity. Hence, a new class of drugs for leukemia can be provided using cAMP efflux inhibitor compounds.

Conflict of Interest

The author declared that they do not have a conflict of interest.

References

1 Riggio, A., Varley, K., and Welm, A. (2021). The lingering mysteries of metastatic recurrence in breast cancer. *Br. J. Cancer* 124: 13–26. https://doi.org/10.1038/s41416-020-01161-4.

2 Aggarwal, A.S., Verma, S.S., Aggarwal, S., and Gupta, S.C. (2021). Drug repurposing for breast cancer therapy: old weapon for new battle. *Semin. Cancer Biol.* 68: 8–20.

3 Yalaza, M., Inan, A., and Bozer, M. (2016). Male breast cancer. *J. Breast Health* 12 (1): 1–8.

4 Onitilo, A., Engel, J., Greenlee, R., and Mukesh, B. (2009). Breast cancer subtypes based on ER/PR and Her2 expression:comparison of clinicopathologic features and survival. *Clin. Med. Res.* 7 (1, 2): 4–13.

5 Waks, A. and Winer, E. (2019). Breast cancer treatment: a review. *JAMA* 321 (3): 288–300.

6 Weigelt, B. and Reis-Filho, J. (2010). Molecular profiling currently offers no more than tumour morphology and basic immunohistochemistry. *Breast Cancer Res.* 12 (4): S5.

7 Inoue, K. and Fry, E. (2016). Novel molecular markers for breast cancer. *Biomark. Cancer* 8: 25–42.

8 Adimule, V., Medapa, S., Rao, P.K., and Kumar, L.S. (2014). Synthesis of Schiff bases of 5-[5-(4-fluorophenyl) thiophen-2-yl]-1,3,4-thiadiazol-2-amine and its anticancer activity. *Int. J. Adv. Pharm. Sci.* 5 (1): 1761–1768.

9 Adimule, V., Yallur, B.C., Bhowmik, D. et al. (2021). Morphology, structural and photoluminescence properties of shaping triple semiconductor $Y_xCoO:ZrO_2$ nanostructures. *J. Mater. Sci.: Mater. Electron.* 32: 12164–12181. https://doi.org/10.1007/s10854-021-05845-2.

10 Nounou, M., Eiamrawy, F., Ahmed, N. et al. (2015). Breast cancer: conventional diagnosis and treatment modalities and recent patents and technologies. *Breast Cancer: Basic Clin. Res.* 9 (2): 17–34.

11 Dhankhar, R., Vyas, S., Jain, A. et al. Advances in novel drug delivery statergies for breast cancer therapy. *Artif. Cells, Blood Substitutes, Immobilization Biotechnol.* 38 (5): 230–249.

12 Adimule, V., Yallur, B.C., and Sharma, K. (2022). Studies on crystal structure, morphology, optical and photoluminescence properties of flake-like Sb doped Y_2O_3 nanostructures. *J. Opt.* 51: 173–183. https://doi.org/10.1007/s12596-021-00746-3.

13 Robson, M., Im, S., Senkus, B. et al. Olaparib for metastatic breast cancer in patients with a Germline BRCA mutation. *N. Engl. J. Med.* 377 (6): 523–533.

14 Telli, M., Hellyer, J., Audeh, W. et al. (2018). Homologous recombination deficiency (HRD) status predicts response to standard neoadjuvant chemotherapy in patients with triple-negative or BRCA1/2 mutation-associated breast cancer. *Breast Cancer Res. Treat.* 168 (3): 625–630.

15 Neuhouser, M., Smith, A., George, S. et al. Use of complementary ans alternative medicine and breast cancer survival in the Health, eating, activity and lifestyle study. *Breast Cancer Res. Treat.* 160 (3): 539–546.

16 Kouchkovsky, I. and Abdul-Hay, M. (2016). Acute myeloid leukemia: a comprehensive review and 2016 update. *Blood Cancer J.* 6 (7): e441.

17 Early on Cancer Trialists' Collaborative Group (1993). Polychemotherapy for early breast cancer: an overview of the randomised trials. *Lancet* 352: 930–942.

18 Adimule, V., Suryavanshi, A., Yallur, B.C., and Nandi, S.S. (2020). A facile synthesis of poly(3-octyl thiophene): $Ni_{0.4}Sr_{0.6}TiO_3$ hybrid nanocomposites for solar cell applications. *Macromol. Symp.* 392: 2000001. https://doi.org/10.1002/masy.202000001.

19 Pedersen-Bjergaard, J., Daugaard, G., Hansen, S. et al. (1991). Increased risk of myelodysplasia and leukaemia after etoposide, cisplatin and bleomycin for germ-cell tumors. *Lancet* 338: 359–363.

20 Chaplain, G., Milan, C., Sgro, C., and Carli, P. (2000). Bonithon-Kopp C (2000) increasd risk of acute leukemia after adjuvant chemotherapy for breast cancer: a population-based study. *J. Clin. Oncol.* 18 (15): 2836–2842.

21 Adimule, V., Nandi, S.S., Jagadeesha, G., and Haramballi, A. (2020). Enhanced power conversion efficiency of the P3BT (poly-3-butyl thiophene) doped nanocomposites of Gd-TiO$_3$ as working electrode. In: *Techno-Societal* (ed. P.M. Pawar, R. Balasubramaniam, B.P. Ronge, et al.). Cham: Springer https://doi .org/10.1007/978-3-030-69925-3_6.

22 Kansal, R. (2016). Acute myeloid leukemia in the era of precision medicine:recent advances in diagnostic classification and risk stratification. *Cancer Biol. Med.* 13 (1): 41–54.

23 Walter, R., Othus, M., Burnett, A. et al. (2014). Resistance prediction in AML: analysis of 4601 patients from MRC/NCRI, HOVON/SAKK, SWOG and MD Anderson cancer center. *Leukemia* 29: 312.

24 Adimule, V., Bowmik, D., and Adarsha, H.J. (2020). A facile synthesis of Cr doped WO$_3$ nanocomposites and its effect in enhanced current-voltage and impedance characteristics of thin films. *Lett. Mater.* 10 (4): 481–485. https://doi.org/10.22226/ 2410-3535-2020-4-481-485.

25 Moorman, A., Ensor, H., Richards, S. et al. (2010). Prognostic effect of chromosomal abnormalities in childhood B-cell precursor acute lymphoblastic leukemia: results from the UK medical Research Council ALL97/99 randomised trial. *Lancet Oncol.* 11 (5): 429–438.

26 Kang, H., Wilson, C., Harvey, R. et al. (2012). Gene expression profiles predictive of outcome and age in infant acute lymphoblastic leukemia: a Children's Oncology Group study. *Blood* 119 (8): 1872–1881.

27 Matlawska-Wasowska, K., Kang, H., Devidas, J. et al. (2016). MLL rearrangement impact outcome in HOXA-deregulated T-lineage acute lymphoblastic leukemia: a children's oncology group study. *Leukemia* 30: 1909.

28 Adimule, V., Yallur, B.C., Challa, M., and Joshi, R.S. (2021). Synthesis of hierarchical structured Gd doped α-Sb$_2$O$_4$ as an advanced nanomaterial for high performance energy storage devices. *Heliyon* 7 (12): e08541. https://doi .org/10.1016/j.heliyon.2021.e08541.

29 Kadan-Lottick, N., Dinu, I., Wasilewski-Masker, K. et al. Osteonecrosis in adult survivors of childhood cancer: a report from the childhood cancer survivor study. *J. Clin. Oncol.* 26 (18): 3038–3045.

30 Te Winkel, M., Pieters, R., Wind, E. et al. (2014). Management and treatment of osteonecrosis in children and adolescents with acute lymphoblastic leukemia. *Haematologica* 99 (3): 430–436.

31 Krull, K., Brinkman, T., Li, C. et al. (2014). Neurocognitive outcomes decades after treatment for childhood acute lymphoblasic leukemia: a report from the St Jude lifetime cohort study. *J. Clin. Oncol.* 31 (35): 4407–4415.

32 Duffner, P., Armstrong, F., Chen, L. et al. (2014). Neurocognitive and neuroradiologic central nervous system late effects in children treated on pediatric Oncology group. *J. Pediatr. Hematol. Oncol.* 36 (1): 8–15.

33 Insel, P., Zhang, L., Murry, F. et al. (2012). Cyclic AMP is both a pro-apoptotic and anti-apoptotic second messenger. *Acta Physiol. (Oxford)* 204 (2): 277–287.

34 Adimule, V., Vageesha, P., Bagihalli, G. et al. (2019). Synthesis characterization of hybrid nanomaterials of strontium, yttrium, copper doped with indole Schiff base derivatives possessing dielectric and semiconductor properties. In: *Emerging Research in Electronics, Computer Science and Technology*, Lecture Notes in Electrical Engineering, vol. 545 (ed. V. Sridhar, M. Padma, and K. Rao). Singapore: Springer https://doi.org/10.1007/978-981-13-5802-9_97.

35 Antoni, F. (2012). New paradigms in cAMP signalling. *Mol. Cell. Endocrinol.* 353 (1, 2): 3–9.

36 Adimule, V., Medapa, S., Adarsha, H.J. et al. (2014). Synthesis, characterization and in vitro anticancer properties of 1-{5-aryl-2-[5-(4-fluoro-phenyl)-thiophen-2-yl]-[1,3,4] oxadiazol-3-yl}-ethanone. *Int. J. Pharma Res. Rev.* 3 (12): 20–25.

37 Acin-Perez, R., Salazar, E., Kamenetsky, M. et al. (2009). Cyckic AMP produced inside mitochondria regulates oxidative phosphorylation. *Cell Metab.* 9 (3): 265–276.

38 Zippin, J., Farrell, J., Huron, D. et al. (2004). Bicarbonate-responsive "soluble" adenylyl cyclase defines a nuclear cAMP microdomain. *J. Cell Biol.* 164 (4): 527–534.

39 Steeborn, C. (2014). Structure, mechanism and regulation of soluble adenylyl cyclases-similarities and differences to transmembrane adenylyl cyclases. *Biochim. Biophys. Acta* 1842: 2535–2547.

40 Chen, Y., Cann, M., Litvin, T. et al. (2000). Soluble adenylyl cyclase as an evolutionarily conserved bicarbonate sensor. *Science* 289 (5479): 625–628.

41 Adimule, V., Yallur, B., and Gowda, A. (2022). Crystal structure, morphology, optical and super-capacitor properties of Srx: α-Sb$_2$O$_4$ nanostructures. *Anal. Bioanal. Electrochem.* 14 (1): 1–17.

42 Kumar, S., Appukuttan, A., Maghnouj, A. et al. (2014). Suppression of soluble adenylyl cyclase protects smooth muscle cells against oxidative stress-induced apoptosis. *Apoptosis* 19 (7): 1069–1079.

43 Husbey, S., Gausdal, G., Keen, T. et al. (2011). Cyclic AMP induces IPC leukemia cell apoptosis via CRE-And CDK-dependent bim transcription. *Cell Death Dis.* 2: e237.

44 Kumar, S., Kostin, S., Flacke, J. et al. (2009). Soluble adenylyl cyclase controls mitochondria-dependent apoptosis in coronary endothelial cells. *J. Biol. Chem.* 284 (22): 14760–14768.

45 Appuluttan, A., Kasseckert, S., Kumar, S. et al. (2013). Oxysterol-induced apoptosis of smooth muscle cells is under the control of a soluble adenylyl cyclase. *Cardiovasc. Res.* 99 (4): 734–742.

46 Adimule, V., Yallur, B.C., and Gowda, A.H.J. (2022). Chapter 14 – Advanced sensors based on carbon nanomaterials. In: *Carbon Nanomaterials-Based Sensors* (ed. J.G. Manjunatha and C.M. Hussain), 259–268. Elsevier https://doi.org/10.1016/B978-0-323-91174-0.00004-4.

47 Davoren, P. and Sutherland, E. (1963). The effect of L-Epinephrine and other agents on the synthesis and release of adenosine 3′,5′-phosphate by whole pigeon erythrocytes. *J. Biol. Chem.* 238: 3009–3015.

48 Mohan Gift, M.D., Pattnaik, B., Nandi, S.S. et al. (2022). Determination of prohibition mechanism of cationic polymer/SiO$_2$ composite as inhibitor in water using drilling fluid. *Mater. Today: Proc.* https://doi.org/10.1016/j.matpr.2022.08.171.

49 Lefkimmiatis, K. and Zaccolo, M. (2014). cAMP signaling in subcellular compartments. *Pharmacol. Ther.* 143 (3): 295–304.

50 Adimule, V., Nandi, S.S., and Yallur, B.C. (2022). Devices and sensors based on additively manufactured shape-memory of hybrid nanocomposites. In: *Shape Memory Composites Based on Polymers and Metals for 4D Printing* (ed. M.R. Maurya, K.K. Sadasivuni, J.J. Cabibihan, et al.). Cham: Springer https://doi .org/10.1007/978-3-030-94114-7_15.

51 Arora, K., Sinha, C., Zhang, W. et al. (2013). Compartmentalization of cyclic nucleotide signaling: a question of when, where and why? *Pflugers Arch.* 465 (10): 1397–1407.

52 Murray, F. and Insel, P. (2013). Targetting cAMP in chronic lymphocytic leukemia: a pathway-dependent approach for the treatment of leukemia and lymphoma. *Expert Opin. Ther. Targets* 17 (8): 937–949.

53 Sapio, L., Di Maiolo, F., Illiano, M. et al. (2020). Targeting protein kinase A in cancer therapy: an update. *EXCLI J.* 13: 843–855.

54 Shaikh, N.M., Bagihalli, G.B., Adimule, V. et al. (2022). A novel silica immobilized acidic ionic liquid [BMIM][AlCl$_4$] as an effective catalyst for biscoumarine synthesis. *Top. Catal.* https://doi.org/10.1007/s11244-022-01591-9.

55 Meyers, J., Su, D., and Lerner, A. (2009). Chronic lymphocytic leukemia and B and T cells differ in their response to cyclic nucleotide phosphodiesterase inhibitors. *J. Immunol.* 182 (9): 5400–5411.

56 Li, C., Xie, J., Lu, Z. et al. (2015). ADCY7 supports development of acute myeloid leukemia. *Biochem. Biophys. Res. Commun.* 465 (1): 47–52.

57 Nandi, S.S., Adimule, V., and Yallur, B.C. (2022). Synthesis, structural and optical properties of co doped Sm$_2$O$_3$ nanostructures. *Adv. Mater. Res.* 1173: 59–69. Trans Tech Publications, Ltd.

58 Adimule, V., Bhat, V.S., Yallur, B.C. et al. (2022). Facile synthesis of novel SrO$_{0.5}$:MnO$_{0.5}$ bimetallic oxide nanostructure as a high-performance electrode material for supercapacitors. *Nanomater. Nanotechnol.* 12: 1–14. https://doi .org/10.1177/18479804211064028.

59 Adimule, V., Yallur, B.C., Batakurki, S., and Nandi, S.S. (2022). Synthesis, morphology and enhanced optical properties of novel Gd$_x$Co$_3$O$_4$ nanostructures. *Adv. Mater. Res.* 1173, Trans Tech Publications, Ltd.: 71–82.

60 van der Sligte, N., Kampen, K., ter Elst, A. et al. (2015). Essential role for cyclic-AMP responsive element binding protein 1(CREB) in the survival of acute lymphoblastic leukemia. *Oncotarget* 6 (17): 14970–14981.

61 Batakurki, S.R., Adimule, V., Pai, M.M. et al. (2022). Synthesis of Cs-Ag/Fe$_2$O$_3$ nanoparticles using *Vitis labrusca* rachis extract as green hybrid nanocatalyst for the reduction of arylnitro compounds. *Top. Catal.* https://doi.org/10.1007/ s11244-022-01593-7.

62 Petrov, I., Suntsova, M., Mutorova, O. et al. (2005). Molecular pathway activation features of pediatric acute myeloid leukemia (AML) and acute lymphoblast leukemia (ALL) cells. *AgingUs* 8 (11): 2936–2947.

63 Mitton, B., Chae, H., Hsu, K. et al. (2016). Small molecule inhibition of cAMP response element binding protein in human acute myeloid leukemia cells. *Leukemia* 30 (12): 2302–2311.

64 Pigazzi, M., Manara, E., Baron, E., and Basso, G. (2008). ICER expression inhibits leukemia phenotype and controls tumor progression. *Leukemia* 22 (12): 2217–2225.

65 Ding, L., Sun, Q., Tn, K. et al. Mutational Landscape of pediatric acute lymphoblastic leukemia. *Cancer Res.* 77 (2): 390–400.

66 Propper, D., Saundres, M., Salisbury, A. et al. (1999). Phase I study of the novel cyclic AMP (cAMP) analogue 8-chloro-cAMP in patients with cancer :toxicity, hormonal and immunological effects. *Clin.Cancer Res.* 5 (7): 1682–1689.

67 Schmidit, O., Harbauer, A., Rao, S. et al. Regualtion of mitochondrial protein import by cytosolic kinase. *Cell* 144 (2): 227–239.

68 Gerbeth, C., Schmidt, O., Rao, S. et al. (2013). Glucose-induced regulation of protein import receptor Tom22 by cytosolic and mitochondria-bound kinases. *Cell Metab.* 18 (4): 578–587.

69 Rao, S., Schmidt, O., Harbauer, A. et al. Biogenesis of the preprotein translocase of the outer mitochondrial membrane: protein kinase A phosphorylates the precursor of Tom40 and impairs its import. *Mol. Biol. Cell* 23 (9): 1618–1627.

70 Ould Amer, Y. and Hebert-Chatelain, E. (2018). Mitochondrial cAMP-PKA signaling: what do we really know? *Biochim. Biophys. Acta Bioenerg.* 1859 (9): 868–877.

71 Adimule, V., Kerur, S.S., Chinnam, S. et al. (2022). Guar Gum and its nanocomposites as prospective materials for miscellaneous applications: a short review. *Top. Catal.* https://doi.org/10.1007/s11244-022-01587-5.

72 Perez, D., Smagley, Y., Garcia, M. et al. (2016). Cyclic AMP efflux inhibitors as potential therapeutic agents for leukemia. *Oncotarget* 7 (23): 33960–33982.

73 Adimule, V., Yallur, B.C., Pai, M.M. et al. (2022). Biogenic synthesis of magnetic palladium nanoparticles decorated over reduced graphene oxide using piper betle petiole extract (Pd-rGO@Fe_3O_4 NPs) as heterogeneous hybrid nanocatalyst for applications in Suzuki-Miyaura coupling reactions of biphenyl compounds. *Top. Catal.* https://doi.org/10.1007/s11244-022-01672-9.

74 Adimule, V., Jagadeesha Gowda, A.H., Nandi, S.S., and Bowmik, D. (2022). Antimalarial activity of novel class of 1,3-benzoxaborole derivatives containing 1,3,4-oxadiazole moiety. In: *Drug Development for Malaria: Novel Approaches for Prevention and Treatment* (ed. P. Kendrekar). Wiley.

75 Opanlinska, M. and Meisinger, C. (2015). Metabolic control via the mitochondrial protein machinery. *Curr. Opin. Cell. Biol.* 33: 42–48.

76 Ventura-Clapier, R., Garnier, A., and Vekser, V. Transcriptional control of mitochondrial biogenesis: the central role of PGC-1 alpha. *Cardiovasc. Res.* 79 (2): 208–217.

77 Scarpulla, R., Vega, R., and Kelly, D. Transcriptional integration of mitochondrial biogenesis. *Trends Endocrinol. Metlab.* 23 (9): 459–466.

78 Xu, D., Li, R., Wu, J. et al. (2016). Drug design targeting the CXCR4/CXCR7/ CXCL12 pathway. *Curr. Top. Med. Chem.* 16 (13): 1441–1451.

79 Monczor, F., Copsel, S., Fernandez, N. et al. (2017). Histamine H2receptor in blood cells: a suitable target for the treatment of acute myeloid leukemia. *Handb. Exp. Pharmacol.* 241: 141–160.

7

Novel Pharmaceutical Nanomaterials to Advance the Current Breast Cancer Treatment – Current Trends and Future Perspective

Steven Mufamadi, Mpho Ngoepe, Aidan Battison and Itumeleng Zosela

Nelson Mandela University, Faculty of Health Sciences, DSI-Mandela Nanomedicine Platform, Gqeberha, 6059, South Africa

7.1 Introduction

The current conventional cancer therapy is limited because it kills cancerous as well as normal cells, leading to side effects such as anemia, discomfort, vomiting, weight fluctuations, and loss of hair. Over the years, research and development of cancer therapy has focused on nanoparticles, receptor targeting molecules, and antibodies for better cancer treatment. Nanocarrier systems such as antibody (Ab)–drug conjugates, micelles, liposomes, dendrimers, nanoemulsions, and polymeric nanoparticles have a clinical advantage over conventional chemotherapy as they aid in decreasing the side effects of anticancer drugs [1]. Designing targeted drug delivery systems enables passive or active targeting of nanoparticles to reach the cancer site. Liposomes (50–200 nm) were first developed in 1964 by Bangham and Horne, followed by FDA approval of the liposomal formulation of doxorubicin (Doxil®) for ovarian cancer in 1965 and later for breast cancer [2].

Since then, several anticancer formulations have been successfully formulated, including DaunoXome®, Depocyt®, Myocet®, Mepact®, Marqibo®, or Onivyde™[3]. Liposomes are spherical vesicles prepared from a naturally derived phospholipid (nontoxic, non-immunogenic, and biodegradable) consisting of one or more lipid bilayers with discrete aqueous spaces [4]. For enhanced permeability and retention effect (EPR), the liposome can be designed to have triggered drug release in response to pH, enzymes, light, electromagnetic field change, and heat (e.g. ThermoDox® – Lysolipid thermosensitive liposomal doxorubicin) [5]. Various organic and inorganic nanomaterials used in cancer are illustrated in Table 7.1. Apart from using nanoparticles as drug delivery systems, the use of inorganic nanoparticles expands the application of nanotechnology in cancer management. The clinically approved Hensify® (hafnium oxide nanoparticle),

Drug and Therapy Development for Triple Negative Breast Cancer, First Edition.
Edited by Pravin Kendrekar, Vinayak Adimule, and Tara Hurst.
© 2023 WILEY-VCH GmbH. Published 2023 by WILEY-VCH GmbH.

Table 7.1   Clinically approved and currently undergoing clinical trials nanoparticle therapies for cancer [6–12].

Name	Drug/active component	Company
Liposomes		
Promitil	Mitomycin C (PEGylated)	Lipomedix Pharmaceuticals
ThermoDox	Doxorubicin (PEGylated)	Celsion
MM-302	Doxorubicin (PEGylated)	Merrimack Pharmaceutical
Vyxeos[a]	Cytarabine: daunorubicin	Jazz Pharmaceutical
Doxil[a]	Doxorubicin (PEGylated)	Janssen
DaunoXome[a]	Doxorubicin	Galen
Onivyde[a]	Irinotecan (PEGylated)	Merrimack
Caelyx[a]	Doxorubicin (PEGylated)	Janssen Pharmaceuticals
Myocet[a]	Doxorubicin	Teva Pharmaceutical Industries Ltd
Marqibo[a]	Vincristine	Spectrum
Onivyde[a]	Irinotecan (PEGylated)	Merrimack
Lipusu[a]	Paclitaxel	Nanjing LuyeSike Pharmaceutical Co., Ltd
Lipodox[a]	Doxorubicin (PEGylated)	Sun Pharma Global FZE
EndoTAG-1	Paclitaxel (cationic)	SynCore Biotechnology
Micelles		
Taxotere[a]	Docetaxel	Sanofi-Aventis
Nanoplatin	Cisplatin	NanoCarrier
Cynviloq	Paclitaxel	Sorrento
Genexol-PM	Paclitaxel	Samyang Biopharmaceuticals
CriPec	Docetaxel	Cristal Therapeutic
NC-6300	Epirubicin	NanoCarrier
NK105	Paclitaxel	Nippon Kayaku
Imx-110	Stat3/NF-κB/polytyrosine kinase inhibitor and low-dose doxorubicin	Immix Biopharma Australia
IT-141	SN-38 (irinotecan analog)	Intezyne Technologies
Dentrimer		
DEP	Docetaxel	Starpharma
Oncaspar	Asparaginase (PEGylated)	Servier Pharmaceuticals, LLC
Prothecan	Camptothecin	Enzon, Inc.
Polymer		
Livatag[a]	Doxorubicin: PiHCA	Onxeo
Apealea[a]	Paclitaxel: XR-17™	Oasmia Pharmaceutical AB
Eligard[a]	Leuprolide: PLGA	Tolmar Pharmaceuticals Inc.

(Continued)

Table 7.1 (Continued)

Name	Drug/active component	Company
Smancs[a]	Neocarzinostatin: SMA	Yamanouchi Pharmaceutical Co., Ltd
AZD2811	Aurora B kinase inhibitor: Accurin™	AstraZeneca & BIND Therapeutic
BIND-014	Docetaxel: PEG-PLGA	BIND Therapeutics
CRLX101	Camptothecin: cyclodextrin	Cerulean
	Protein	
Abraxane[a]	Paclitaxel: albumin	Celgene Pharmaceutical Co. Ltd.
ABI-009	Rapamycin: albumin	Aadi with Celgene
ABI-011	Thiocolchicine analog: albumin	NantBioscience, Inc.
Ontak[a]	Diphtheria toxin: interleukin-2	Eisai Co., Ltd.
	Inorganic	
Hensify[a]	Hafnium oxide nanoparticles	Nanobiotix
NanoTherm[a]	Iron oxide nanoparticles	MagForce AG
Aurolase[a]	PEG-coated silica-gold	Nanospectra Biosciences

SMA, poly(styrene-*co*-maleic acid), PiHCA, poly(isohexyl cyanoacrylate), PLG, poly(lactic-*co*-glycolic acid), PEG-PLGA, polyethylene glycol–poly lactic acid-*co*-glycolic acid.
[a] Clinically approved for use by Food and Drug Administration (FDA) or European Medicines Agency (EMA).

MagForce® (iron oxide nanoparticles), or Aurolase® (gold nanoparticles) can be used as indirect therapeutic agents by producing local heat, leading to cancer cell death [13].

Nanotechnology-based breast cancer management strategies should focus on emerging nanomaterials to advance existing treatments and diagnostics. Current trends and emerging research using nanotechnology to advance breast cancer treatment include graphene and graphene oxide (GO)-based nanosystems, light-based nanotechnology (photodynamic therapy [PDT] and photothermal therapy [PTT]), and green synthetic metal nanoparticles. In addition, nucleic acid-based gene therapy and immunotherapy for breast cancer have also received attention in recent years.

7.2 Graphene-Based Nanomaterials for Breast Cancer

For efficient cancer treatment, simultaneous detection and treatment of cancer cells can aid in ensuring complete management of cancer. Graphene-based nanosystems (GrNSs) have emerged as ideal platforms due to their unique physicochemical properties, which allow for easy surface functionalization with various targeting and imaging ligands, as well as PTT and PDT due to optical properties (e.g. GO and reduced graphene oxide [rGO]) and use as drug delivery vehicles [14]. Even though GO and rGO have many limitations, including a lack of cellular imaging potential, cellular entry, and functionalization, as well as economic issues, converting them into graphene quantum dots (GQDs) produces nanomaterials with inherent fluorescence properties, high aqueous dispersity, low cytotoxicity, stable photoluminescence (PL), and good biocompatibility [15]. A study by de Melo-Diogo et al. has shown that surface modification of GO nanoparticles with poly(2-ethyl-2-oxazoline) (PEtOx) and then combined incorporation of doxorubicin (DOX) and D-α-Tocopherol succinate (TOS) into nanomaterials enabled chemo-phototherapy against breast cancer cells [16]. Similar studies were also conducted by Prasad et al. whereby GQDs and DOX were loaded inside mesoporous silica nanoparticles (carbanosilica) and enabled enhanced ultrahigh penetration and retention into solid tumors [17]. The benefit of using GQDs can enable simultaneous tumor imaging, drug delivery, and phototherapy, which cannot be obtained by other nanomaterials alone.

7.3 Light-Based Nanotechnology for Breast Cancer

Photo-triggered therapy has emerged as a powerful tool to treat and diagnose different types and stages of cancer. Moreover, the use of nanomaterials as effective photoactive cancer theranostic systems has gained increasing focus due to their advantageous chemical and physical properties. Phototherapy may be divided into four main classes, namely photothermal therapy (PTT), photodynamic therapy (PDT), photoimmunotherapy (PIT), and image-guided gene therapy (IGGT) [18].

PTT exploits photothermal effects using photothermal agents (PTAs) that can convert light energy into heat, thereby increasing the temperature of surrounding tissue and causing cell death [19]. PDT is based on the activation of photosensitizers (PS) to produce cytotoxic reactive oxygen species (ROS) by the irradiation of light of a suitable wavelength [20].

7.3.1 Photodynamic Therapeutic Nanomaterials

PDT is an ideal cancer treatment method to kill cancer cells through the generation of ROS by a PS under light irradiation in the presence of molecular oxygen [21]. PDT consists of three essential components: the PS, excitation radiation, and oxygen (O_2) [22]. These components alone are not toxic, but when combined will trigger a photochemical reaction to produce cytotoxic ROS. Under the irradiation of light of a specific wavelength, the PS can be excited and then react with substrates and molecular oxygen to generate free radicals, such as the superoxide radical ($^{\cdot}O_2$) and hydroxyl radical ($^{\cdot}OH$). Alternatively, the excited PS can directly transfer its energy to oxygen to form a highly reactive singlet oxygen ($^1O_2$), resulting in the significant cellular toxicity (Type II reaction) [23]. Apoptosis, necrosis, and autophagy are the main cell death pathways by PDT [24].

In recent years, the introduction of nanoparticles into PDT has become a promising strategy for overcoming some of these problems and offers certain advantages over molecular-based PS. The introduction of PS into nanoparticles can increase the water solubility of the drugs, thereby extending their blood circulation time thereby improving their tumor uptake; they may improve the pharmacokinetics and biodistribution; and can protect the PS against metabolic species [25]. Nanomaterial carriers currently used in PDT are shown in Figure 7.1 [26, 27].

Guan et al. have reported laser-responsive multifunctional nanoparticles for the efficient combinational chemo-photodynamic therapy against breast cancer [28]. They developed an injectable, laser-responsive multifunctional nanodrug delivery system, which was composed of both liposomal bilayers and nanoparticles. The chemotherapeutic drug paclitaxel (PTX) and P-gp inhibitor cyclosporin A (CyA) were encapsulated in bovine serum albumin (BSA)-based nanoparticles (PTX/CyA NPs), which constituted the core aqueous phase of the nanostructure. PTX/CyA NPs were then coated by a liposomal bilayer encapsulating the PS agent Chlorin e6 (Ce6), which was subsequently surface modified by the active targeting moiety transferrin (Tf). The nanomaterial possessed laser-responsive ROS release profiles upon irradiation at 660 nm, which was shown to be an efficient combinatorial chemo-photodynamic therapy method and could effectively reverse multiple drug resistance (MDR). Under laser irradiation, the nanomaterial demonstrated appreciable intracellular ROS productivity and noteworthy *in vitro* and *in vivo* anticancer therapy effects without the induction of systemic toxicity. Other organic-, carbon-, and inorganic-based nanomaterials applied with PDT that have been studied for breast cancer theranostics have been reported [29–31].

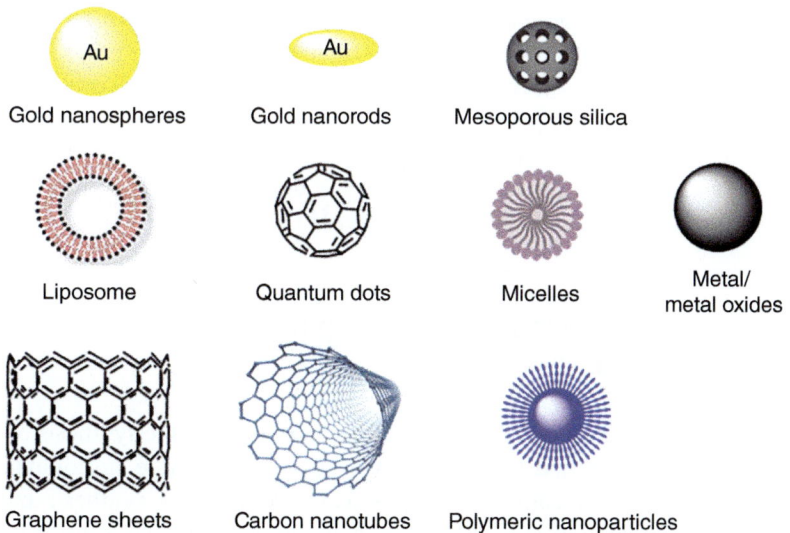

Gold nanospheres Gold nanorods Mesoporous silica

Liposome Quantum dots Micelles Metal/metal oxides

Graphene sheets Carbon nanotubes Polymeric nanoparticles

Figure 7.1 Inorganic, organic, and carbon-based nanomaterials currently used in PDT.

7.3.2 Photothermal Therapeutic Nanomaterials

The use of hyperthermia as a cancer treatment method works by elevating the body's temperature thereby inducing hyperthermic stress, which in turn suppresses cancer cell growth and induces hyperthermia-associated genes. However, the inherent disadvantages of hyperthermia treatment, including poor selectivity for cancer tissues and the requirement of advanced facilities and devices for temperature control, have significantly limited its application to all cancer patients [18]. Therefore, PTT has become a widely studied therapeutic method in hyperthermia due to the advantages of simplicity, high spatiotemporal precision, minimal invasiveness and side effects, negligible toxicity to normal tissue, and remote activation [26, 32]. PTT employs a PTA to convert light energy into heat, as opposed to ROS in PDT, thereby inducing the thermal ablation of cancer cells [33]. PTAs are used to optimize the efficacy of cancer treatment. They can include inorganic and organic materials [34]. As a result of the generated heat, cellular effects such as protein aggregation, denaturation, and cell lysis can occur [35]. Owing to the unique characteristics of photothermal conversion, these functional materials are preferred in biomedical applications [36].

Nanomaterials have remarkable properties such as a very high absorbance effect with NIR and the ability to effectively convert laser energy into heat. Moreover, some of the advantageous properties of nanoparticles promoting their application in PTT include longer circulation time in the bloodstream, their ability to disperse in biological fluids, negligible adverse effects on the natural behavior of the host biosystem, and efficient heat production during external excitation [37]. The heat generated by the vibrational relaxation of stimulated PTT agents ($>42\,°C$) induces the photothermal effect that causes cell death. According to recent findings, PTT causes cell death through apoptosis internal pathway rather than necrosis [38]. In general, photothermal heating to temperatures around $41–45\,°C$ has been reported to induce reversible damage to cells such as affecting DNA repair mechanisms or metabolic pathways or sensitizing them to the action of other agents like chemotherapeutic drugs. In turn, temperatures of $50\,°C$ (or above) cause irreversible damage in cells in the primary tumor (e.g. mitochondrial/enzymatic dysfunctions, protein denaturation, membrane destruction) [34, 39]. In recent approach, PTT together with chemotherapy have been combined to provide a promising synergistic therapeutic approach for breast cancer [37].

7.4 Green Synthesis of Gold Nanoparticles for Breast Cancer

Green synthesis is the use of plants and microorganisms for the synthesis of metal nanoparticles. This method is a safer, less expensive, and more sustainable technique compared to the conventional physical and chemical synthesis methods [40, 41]. When it comes to the synthesis of gold nanoparticles, conventional methods have various drawbacks, including the use of hazardous compounds, excessive power consumption, high cost, and specialized apparatus [42]. Plant extracts have been preferred over microbes for green synthesis of gold nanoparticles because they are generally available, found all over the world, easily digestible, largely nontoxic, and cost-effective. Whereas using microorganisms involves careful handling and maintenance, which increases synthesis costs [43, 44]. A wide range of plant extracts and bioactive components are employed in the manufacture of gold nanoparticles. During the synthesis of gold nanoparticles (Figure 7.2), plant bioactive components such as polyphenols, phenolic acids, flavonoids, terpenes, and alkaloids act as reducing and stabilizing agents [45, 46]. The advantage of gold nanoparticles is their small size and large surface area [47]. Their resistance to oxidation by air, moisture, and mild acids, as well as their biocompatibility, makes gold nanoparticles ideal candidates for biological applications, particularly in cell targeting, drug adminis-tration, tumor detection, antimicrobial, cancer therapy, chemical, and biological sensing [48, 49].

A study done by Vemuri et al. investigated the synthesis and testing of several bio-synthesized gold nanoparticles (b-AuNPs) using naturally derived phytochemicals (Curcumin: Cur, Turmeric: Tur, Quercetin: Qu, and Paclitaxel: Pacli) [50]. The pro-duced b-AuNPs were tested in several breast cancer cells, either individually or in combination. In comparison to their pure administration, a combination of all four forms of b-AuNPs (AuNPs-Cur, AuNPs-Tur, AuNPs-Qu, and AuNPs-Pacli) demon-strated the highest therapeutic activity. Furthermore, mechanistic studies of these compounds revealed that combinations of AuNPs-Cur, AuNPs-Tur, AuNPs-Qu, and AuNPs-Pacli were significantly effective in inhibiting cell proliferation, apoptosis, and angiogenesis, indicating a synergistic effect when compared to individual treat-ment against various breast cancer cell lines (MCF-7 and MDA-MB 231). Li et al. synthesized gold nanoparticles from *Mentha longifolia* leaf extract. The biosynthe-sized nanoparticles were effective against breast adenocarcinoma (MCF7), breast carcinoma (Hs 578Bst), breast infiltrating ductal cell carcinoma (Hs 319.T), and

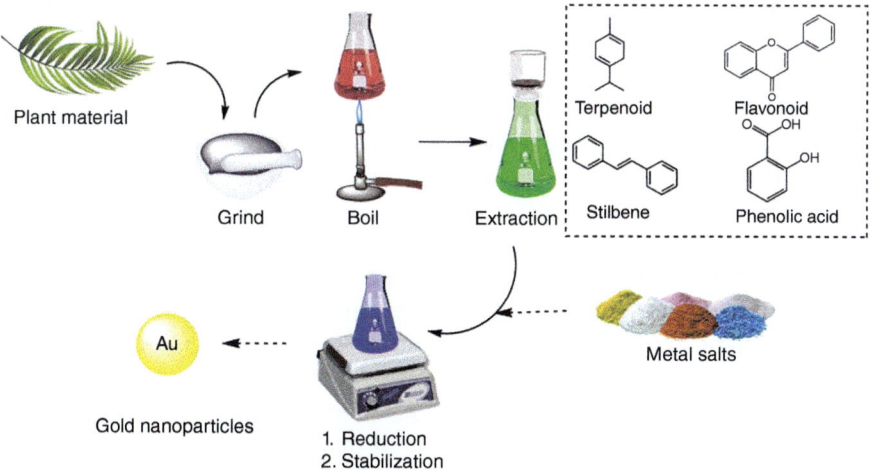

Figure 7.2 Synthesis of gold nanoparticles.

breast infiltrating lobular carcinoma (UACC-3133) cell lines, with no cytotoxicity activity against normal cell lines, such as HUVEC [51]. In a study by Yas et al., the orchid plant was used as a reducing agent for the synthesis of gold nanoparticles [49]. The cytotoxicity test demonstrated the anticancer activity of produced gold nanoparticles in the breast cancer AMJ-13 cell line [49]. AuNPs reduced cell growth and produced cytotoxicity in cancer cells at dose-dependent concentrations. Gold nanoparticles were synthesized using *Curcuma longa* extract in a study done by Hendi et al., influenced breast cancer DMBA-induced in female Swiss albino mice, which reduced the incidence of tumor in the breast and lymph nodes; additionally, the AuNPs improved biological functions such as liver and kidney functions that were disrupted by tumor existence [52]. As a result, the green synthesis of AuNPs remains a potential topic that requires additional research in the future.

7.5 Nanocarriers for Gene Therapy and Immunotherapy

Over the past few years, research focusing on the use of gene therapy and immunotherapy with nanosystems has emerged and it has become an innovative strategy for breast cancer treatment. Gene therapy involves sending nucleic acid modalities through a viral vector (e.g. adenovirus) and nonviral vectors (e.g. nanosystems) into target cells, where gene expression can be corrected, increased, and downregulated [53]. Small interfering RNA (siRNA) and microRNA (miRNA) therapeutics are among a promising therapeutic modality employed for inhibition of gene expression (gene silencing) for breast cancer therapy [54]. While small activating RNA (saRNA), and clustered regularly interspaced short palindromic repeats (CRISPR) gene editing systems are usually employed to increase, activate, and correct target gene expression for breast cancer [55]; Onpattro (Patisiran), a lipid nanoparticle containing siRNA, is the first FDA-approved gene therapy consisting of a nanoformulation in 2018 [56, 57].

Yu et al. reported gene silencing *in vitro* and *in vivo* using siRNA-functionalized lanthanide nanoparticles, and successfully represses the tumor growth in CT26 tumor model and 4T1 orthotopic mouse model of spontaneous breast cancer metastasis [58]. Li et al., used breast cancer stem cells-derived extracellular vesicles (BCSCs-EVs) to activate epithelial-mesenchymal transition of breast cancer cells [59]. To achieve this, microRNA 197 was delivered into breast cancer cells using the BCSCs-EVs and inhibiting peroxisome proliferator-activated receptor γ (PPARG) expression, thereby promoting growth and metastasis of breast cancer cells. Xiong et al. delivered saRNA-employed dendrimer nanoparticles in breast, ovarian, and pancreatic cancer cells for enhancing Mas receptor (MAS1) gene expression [60]. Shaikh et al. reported an efficient delivery of CRISPR/Cas9 and single guide RNA (sgRNA) into MDA-MB-231 human breast cancer cell line employing polyethylenimine-coated-BSA nanoparticles as a delivery vehicle [61].

Breast cancer immunotherapy is a form of treatment that uses the body's own immune system to eliminate cancer. A study done by Cao et al. reported an increase in cellular uptake of the drug-carrying liposome in a metastatic 4T1 breast cancer cells and *in vivo* it inhibits the effect on lung metastasis of breast cancer. Liu et al. combined immunotherapy of the transmembrane glycoprotein mucin 1 (MUC1) mRNA nano-vaccine and cytotoxic T-lymphocyte-associated protein 4 (CTLA-4) to inhibit growth of triple-negative breast cancer. The use of mRNA vaccines for cancer immunotherapy has become an innovative ground for types of cancer, such as brain, breast, melanoma, lung, ovarian, prostate cancer, and solid tumors.

7.6 Conclusion and Recommendations

Considerable progress has been reported on the use of nanotechnology-based breast cancer therapy and some of them had already received FDA approval and others were already in our clinic. Nonviral vectors (e.g. nanosystems) for gene and immunotherapy delivery are safer compared to viral vectors that are associated with immunogenicity and cytotoxicity. Nanomaterial-based vectors are biocompatible, nontoxic, can improve the half-life of genetic material in the bloodstream, target delivery, and enhance intracellular uptake, followed by gene knockdown and/or activation. Despite the recent research activities and the rapid growth of the GrNSs, green synthesized metal-based nanoparticles using plant extracts, and light-based nanotechnology (PDT and PTT) as innovative strategies for advanced breast cancer treatment, most of the studies were conducted *in vitro*. To understand the long-term effect of these emerging nanomaterials on human health, intensive *in vivo* studies are needed to confirm safety and clinical significance. This chapter provides some details on how nanotechnology-based strategies can advance research in drug and gene delivery systems and breast cancer treatment. Overall, the advancement of nanosystems and cancer therapy development could provide a promising synergistic therapeutic approach for breast cancer.

References

1 Chaturvedi, V.K., Singh, A., Singh, V.K., and Singh, M.P. (2019). Cancer nanotechnology: a new revolution for cancer diagnosis and therapy. *Curr. Drug Metab.* 20: 416–429.

2 Tran, P., Lee, S.-E., Kim, D.-H. et al. (2020). Recent advances of nanotechnology for the delivery of anticancer drugs for breast cancer treatment. *J. Pharm. Invest.* 50: 261–270.

3 Filipczak, N., Pan, J., Yalamarty, S.S.K., and Torchilin, V.P. (2020). Recent advancements in liposome technology. *Adv. Drug Delivery Rev.* 156: 4–22.

4 Keri, R.S., Adimule, V., Kendrekar, P. et al. (2022). The nano-based catalyst for the synthesis of benzimidazoles. *Top. Catal.* https://doi.org/10.1007/s11244-022-01562-0.

5 Chaudhry, M., Lyon, P., Coussios, C., and Carlisle, R. (2022). Thermosensitive liposomes: a promising step toward localised chemotherapy. *Expert Opin. Drug Delivery* 19: 899–912.

6 Halwani, A.A. (2022). Development of pharmaceutical nanomedicines: from the bench to the market. *Pharmaceutics* 14: 106.

7 Anselmo, A.C. and Mitragotri, S. (2021). Nanoparticles in the clinic: an update post COVID-19 vaccines. *Bioeng. Transl. Med.* 6: 1–20.

8 Mignani, S., Shi, X., Guidolin, K. et al. (2021). Clinical diagonal translation of nanoparticles: case studies in dendrimer nanomedicine. *J. Controlled Release* 337: 356–370.

9 Mitchell, M.J., Billingsley, M.M., Haley, R.M. et al. (2021). Engineering precision nanoparticles for drug delivery. *Nat. Rev. Drug Discovery* 20: 101–124.

10 Anselmo, A.C. and Mitragotri, S. (2019). Nanoparticles in the clinic: an update. *Bioeng. Transl. Med.* 4: 1–16.

11 Alshehri, S., Imam, S.S., Rizwanullah, M. et al. (2021). Progress of cancer nanotechnology as diagnostics, therapeutics, and theranostics nanomedicine: preclinical promise and translational challenges. *Pharmaceutics* 13: 24.

12 Martinelli, C., Pucci, C., and Ciofani, G. (2019). Nanostructured carriers as innovative tools for cancer diagnosis and therapy. *APL Bioeng.* 3: 1–13.

13 Maksoudian, C., Saffarzadeh, N., Hesemans, E. et al. (2020). Role of inorganic nanoparticle degradation in cancer therapy. *Nanoscale Adv.* 2: 3734–3763.

14 Dolatkhah, M., Hashemzadeh, N., Barar, J. et al. (2020). Graphene-based multifunctional nanosystems for simultaneous detection and treatment of breast cancer. *Colloids Surf., B* 193: 1–15.

15 Tade, R.S. and Patil, P.O. (2020). Theranostic prospects of graphene quantum dots in breast cancer. *ACS Biomater. Sci. Eng.* 6: 5987–6008.

16 de Melo-Diogo, D., Costa, E.C., Alves, C.G. et al. (2018). POxylated graphene oxide nanomaterials for combination chemo-phototherapy of breast cancer cells. *Eur. J. Pharm. Biopharm.* 131: 162–169.

17 Prasad, R., Jain, N.K., Yadav, A.S. et al. (2021). Ultrahigh penetration and retention of graphene quantum dot mesoporous silica nanohybrids for image guided tumor regression. *ACS Appl. Bio Mater.* 4: 1693–1703.

18 Yin, X., Cheng, Y., Feng, Y. et al. (2022). Phototheranostics for multifunctional treatment of cancer with fluorescence imaging. *Adv. Drug Delivery Rev.* 189: 114483.

19 Han, H.S. and Choi, K.Y. (2021). Advances in nanomaterial-mediated photothermal cancer therapies: toward clinical applications. *Biomedicines* 9: 305.

20 Arbeloa, E.M., Militello, M.P., Bertolotti, S.G., and Previtali, C.M. (2021). Photosensitizer-dendrimer systems in anticancer treatments: from photophysics to PDT applications. In: *Dendrimer-Based Nanotherapeutics* (ed. P. Kesharwani), 311–326. Elsevier.

21 Dunkel, P. and Ilaš, J. (2021). Targeted cancer therapy using compounds activated by light. *Cancers* 13: 3237.

22 Li, J.-Q., Zhao, R.-X., Yang, F.-M. et al. (2022). An erythrocyte membrane-camouflaged biomimetic nanoplatform for enhanced chemo-photothermal therapy of breast cancer. *J. Mater. Chem. B* 10: 2047–2056.

23 Yang, Y., Zeng, Z., Almatrafi, E. et al. (2022). Core–shell structured nanoparticles for photodynamic therapy-based cancer treatment and related imaging. *Coord. Chem. Rev.* 458: 214427.

24 Mishchenko, T., Balalaeva, I., Gorokhova, A. et al. (2022). Which cell death modality wins the contest for photodynamic therapy of cancer? *Cell Death Dis.* 13: 455.

25 del Valle, C.A., Hirsch, T., and Marín, M.J. (2022). Recent advances in near infrared upconverting nanomaterials for targeted photodynamic therapy of cancer. *Methods Appl. Fluoresc.* 10: 034003.

26 Nandi, S.S., Adimule, V., Kadapure, S.A., and Kerur, S.S. (2022). Rare earth based nanocomposite materials for prominent performance supercapacitor: a review. *AMM* 908: 3–18. https://doi.org/10.4028/p-rff302.

27 de Freitas, L.F. (2020). Nanomaterials for enhanced photodynamic therapy. In: *Photodynamic Therapy – From Basic Science to Clinical Research*. IntechOpen.

28 Guan, Q., Li, Y., Zhang, H. et al. (2022). Laser-responsive multi-functional nanoparticles for efficient combinational chemo-photodynamic therapy against breast cancer. *Colloids Surf., B* 216: 112574.

29 Bekmukhametova, A., Uddin, M.M.N., Houang, J. et al. (2022). Fabrication and characterization of chitosan nanoparticles using the coffee-ring effect for photodynamic therapy. *Lasers Surg. Med.* 54: 758–766.

30 Yuan, X., Cen, J., Chen, X. et al. (2022). Iridium oxide nanoparticles mediated enhanced photodynamic therapy combined with photothermal therapy in the treatment of breast cancer. *J. Colloid Interface Sci.* 605: 851–862.

31 Yang, Y., Chen, F., Xu, N. et al. (2022). Red-light-triggered self-destructive mesoporous silica nanoparticles for cascade-amplifying chemo-photodynamic therapy favoring antitumor immune responses. *Biomaterials* 281: 121368.

32 Gao, Y., Gao, D., Shen, J., and Wang, Q. (2020). A review of mesoporous silica nanoparticle delivery systems in chemo-based combination cancer therapies. *Front. Chem.* 8: https://doi.org/10.3389/fchem.2020.598722.

33 Zhi, D., Yang, T., O'hagan, J. et al. (2020). Photothermal therapy. *J. Controlled Release* 325: 52–71.

34 Zhang, L., Jia, H., Liu, X. et al. (2022). Heptamethine cyanine-based application for cancer theranostics. *Front. Pharmacol.* 12: 764654. https://doi.org/10.3389/fphar.2021.764654.

35 Sargazi, S., Simge, E., Gelen, S.S. et al. (2022). Application of titanium dioxide nanoparticles in photothermal and photodynamic therapy of cancer: an updated and comprehensive review. *J. Drug Delivery Sci. Technol.* 75: 103605.

36 Shukla, A. and Maiti, P. (2022). Nanomedicine and versatile therapies for cancer treatment. *MedComm* 3: e163.

37 Alamdari, S.G., Amini, M., Jalilzadeh, N. et al. (2022). Recent advances in nanoparticle-based photothermal therapy for breast cancer. *J. Controlled Release* 349: 269–303.

38 Eskiizmir, G., Baskın, Y., and Yapıcı, K. (2018). Graphene-based nanomaterials in cancer treatment and diagnosis. In: *Fullerens, Graphenes and Nanotubes* (ed. A.M. Grumezescu), 331–374. Elsevier.

39 Alves, C.G., Lima-Sousa, R., Melo, B.L. et al. (2022). Heptamethine cyanine-loaded nanomaterials for cancer immuno-photothermal/photodynamic therapy: a review. *Pharmaceutics* 14: 1015.

40 Chellapandian, C., Ramkumar, B., Puja, P. et al. (2019). Gold nanoparticles using red seaweed *Gracilaria verrucosa*: green synthesis, characterization and biocompatibility studies. *Process Biochem.* 80: 58–63.

41 Elbialy, N.S., Abdelfatah, E.A., and Khalil, W.A. (2019). Antitumor activity of curcumin-green synthesized gold nanoparticles: in vitro study. *BioNanoScience* 9: 813–820.

42 Soshnikova, V., Kim, Y.J., Singh, P. et al. (2017). Cardamom fruits as a green resource for facile synthesis of gold and silver nanoparticles and their biological applications. *Artif. Cells Nanomed. Biotechnol.* 46: 108–117.

43 Hosny, M., Fawzy, M., Abdelfatah, A.M. et al. (2021). Comparative study on the potentialities of two halophytic species in the green synthesis of gold nanoparticles and their anticancer, antioxidant and catalytic efficiencies. *Adv. Powder Technol.* 32: 3220–3233.

44 Dey, A., Yogamoorthy, A., and Sundarapandian, S. (2018). Green synthesis of gold nanoparticles and evaluation of its cytotoxic property against colon cancer cell line. *Res. J. Life Sci. Bioinform. Pharm. Chem. Sci.* 4: 1–17.

45 Vaid, P., Raizada, P., Saini, A.K., and Saini, R.V. (2020). Biogenic silver, gold and copper nanoparticles – a sustainable green chemistry approach for cancer therapy. *Sustainable Chem. Pharm.* 16: 100247.

46 Aljabali, A., Akkam, Y., Al Zoubi, M. et al. (2018). Synthesis of gold nanoparticles using leaf extract of *Ziziphus zizyphus* and their antimicrobial activity. *Nanomaterials* 8: 174.

47 Groysbeck, N., Stoessel, A., Donzeau, M. et al. (2019). Synthesis and biological evaluation of 2.4 nm thiolate-protected gold nanoparticles conjugated to Cetuximab for targeting glioblastoma cancer cells via the EGFR. *Nanotechnology* 30: 184005.

48 Satpathy, S., Patra, A., Ahirwar, B., and Hussain, M.D. (2020). Process optimization for green synthesis of gold nanoparticles mediated by extract of *Hygrophila spinosa* T. Anders and their biological applications. *Physica E* 121: 113830.

49 Yas, R.M., Ghafoor, A., and Saeed, M.A. (2021). Anticancer effect of green synthesized gold nanoparticles using orchid extract and their characterizations on breast cancer AMJ-13 cell line. *Syst. Rev. Pharm.* 12: 500–505.

50 Vemuri, S.K., Banala, R.R., Mukherjee, S. et al. (2019). Novel biosynthesized gold nanoparticles as anti-cancer agents against breast cancer: synthesis, biological evaluation, molecular modelling studies. *Mater. Sci. Eng., C* 99: 417–429.

51 Li, S., Al-Misned, F.A., El-Serehy, H.A., and Yang, L. (2021). Green synthesis of gold nanoparticles using aqueous extract of *Mentha longifolia* leaf and investigation of its anti-human breast carcinoma properties in the in vitro condition. *Arabian J. Chem.* 14: 102931.

52 Hendi, A.A., El-Nagar, D.M., Awad, M.A. et al. (2020). Green nanogold activity in experimental breast carcinoma *in vivo*. *Biosci. Rep.* 40: https://doi.org/10.1042/BSR20200115.

53 Ghanbarian, H., Aghamiri, S., Eftekhary, M. et al. (2021). Small activating RNAs: towards the development of new therapeutic agents and clinical treatments. *Cells* 10: 591.

54 Ahmadzada, T., Reid, G., and McKenzie, D.R. (2018). Fundamentals of siRNA and miRNA therapeutics and a review of targeted nanoparticle delivery systems in breast cancer. *Biophys. Rev.* 10: 69–86.

55 Karn, V., Sandhya, S., Hsu, W. et al. (2022). CRISPR/Cas9 system in breast cancer therapy: advancement, limitations and future scope. *Cancer Cell Int.* 22 (1): 234. https://doi.org/10.1186/s12935-022-02654-3.

56 Akinc, A., Maier, M.A., Manoharan, M. et al. (2019). The Onpattro story and the clinical translation of nanomedicines containing nucleic acid-based drugs. *Nat. Nanotechnol.* 14: 1084–1087.

57 Yoon, S. and Rossi, J.J. (2018). Therapeutic potential of small activating RNAs (saRNAs) in human cancers. *Curr. Pharm. Biotechnol.* 19: 604–610.

58 Yu, C., Li, K., Xu, L. et al. (2022). siRNA-functionalized lanthanide nanoparticle enables efficient endosomal escape and cancer treatment. *Nano Res.* 15: 9160–9168.

59 Li, R.-T., Zhu, Y.-D., Li, W.-Y. et al. (2022). Synergistic photothermal-photodynamic-chemotherapy toward breast cancer based on a liposome-coated core–shell AuNS@ NMOFs nanocomposite encapsulated with gambogic acid. *J. Nanobiotechnol.* 20 (1): 212. https://doi.org/10.1186/s12951-022-01427-4.

60 Xiong, Y., Ke, R., Zhang, Q. et al. (2022). Small activating RNA modulation of the G protein-coupled receptor for cancer treatment. *Adv. Sci.* 9: 2270162.

61 Shaikh, N.M., Sawant, A.D., Bagihalli, G.B. et al. (2022). Highly active mixed Au–Pd nanoparticles supported on RHA silica through immobilised ionic liquid for Suzuki coupling reaction. *Top. Catal.* https://doi.org/10.1007/s11244-021-01547-5.

Part III

Advanced Technologies in Breast Cancer Therapy

8

Artificial Intelligence-Driven Decisions in Breast Cancer Diagnosis

Amit Gangwal[1] and Rupesh K. Gautam[2]

[1] *Shri Vile Parle Kelavani Mandal's Institute of Pharmacy, Department of Pharmacognosy, SVKM Campus, AB Road, Behind Gurudwara, 424001 Dhule, Maharashtra, India*
[2] *Indore Institute of Pharmacy, Department of Pharmacology, IIST Campus, Opposite IIM Indore, Rau-Pithampur Road, 453331 Indore, Madhya Pradesh, India*

8.1 Introduction

Breast cancer is the most prevalent cancer in females across the world, and in early-stage, non-metastatic illness is treatable in almost 70–80% of cases. With current treatment choices, advanced breast tumor having distant organ metastases is regarded as untreatable. Because there is currently no cure for these people, the condition progresses to a chronic stage that requires continuous therapy besides monitoring [1]. Several nations have adopted mammography-based screening program for before-time identification as well as the cure of breast tumors, with an aim to reduce death and other negative repercussions. Breast imaging appears to show an effect on death; randomized controlled studies have demonstrated that including mammography in breast cancer screening reduces breast cancer-related mortality by around 20% [2, 3]. Several nations developed country-wide advisories including breast scans for identification of breast tumors.

The effectiveness of this screening has been proved; nevertheless, there are some possible downsides. Mammography screening appears to reduce mortality; randomized controlled studies have found a decrease of 20% in breast cancer-associated death having added mammography to breast cancer diagnosis. Several nations developed national advice or guidelines that incorporated mammography in breast cancer screening as a result of these findings. Although this screening has shown to be useful, a few of the downsides are: (i) false-positive recalls, which result in extra imaging work, increasing medical costs and anxiety for the women; (ii) false negatives, which occur when breast tumors are beyond detection by mammography alone or when explanation errors occur, which further delays the verdict; (iii) radiation exposure; and (iv) diagnosis of cancers that are not serious, such as low-risk ductal carcinoma in situ [4].

X-ray pictures of each breast are obtained from two perspectives during a normal mammographic screening test. One or two expert radiologists examine these pictures for malignant tumors. Suspicious instances are reexamined for a more thorough diagnosis. Human readers examine mammograms for screening. The reading process is tedious, exhausting, time-consuming, expensive, and most prominently, prone to mistakes. Manifold investigations have indicated that blinded reviewers can find 20–30% of confirmed malignancies by looking back at prior negative screening exams [5–7]. Despite a current full-field digital mammography (FFDM), the problem of undetected tumors continues. Screening mammography is said to have a sensitivity and specificity of 77–87% and 89–97%, respectively. The stated false-positive rates range from 1% to 29%, and sensitivities range from 29% to 97% [8–10]. Double reading has been demonstrated to increase mammographic assessment performance, and it is now used in many countries. Several readings can increase prognosis performance by up to ten readers, demonstrating that mammography assessment can be improved beyond double reading [11, 12].

Breast magnetic resonance imaging (MRI) is now employed in the treatment of breast cancer for a variety of reasons. The most common uses for MRI include screening individuals at high risk for breast cancer, evaluating the extent of illness, detecting positive margins, and monitoring the response of neoadjuvant treatment. An MRI of the breast is a multi-parametric examination. A common approach in this regard involves dynamic T1-weighted gradient echo pictures recorded before and after intravenous gadolinium chelate delivery, T2-weighted images or short-tau inversion recovery, and diffusion-weighted images [13] Decision fatigue can affect radiologists and other medical imaging experts, resulting in a high rate of medical errors such as missing, inaccurate, or delayed diagnoses. Furthermore, radiologists' interpretations are subject to a lot of intra and interindividual variation. As the quantity of radiological tests grows, so does the difficulty of interpreting them and the responsibilities placed on providers. Notwithstanding substantial developments in genetics and contemporary imaging, a lot of breast tumor victims are not happy with the diagnosis ecosystem. It may be too late for some. Later diagnosis involves more intensive treatments, unknown results, and higher medical costs for the afflicted. As a result, patient identification has become a critical component of breast cancer research and early diagnosis. Breast cancer has a good long-term prognosis when caught early. Despite this, breast cancer is a common disease, with around 10% of patients relapsing after the first therapy each year. Here comes the role of artificial intelligence (AI) and machine learning (ML). AI enables medical experts like radiologists, and doctors to uncover precise insights and identify complex relationships from immense amounts of data. ML is the knowledge of algorithms that learn and refine on their own through repeated exposure to data and fine-tuning. It is considered a part of AI. ML algorithms shape a model basis sample input dataset [14]. This is known as ML model training data, to make forecasts sans being openly or very clearly programmed to do so. ML procedures are deployed in various domains like email sifting and computer vision, where it is tough or infeasible to design conformist algorithms to achieve the required output. AI and ML have made noteworthy growth in the level of sophistication in the past decade, resulting in passionate acceptance across many industries. Deep learning (DL) is a subcategory in ML, which has systems of layers of nodes or neurons. This system of layers of nodes is called neural networks and has established greater performance versus normal computer vision procedures. This neural network in medical imaging has the power to considerably change almost every stage of the pipeline of medical imaging. Approximations for the value that AI and ML could bring to the world economy by 2030 are as high as US$13-US$16T1. Biopharmaceutical companies (pharmaceutical, biotechnological, and biologicals mainly) continue to make significant investments in AI and ML to both improve their decision-making across R&D and commercialization, and to deliver better outcomes for patients, physicians, and payers. This is no more a sci-fi movie. AI is very much now involved in our routine activities and decisions. The simplest example may be when you log on to e-commerce website, it shows you results based on your past purchase history and browsing habits. There are now ample evidences of success stories of AI, as DL models are being used to distinguish between malignant and benign breast tumors. There are DL models that have been trained, which concurrently get trained to notice lesions, and describe them with an accuracy that is in no way inferior to human decisions.

This chapter provides an overview of the traditional approach to diagnose the causes of breast cancer along with its limitations. This is followed by the fundamentals of AI and ML, before moving on to the application and advantages of these technologies in the early detection of cases of breast cancer versus traditional human-involved decisions. In recent years, incorporation of such computational techniques has been observed in screening patients after analyzing medical images like CT scans, MRI, X-rays, and others. Basics of ML models like supervised and unsupervised; basic functioning of the deep neural networks have also been discussed.

8.2 Breast Cancer

The most common cancer among women globally is breast cancer, and 70–80% of individuals with the initial stage, nonmetastatic illness can be cured. With current therapy, late-stage cases of breast cancer with remote organ metastases are said to be untreatable. At the molecular level breast cancer is an assorted disease involving the triggering of human epidermal growth factor receptor 2 and activation of hormone receptors BRCA alteration is the most prevalent molecular marker. Breast cancer management is an inclusive approach, covering surgery and radiation therapy besides the systemic types of therapies. Customization of treatment, treatment reduction, and intensification basis tumor biochemistry, and sensitivity to primary therapy, are among the goals of future therapeutic approaches in breast cancer. Equal worldwide access to therapeutic improvements, in addition to new treatment innovations, will continue to be a global concern in breast cancer care in the future [1]. Certain catalysts were reported in the literature [15–22] for effecient convertion of organic molecules which are essential for the synthesis of novel molecules for breast cancer.

Screening of breast cancer resulted in decrease in death by a substantial amount. With the rising use of screening exams [23], there has been a rise in the requirement for quick and reliable diagnostic reporting. Characteristic equivalent mammography in modern breast imaging has been substituted by FFDM. Moreover, this approach has lent newer promises for generating computational stages to automate detection. When compared to prior computer-based technologies [24–27], such as computer-aided findings besides diagnosis, AI, and DL are proving quite effective in diagnostic correctness [28].

8.3 Diagnosis of Breast Cancer

Breast cancer screening and diagnosis continue to rely on mammography. Mammography screening proponents point to its well-documented role in lowering breast cancer death rates [29–31]. Whilst mammographic screening has been shown to reduce mortality, it does come with a false-positive rate, much like any other test. While approximately 7–12% of women are wrongly recalled after a single mammogram, more than 45% of women who have had yearly mammography screening for 10 years will be incorrectly recalled [32]. These false positives result in more benign biopsies, more money spent, and poor psychological repercussions for the patients [33]. Similarly, because of their tiny size or the abundant fibro glandular tissue around them, possibly malignant neoplasms are in danger of being ignored [34, 35]. Women aged 50–89 who have had benign biopsies are more likely to have false-negative mammograms. However, false-negative findings are still uncommon [22], with rates ranging from 1.0 to 1.5 per 1000 women [36, 37]. Diagnostic mammography remains the widely accepted criterion for evaluating breast cancer, even though fact that its accuracy is improving with technological advancements [35]. Recent breakthroughs in AI have made it possible to design software, which can make job of radiologists easier in routine practice [38]. By increasing more scan experts, escalating screening numbers [39], or integrating supplemental imaging tools with traditional breast scanning, significant advancements have been achieved in the previous decade interms of novel material synthesis [40, 41] to address these shortcomings of mammographic screening [3]. The European validations for breast tumor identification and investigation, for instance, proposed twofold reading in which mammograms are examined separately by two different radiologists to improve sensitivity [4]. To improve breast cancer screening results, several imaging techniques like digital breast tomosynthesis (DBT), ultrasonography (US), and MRI are part of traditional four-point mammography [42]. Although increasing screening and leveraging additional imaging tools may hone the detection of breast cancer, guaranteeing adequate materials seems to be difficult since the responsibility of scan analysis by human experts will jump twofold setting beside an unavoidable jump in treatment costs as sophisticated machines would be required essentially [43]. DBT has been linked to concerns such as higher radiation exposure among other imaging modalities, and data to prove reduced mortality is scarce [44]. Finally, as community breast tumor identification programs have grown more popular, demand for screening of breast cancer has increased, as has the number of associated tests [45]. Medical resources, on the other hand, are still scarce. As a result, it's vital to improve and streamline present screening procedures [46]. Computer-assisted medical image analysis has sparked attention due to advancements in techniques and software, plus a pressing necessity to increase the speed and correctness of imaging interpretation workflows [47]. A few important reasons to leverage software programs to aid in scan analysis are: (i) computer-aided detection (CADe), which focuses on the localization of dubious irregularities in a mammogram, and/or (ii) computer-aided diagnosis, which focuses on the identification of irregularities identified by either the human expert or the computer (CADx). The radiologist decides the medical relevance of the discovered aberration and if it requires additional study basis on CADe. Although the name CAD refers to computer-aided design, it may be utilized for a variety of reasons based on the requirements of the radiologist. System algorithms may give numerical data from scans like parenchymal density from breast tissues, which were previously judged subjectively by the human eye [48]. At particular moments in the entire process of mammography interpretation, computer intervention is thought to be most useful. Initial CAD tools [49, 50], known as conventional CAD, were dependent on numerical frameworks that discovered forms linked [51, 52] to breast tumors and presented such designs as some markings on breast scans [53]. In a nutshell, such highlighted regions reflected what doctors needed to look into following the screening [54].

Manifold randomized type clinical trials have shown that mammography investigation reduces breast cancer death by 20–22% [55]. Owing to this, mammography has become the gold standard for breast tumor detection [56]. Breast scanning is the first test for numerous women who have breast complaints [57]. In the United States alone, more than 22 million mammograms were done in 2015 [57]. As a result, mammogram evaluation necessitates a good number of radiologists. Regrettably, many countries are experiencing a growing lack of qualified people [58]. Even among women who have had their mammographers, one in every three malignancies appears to be intermission cancer; a high fraction of such cases was apparent from earlier scans [59]. As a result, missed malignancies on mammography are among the most usual causes of radiography-related suits in courts [60, 61]. The new invention of DBT, which reconstructs a 3D dataset of mammography pictures by obtaining several angles of the breast is only a partial answer [62]. Although DBT detects 30–40% more tumors than FFDM, scanning time is nearly twofold and mental and perception mistakes still occur. As a result, support with mammography and DBT evaluation is required, both to maximize cancer detection rates and to solve workload concerns. These requirements could be met by automated analysis (through AI) of breast scans as well as DBT pictures. Since the late 1960s, CAD for mammograms has been in progress. Its major goal is to help radiologists in detecting tumors that might otherwise go undetected. CAD programs identify high-density areas and microcalcifications [63]. In 1998, the US Food and Drug Administration (FDA) approved the first CAD software for mammography screening. Initial results were good and CAD has been extensively incorporated into medical practice, with CAD being used in about 92% of all breast scanning facilities in the US by 2016. But its clinical utility is debatable owing to the high frequency of false-positive results [64]. The victory of convolutional neural networks (CNNs) in the

2012 ImageNet Large Scale Visual Recognition Challenge sparked renewed interest in developing more automated image analysis methods [65]. Parallel deep neural networks are extremely effective in a variety of jobs in recent years, ranging from face detection (computer vision) to driverless cars [66]. Recent research has demonstrated that CNNs can be very effective in a variety of activities in the healthcare business, extending from retina scanning assessment to computer-aided pathology, as well as in a variety of radiology applications.

8.4 Artificial Intelligence

AI refers to the capacity of computer programs to learn and solve issues by imitating human brain functioning [67–69]. More than around 60 years ago the concept of AI was proposed by John McCarthy. AI technology has advanced rapidly in the last 10 years. An important arm of computer technology, AI tools attempt to mimic human intelligence. Currently, accepted fields where AI is playing a crucial role are computer vision, natural language processing, and robotics among others [70]. In the medical range, AI is being used for management of health, clinical decision-making, radiology, and reading and extracting data from vast medical data [71–73], documents, literature, etc. AI can analyze medical pictures and data for illness screening and prediction, as well as help clinicians [74, 75] in establishing diagnoses. AI may be used in mammography to identify, section, and categorize masses. In one of the experiments, its accuracy in all aspects was greater than 92% [76]. Researchers collected 2654 exams and readings from around a hundred radiologists and used a skilled AI tool to rank the likelihood of tumors on a scale of 1–10. It was discovered that an AI rank of 2 as the threshold can decrease the assignment by 17%, proving that automation owing to AI can considerably lessen radiologists' burden [77]. The training of a mathematical system that studies structures and related parameters, which probably reflect or define the given data is referred to as ML. DL is one such ML tool based on an artificial neural network that works the way human neurons do (Figure 8.1)

ML uses a data-guided technique to understand a mathematical system from a training dataset. The objective is to use the trained model to generate predictions for the new test dataset. In a typical ML setup, training means repeated rounds of calculating the difference between machine prediction and the ground truth (that is actual value or true label). The ultimate goal is to adjust to narrow this difference in a clinically meaningful way, i.e. until the machine-generated output exactly matches with ground truth. Supervised learning is a type of ML learning where input data, such as mammograms, and results that are benign or malignant type are available for training the algorithms [78]. In mammography, for example, a tumor is precisely drawn on the picture, letting the computer learn the characteristics of a malignant type tumor from the given information unsupervised learning is when no diagnostic or normal/abnormal labels are available or included in the training data. Semi-supervised learning, also known as weakly supervised learning, gives the algorithm some information, but not for all of the training samples. In picture classification problems, supervised learning appears to be the most common method. Traditional

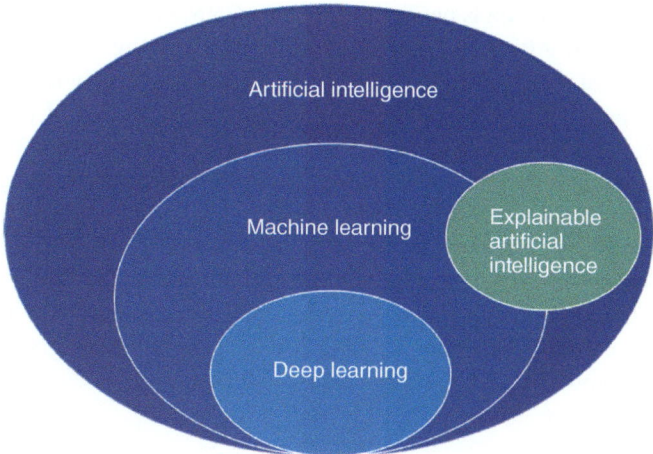

Figure 8.1 The relationship between AI, ML, DL, and explainable artificial intelligence.

ML methods like decision trees, support vector machines, random forests, and dimensionality reduction, have lesser requirements from the viewpoint of computational tools compared to DL tools [79, 80]. These methods are frequently employed in object identification for hand-crafted feature filtration and assortment. DL takes longer to train models since it has thousands of variables and hence demands high-performance computer tools like graphics processing units (GPUs). Although there are a variety of DL designs in the literature, the bulk of these networks are based on certain elementary and comparable neural network structure elements called layers. A neural network is made up of layers that include an input layer, hidden layer(s), and finally output layer, which gives the prediction. Early layers in a neural network are thought to behave like cells in the human primary visual cortex, which acquire low-level information like borders in certain orientations in the picture. Multiple stacked layers result in higher degrees of abstraction [81]. The DL architecture propagates information, and increasingly complicated characteristics are extracted. Finally, these characteristics are put into the network architecture's last layer for prediction or segmentation [82]. CNNs are a type of network used in image-related tasks. CNNs include distinct layers in which a convolution operation (kernel) serves as a filter on the pixel image matrix to extract spatially correlated characteristics from the input picture, resulting in some shift invariance. The output of the convolutional layer is linear (feature map). The pooling and completely linked layers are among the other layers. The convolutional layer's feature map is first processed via a nonlinear activation function, such as a rectified linear unit (ReLU), which reduces all negative values to 0 and maps it to an output. This is then sent to the pooling layer, which allows the feature map to be down sampled. Finally, the output is sent to the fully linked layer, which classifies the total result. The hidden layers of a CNN are composed of these layers. There can be hundreds of hidden levels, and the intricacy of feature extraction increases as the number of layers grows.

A large quantity of data is necessary to train a CNN, far more than is required for other forms of ML. Although radiology is rapidly welcoming digitization of routine tasks, the bulk of the data isn't open to the public, isn't properly labeled, or isn't of the needed quality from the DL learning model training point of view. In such a scenario, method of transfer learning can be used to circumvent this necessary training constraint. This method involves pretraining a model to recognize several essential characteristics, such as edges, and then applying that knowledge to a relevant picture job, such as identifying marks on a mammogram [83]. Following that, feature filtration may be used to modify the neural network for the new feature filtration job. Other approaches to overcoming data restrictions are an augmentation of data unsupervised learning, and the use of generative adversarial networks invented by renowned AI expert Ion Goodfellow [84].

8.4.1 Artificial Intelligence and Medical Imaging

AI is an arena that combines computer science, engineering [85–87], and mathematics (statistics) to create intelligent computing tools that can accomplish jobs that would normally need human intelligence [88, 89]. AI is in extensive use in the medical fields like imaging, retina scan, and others. Though AI applications in medicine are still restricted to research, radiologists are utilizing commercial algorithms in clinical practice to assist reduce flawed conclusions [54, 90]. Most of the AI-driven solutions for assisting imaging experts and analyzing scans are now being established using ML. ML is a branch of AI, which is mainly for developing algorithms that allow computers to do tasks without explicit instructions, relying instead on inference and patterns, and improving their performance over time [91].

The fast growth of ML, particularly DL, continues to attract the consideration of the medical radiology fraternity using these approaches to increase the accuracy of cancer detection. Cancer of breast tissues is the second highest cause of cancer deaths among American women, according to research, and screening breast scans has been shown to lower mortality. Despite its advantages, mammography screening is coupled with a significant risk of incorrect positives and untrue negatives. In the United States, digital screening mammography has an average sensitivity of 86.9% and an average specificity of 88.9%. For almost the last three decades CADe and diagnosis tools have been software has been generated to assist radiologists in bettering the prediction score of medical scans [92, 93]. Regrettably, evidence showed that early commercial CAD systems did not result in considerable performance improvements and that development remained stagnant for more than a decade after their introduction [94]. With DL's spectacular performance in vision-based recognition and detection, as well as many other domains there is a lot of curiosity in creating DL tools to aid imaging experts besides increasing screening mammography accuracy. Recent studies found that a DL-based CAD system fared well as radiologists in solo and even outperformed radiologists in assistance approach. Since the tumors occupy just a minor percentage of the picture of the whole breast, detecting subclinical tumors of the breast through scan screening is difficult as a picture image cataloging job [54], [95]. FFDM picture, for example, is generally 4000 × 3000 pixels wide, yet a possibly malignant region of interest (ROI) might be as little as 100 × 100 pixels wide. As a result, several studies have focused only on the categorization of marked lesions. Though physically labeling ROIs is a necessary initial stage, a completely automatic software solution should be capable enough to function on the complete scan to give extra information outside recognized lesions and to supplement scientific clarifications. Established object identification and classification algorithms like region-based convolutional neural network (R-CNN) and its variants might be easily employed if ROI explanations were generally accessible in mammography databases. However, methods that need ROI annotations are frequently incompatible with large datasets of mammography without ROI remarks, which are time-consuming and expensive to create. Only a small percentage of public imaging databases are completely annotated. Some studies have aimed at training neural networks without using any annotations by utilizing entire mammograms. However, it's difficult to say whether these neural networks were competent to find medically relevant lesions and make forecasts based on the related mammography segments. DL is widely known for requiring big training data to be effective. To further hone the correctness of segregation of breast cancer tumor models, it is unavoidable to use a few fully annotated datasets and bigger datasets labeled with the cancer status of each picture [96, 97]. When the desired vast and full training datasets are not available, pretraining is a potential way to handle the challenge of training a classifier. A layer-wise strategy is used in such cases to establish the weight parameters of a deep belief net (DBN) with hidden layers during pretraining, and then fine-tune it for classification. Such pretraining was shown to increase both the training speed and the correctness of handwritten digit identification. Another common training strategy is to build a DL model on a big database, such as ImageNet32, before fine-tuning it for a different goal [77, 98, 99]. Though the exact aim might not be matching with training set, the model's weight values are set for identifying very basic features like corners, lines, edges, angles, etc., and these can be easily used for different tasks. This saves training time and also improves the output of the mode. When it comes to mammography, the software normally focuses on clean-tagged mammograms with labels for faulty diagnosis and normal anatomy. The algorithms utilized by the software are adjusted as the program trains with more and more diverse images, with the decisive aim to be precise in interpreting scans and anticipating outlines. Because current improvements have expanded the manageability of AI software in clinical practice, it is critical to assess the present state of AI in mammography. One of the key benefits of AI in mammography is that it can let radiologists focus on more difficult cases by reducing effort by assisting with the assessment of more obvious situations. DL is used in the majority of image classification. It comes under the purview of ML where algorithms are designed in a similar fashion as that of ML. But DL is different from ML in the sense that the former consists of a number of stacked layers with various nodes or neurons for extracting and processing the features or information from inputs.

8.5 Conclusion

Presently radiologists are under the extreme burden of correctly interpreting mammograms in a timely fashion for the comfort of patients. Though the latest technologies are being used to scan breasts and subsequently interpretation of mammograms, in the current scenario if faster, smarter, and more accurate solutions are required for the interpretation of mammograms, then AI has no alternative. More specifically DL arm of AI has tremendous potential in making the entire process of interpretation fast based on the deployment of potent CNNs with multiple hidden layers. AI systems can be made more accurate and bias-free depending on the type and size of the database, which is used to train and test the DL algorithms (artificial neural network) before floating the model for real-world practice. To sum up, until there is full regulatory control on AI technology (especially from a training point of view) and some federal yardstick to check the bias-free approach of AI systems to analyze the mammograms, and data from millions of subjects to train the model, AI's full potential cannot be realized by a human.

8.6 Future Challenges

AI is an area where no one method or solution can suffice. There is a plethora of algorithms to choose from, as well as different methods for training, testing, and verifying the models. It makes selecting the correct outline even extra challenging, still, it offers you a lot of flexibility when it comes to integrating AI into mammography. Yet, with AI prepared to join the area at a rapid pace, there are several challenges to be aware of. Validation is necessary after a model has been trained. The validation stage involves assessing how successfully the model has trained or fine-tuned hyperparameters. When constructing a clinically useful ML earning model with the best predictive capacity, the training and validation phases are crucial [100, 101]. The creation of unbiased AI methods, which encourage parity in the medical imaging field is becoming a concern [55].

As a result, developers will have to demonstrate that their model works for a vast and varied population with different groups like ethnic, racial, minority groups, and individuals having lower prevalent risk issues. To adequately train and evaluate AI models, collaboration with large hospital networks and other organizations with diverse imaging data would be required [102]. Indeed, AI systems that can be applied to huge groups of people are currently in testing. FFDM is now the most often used imaging modality, and other screening methods like DBT are fast gaining prominence. Coming AI algorithms should be capable to manage the transition from 2D to 3D data, which will necessitate more storage and computational resources. To stay clinically useful, these models, like FFDM, will need to be trained on vast and heterogeneous DBT datasets [80].

AI researchers must adapt to any new mammographic screening procedures in the future. Owing to the specialized system of AI and breast scans collaboration between AI programmers and radiologists is critical to closing information gorge gaps in the two fields. They should be open to controlling bodies like the US FDA, which decides if an AI tool can be utilized as a valid medical tool. AI research is mainly retroactive and comparatively tiny expert investigation designed at demonstrating non-inferiority to scanning specialists [103].

The volume of datasets held in health databases, especially medical scanning data, grows, and there has been a rise in worries about cybersecurity [104]. Only mammograms connected to clinically proven outcomes are utilized to establish a model's genuine performance. As a result, researchers need not just a broad set of mammography registries to work with, but also a thorough understanding of the patient's medical record – particularly if the tool is going to integrate both scans and health checkup output. Blockchain [67] is one of the new technologies being used to disseminate medical imaging data. Nonetheless, without sophisticated data use agreements, imaging data firms may be hesitant to share thousands of mammograms and the history of the patient with research groups.

Even after the AI tool has been authorized and implemented, the AI algorithms must be constantly reviewed for potential enhancements or corrections and security vulnerabilities to reduce the risk of tampering or theft of image databases [105]. Three research demonstrated AI's utility in lowering the workload. Six studies found that AI can help in diagnosis, with a reduction in false positives of up to 69% and an increase in sensitivity of 84–91%. Five studies demonstrate how AI models can automatically flag and categorize worrisome abnormalities on traditional scans, with abilities equivalent to radiologists. Seven research looked into AI's ability to forecast breast cancer risk and calculate risk scores. One of the most significant roadblocks to the advancement of AI in clinical care is the difficulty in developing and implementing complicated algorithms capable of managing multivariable data that influences physicians' decision-making. When reviewing complicated medical issues, this also raises the question of accountability. Furthermore, the use of AI necessitates the establishment of an ethical code that promotes openness and trustworthiness, underline the need of condensing current viewpoints on the issue to focus on radiology particularly, pointing out that various AI models are reasonably simple to construct and develop. Few present studies adequately analyze how AI may be used to maximize effect while remaining responsible, and further research into the accuracy and dependability of AI is needed before this technology can be reliably employed in clinical settings. Despite these limitations, it is believed largely that AI has the potential to play a key role in mammography.

Even with recent advancements, additional research is needed to demonstrate AI's efficacy in assisting scanning radiologists with mammography in a natural setup. Sufficient outside justification is essential [106]. Different ML approaches have already been employed to build CAD algorithms for breast imaging modalities in recent years only a few have been commercially exploited for clinical usage, owing to a lack of clinical validation [107]. We need research that shows how AI will perform in the actual world while taking into account generalization, efficiency, user variability, and approaches to improve algorithms for consistent results [108]. In addition, the mechanism for assessing an AI algorithm's genuine performance has to be improved. To assist the radiologists, who can select to follow or ignore the

CAD instructions when making a final diagnosis is the role of traditional CAD. Characteristically, radiologists make concluding diagnoses, taking into account the CAD data as well.

When evaluating CAD performance under these conditions circumstances that represent real-world practice since radiologists, not CAD, are legally liable for their interpretation – the benefit gained by deploying a competent AI-CAD may be over-estimated. Furthermore, while estimating relevant data is challenging, we must investigate the way to quantify the material and number of hours spent on under-standing to determine if including the AI models increases the competence of over-all understanding. Moreover, in designing AI tools for DBT, a few technological issues need to be addressed. For instance, DL models are trained for 2D scans of breasts, therefore the use of transfer learning is done by using already trained CNNs to make for interpretation of DBT. Additionally, the meager spatial resolution fur-ther restricts the finding correctness of tomosynthesis, which lead to an inferior show of DL tools. Furthermore, various vendors' DBTs have varying angle ranges, acquisition procedures, pixel binning, and reconstruction strategies, all of which have an impact on mammography pictures. Finally, before introducing AI into our interpretation pipeline, there are ethical and legal problems to consider. Should AI be regarded as a stand-alone reader? Should medical records include the analytic data given by AI algorithms? Currently, regardless of whether or not the interpret-ing radiologist uses AI markings, the investigating image expert is legally account-able for her/his consequences, but the lawful implications of employing AI tool and the extent to which it is used should be thoroughly examined. If AI-CAD becomes commonplace, a legal or ethical structure for the use of AI algorithms will be needed, and this framework must represent the consensus of all stakeholders in a real-world mammography context, from patients to radiologists. Furthermore, we must be prepared for the unexpected effects of implementing AI algorithms, such as the diagnosis of numerous in situ cancers rather than invasive malignancies and the loss of interpretative abilities as a result of radiologists' overreliance on AI.

References

1 Harbeck, N., Penault-Llorca, F., Cortes, J. et al. (2019). Breast cancer. *Nat. Rev. Dis. Primers* 5: 66.

2 Yoon, J.H. and Kim, E.K. (2021). Deep learning-based artificial intelligence for mammography. *Korean J. Radiol.* 22 (8): 1225.

3 Myers, E.R., Moorman, P., Gierisch, J.M. et al. (2015). Benefits and harms of breast cancer screening: a systematic review. *JAMA* 314 (15): 1615–1634.

4 Lauby-Secretan, B., Scoccianti, C., Loomis, D. et al. (2015). Breast-cancer screening – viewpoint of the IARC Working Group. *N. Engl. J. Med.* 372 (24): 2353–2358.

5 Ribli, D., Horváth, A., Unger, Z. et al. (2018). Detecting and classifying lesions in mammograms with deep learning. *Sci. Rep.* 8 (1): 1–7.

6 Bae, M.S., Moon, W.K., Chang, J.M. et al. (2014). Breast cancer detected with screening US: reasons for nondetection at mammography. *Radiology* 270 (2): 369–377.

7 Hoff, S.R., Abrahamsen, A.L., Samset, J.H. et al. (2012). Breast cancer: missed interval and screening-detected cancer at full-field digital mammography and screen-film mammography – results from a retrospective review. *Radiology* 264 (2): 378–386.

8 Martin, J.E., Moskowitz, M., and Milbrath, J.R. (1979). Breast cancer missed by mammography. *AJR* 132: 737–739.

9 Banks, E., Reeves, G., Beral, V. et al. (2004). Influence of personal characteristics of individual women on sensitivity and specificity of mammography in the Million Women Study: cohort study. *BMJ* 329 (7464): 477. https://doi.org/10.1136/bmj.329.7464.477.

10 Smith-Bindman, R., Chu, P., Miglioretti, D.L. et al. (2005). Physician predictors of mammographic accuracy. *J. Natl. Cancer Inst.* 97 (5): 358–367. https://doi.org/10.1093/jnci/dji060.

11 Blanks, R.G., Wallis, M.G., and Moss, S.M. (1998). A comparison of cancer detection rates achieved by breast cancer screening programmes by number of readers, for one and two view mammography: results from the UK National Health Service breast screening program. *J. Medical Screening* 5 (4): 195–201.

12 Karssemeijer, N., Otten, J.D., and Roelofs, A.A. (2004). Effect of independent multiple reading of mammograms on detection performance. In: *Medical Imaging 2004: Image Perception, Observer Performance, and Technology Assessment* (4 May 2004), vol. 5372, pp. 82–89. SPIE.

13 Herent, P., Schmauch, B., Jehanno, P. et al. (2019). Detection and characterization of MRI breast lesions using deep learning. *Diagn. Interventional Imaging* 100 (4): 219–225.

14 Kadapure, S.A., Kadapure, P., Nandi, S.S., and Shet, A. (2022). Overview on catalyst and co-solvents for sustainable biodiesel production. *Proc. Inst. Civ. Eng. Energy* 1–9. https://doi.org/10.1680/jener.21.00092.

15 Adimule, V., Yallur, B.C., Kamat, V., and Krishna, P.M. (2021). Characterization studies of novel series of cobalt(II) nickel(II) and copper(II) complexes: DNA binding and antibacterial activity. *J. Pharm. Invest.* 51 (3): 347–359.

16 Adimule, V.M., Nandi, S.S., Kerur, S.S. et al. (2022). Recent advances in the one-pot synthesis of coumarin derivatives from different starting materials using nanoparticles: a review. *Top. Catal.* 1–31.

17 Nandi, S.S., Adimule, V., and Yallur, B.C. (2022). Synthesis, structural and optical properties of co doped Sm_2O_3 nanostructures. *Adv. Mater. Res.* 1173: 59–69. Trans Tech Publications, Ltd.

18 Adimule, V., Yallur, B.C., Batakurki, S., and Nandi, S.S. (2022). Synthesis, morphology and enhanced optical properties of Novel $Gd_xCo_3O_4$ nanostructures. *Adv. Mater. Res.* 1173: 71–82. Trans Tech Publications, Ltd.

19 Adimule, V., Batakurki, S., Yallur, B.C. et al. (2022). Enhanced photoluminescence, optical, structural properties of ZrO_2-incorporated Sm_2O_3:

Co_3O_4 nanocomposite and their applications in photocatalytic degradation of methylene blue. *J. Mater. Res.* 37: 2396–2405.

20 Adimule, V., Yallur, B.C., Pai, M.M. et al. (2022). Biogenic synthesis of magnetic palladium nanoparticles decorated over reduced graphene oxide using *Piper betle* petiole extract (Pd-rGO@Fe_3O_4 NPs) as heterogeneous hybrid nanocatalyst for applications in Suzuki-Miyaura coupling reactions of biphenyl compounds. *Top. Catal.* 1–14.

21 Adimule, V., Jagadeesha Gowda, A.H., Nandi, S.S., and Bowmik, D. (2022). Antimalarial activity of novel class of 1,3-benzoxaborole derivatives containing 1,3,4-oxadiazole moiety. In: Kendrekar, P. (ed) *Drug Development for Malaria*. Wiley.

22 Adimule, V., Nandi, S.S., and Yallur, B.C. (2022). Devices and sensors based on additively manufactured shape-memory of hybrid nanocomposites. In: *Shape Memory Composites Based on Polymers and Metals for 4D Printing* (ed. M.R. Maurya, K.K. Sadasivuni, J.J. Cabibihan, et al.). Cham: Springer https://doi .org/10.1007/978-3-030-94114-7_15.

23 Nelson, H.D., O'meara, E.S., Kerlikowske, K. et al. (2016). Factors associated with rates of false-positive and false-negative results from digital mammography screening: an analysis of registry data. *Ann. Intern. Med.* 164 (4): 226–235.

24 Zhang, Y., Weng, Y., and Lund, J. (2022). Applications of explainable artificial intelligence in diagnosis and surgery. *Diagnostics* 12 (2): 237.

25 Swartout, W.R. (1985). Rule-based expert systems: The mycin experiments of the stanford heuristic programming project. In: *Artificial Intelligence*, vol. 26(3) (ed. B.G. Buchanan and E.H. Shortliffe), 364–366. Reading, MA: Addison-Wesley.

26 Rodriguez-Ruiz, A., Lång, K., Gubern-Merida, A. et al. (2019). Stand-alone artificial intelligence for breast cancer detection in mammography: comparison with 101 radiologists. *J. Natl. Cancer Inst.* 111 (9): 916–922.

27 Berkman, A., Cole, B.F., Ades, P.A. et al. (2014). Racial differences in breast cancer, cardiovascular disease, and all-cause mortality among women with ductal carcinoma in situ of the breast. *Breast Cancer Res. Treat.* 148 (2): 407–413. https:// doi.org/10.1007/s10549-014-3168-3.

28 Tran, W.T., Sadeghi-Naini, A., Lu, F.I. et al. (2021). Computational radiology in breast cancer screening and diagnosis using artificial intelligence. *Can. Assoc. Radiol. J.* 72 (1): 98–108.

29 Batchu, S., Liu, F., Amireh, A. et al. (2021). A review of applications of machine learning in mammography and future challenges. *Oncology* 99 (8): 483–490.

30 Nyström, L., Andersson, I., Bjurstam, N. et al. (2002). Long-term effects of mammography screening: updated overview of the Swedish randomised trials. *Lancet* 359 (9310): 909–919.

31 Duffy, S.W., Tabar, L., Vitak, B. et al. (2003). The relative contributions of screen-detected in situ and invasive breast carcinomas in reducing mortality from the disease. *Eur. J. Cancer* 39 (12): 1755–1760.

32 Tabár, L., Vitak, B., Chen, H.H. et al. (2001). Beyond randomized controlled trials: organized mammographic screening substantially reduces breast carcinoma mortality. *Cancer* 91 (9): 1724–1731.

33 Hubbard, R.A., Kerlikowske, K., Flowers, C.I. et al. (2011). Cumulative probability of false-positive recall or biopsy recommendation after 10 years of screening mammography: a cohort study. *Ann. Intern. Med.* 155 (8): 481–492.

34 Braithwaite, D., Zhu, W., Hubbard, R.A. et al. (2013). Screening outcomes in older US women undergoing multiple mammograms in community practice: does interval, age, or comorbidity score affect tumor characteristics or false positive rates? *J. Natl. Cancer Inst.* 105 (5): 334–341.

35 Nelson, H.D., Pappas, M., Cantor, A. et al. (2016). Harms of breast cancer screening: systematic review to update the 2009 US Preventive Services Task Force recommendation. *Ann. Intern. Med.* 164 (4): 256–267.

36 Huynh, K.T., Chong, K.K., Greenberg, E.S. et al. (2012). Epigenetics of estrogen receptor-negative primary breast cancer. *Expert Rev. Mol. Diagn.* 12: 371–382.

37 Tan, J., Ong, C.K., Lim, W.K. et al. (2s of breast fibroepithelial tumors. *Nat. Genet.* 47 (11): 1341–1345. https://doi.org/10.1038/ng.3409.

38 Linden, O.E., Hayward, J.H., Price, E.R. et al. (2020). Utility of diagnostic mammography as the primary imaging modality for palpable lumps in women with almost entirely fatty breasts. *Am. J. Roentgenol.* 214 (4): 938–944.

39 Keri, T., R.S., Adimule, V., Kendrekar, P. et al. (2022). The nano-based catalyst for the synthesis of benzimidazoles. *Top. Catal.* 1–21.

40 Adimule, V., Nandi, S.S., Yallur, B.C., and Shaikh, N. (2021). CNT/graphene-assisted flexible thin-film preparation for stretchable electronics and superconductors. In: Pal, K. (ed) *Sensors for Stretchable Electronics in Nanotechnology*, 89–103. CRC Press.

41 Nandi, S.S. Suryavanshi, A., Adimule, V., and Maradur, S.R. (2020). Semiconductor current-voltage characteristics of some novel perovskite ionic nanocomposites of $Sr_{0.5}Cu_{0.4}Y_{0.1}$ and $Sr_{0.5}Mn_{0.5}$ and their electronic sensor applications. In: *AIP Conference Proceedings*, vol. 2274, p. 020006. AIP Publishing LLC.

42 Taylor-Phillips, S. and Stinton, C. (2020). Double reading in breast cancer screening: considerations for policy-making. *Br. J. Radiol.* 93 (1106): 20190610.

43 Shaikh, N.M., Adimule, V., Bagihalli, G.B. et al. (2022). A novel mixed Ag–Pd nanoparticles supported on SBA silica through [DMAP-TMSP-DABCO]OH basic ionic liquid for Suzuki coupling reaction. *Top. Catal.* 1–10

44 Houssami, N., Lee, C.I., Buist, D.S., and Tao, D. (2017). Artificial intelligence for breast cancer screening: opportunity or hype? *Breast* 36: 31–33.

45 Reporting, B.R. (2005). Computer-aided detection with screening mammography in a university hospital setting. *Radiology* 236: 451–457.

46 Bi, W.L., Hosny, A., Schabath, M.B. et al. (2019). Artificial intelligence in cancer imaging: clinical challenges and applications. *CA Cancer J. Clin.* 69 (2): 127–157.

47 Abbasi-Sureshjani, S., Yüce, A., and Schönenberger, S. (2021). Molecular subtype prediction for breast cancer using H&E specialized backbone. In: *MICCAI Workshop on Computational Pathology*, pp. 1–9. PMLR.

48 Birdwell, R.L., Ikeda, D.M., O'Shaughnessy, K.F., and Sickles, E.A. (2001). Mammographic characteristics of 115 missed cancers later detected with screening mammography and the potential utility of computer-aided detection. *Radiology* 219 (1): 192–202. https://doi.org/10.1148/radiology.219.1.r01ap16192.

49 Sardanelli, F., Fallenberg, E.M., Clauser, P. et al. (2017). Mammography: an update of the EUSOBI recommendations on information for women. *Insights Imaging* 8 (1): 11–18.

50 Adimule, V., Yallur, B.C., Bhowmik, D., and Gowda, A.H. (2021). Dielectric properties of P3BT doped $ZrY_2O_3/CoZrY_2O_3$ nanostructures for low cost optoelectronics applications. *Trans. Electr. Electron. Mater.* 23(3): 288–303.

51 Adimule, V., Nandi, S.S., and Adarsha, H.J. (2021). A facile synthesis of Cr doped WO_3 nanostructures study of their current-voltage power dissipation and impedance properties of thin films. *J. Nano Res.* 67: 33–42.

52 Adimule, V., Nandi, S.S., and Gowda, A.H.J. (2021). Enhanced power conversion efficiency of the P3BT (poly-3-butyl thiophene) doped nanocomposites of $GdTiO_3$ as working electrode. In: Prashant, M. P., Balasubramaniam, R., Ronge, B.P. et al. (eds) *Techno-Societal 2020*. Springer Nature. 55–68.

53 Fenton, J.J., Abraham, L., Taplin, S.H. et al. (2011). Breast cancer surveillance consortium. Effectiveness of computer-aided detection in community mammography practice. *J. Natl. Cancer Inst.* 103 (15): 1152–1161.

54 Lehman, C.D., Wellman, R.D., Buist, D.S. et al. (2015). Breast cancer surveillance consortium. Diagnostic accuracy of digital screening mammography with and without computer-aided detection. *JAMA Intern. Med.* 175 (11): 1828–1837.

55 Geras, K.J., Mann, R.M., and Moy, L. (2019). Artificial intelligence for mammography and digital breast tomosynthesis: current concepts and future perspectives. *Radiology* 293 (2): 246.

56 Tabár, L., Yen, A.M., Wu, W.Y. et al. (2015). Insights from the breast cancer screening trials: how screening affects the natural history of breast cancer and implications for evaluating service screening programs. *Breast J.* 21 (1): 13–20.

57 Sardanelli, F., Aase, H.S., Álvarez, M. et al. (2017). Position paper on screening for breast cancer by the European Society of Breast Imaging (EUSOBI) and 30 national breast radiology bodies from Austria, Belgium, Bosnia and Herzegovina, Bulgaria, Croatia, Czech Republic, Denmark, Estonia, Finland, France, Germany, Greece, Hungary, Iceland, Ireland, Italy, Israel, Lithuania, Moldova, The Netherlands, Norway, Poland, Portugal, Romania, Serbia, Slovakia, Spain, Sweden, Switzerland and Turkey. *Eur. Radiol.* 27 (7): 2737–2743.

58 Wing, P. and Langelier, M.H. (2009). Workforce shortages in breast imaging: impact on mammography utilization. *Am. J. Roentgenol.* 192 (2): 370–378.

59 Weber, R.J., van Bommel, R.M., Louwman, M.W. et al. (2016). Characteristics and prognosis of interval cancers after biennial screen-film or full-field digital screening mammography. *Breast Cancer Res.Treat.* 158 (3): 471–483.

60 Whang, J.S., Baker, S.R., Patel, R. et al. (2013). The causes of medical malpractice suits against radiologists in the United States. *Radiology* 266 (2): 548–554.

61 Arleo, E.K., Saleh, M., and Rosenblatt, R. (2016). Lessons learned from reviewing breast imaging malpractice cases. *J. Am. Coll. Radiol.* 13 (11): R58–R60.

62 Vedantham, S., Karellas, A., Vijayaraghavan, G.R., and Kopans, D.B. (2015). Digital breast tomosynthesis: state of the art. *Radiology* 277 (3): 663.

63 Sechopoulos, I. (2013). A review of breast tomosynthesis. Part I. The image acquisition process. *Med. Phys.* 40 (1): 014301.

64 Ciatto, S., Houssami, N., Bernardi, D. et al. (2013). Integration of 3D digital mammography with tomosynthesis for population breast-cancer screening (STORM): a prospective comparison study. *Lancet Oncol.* 14 (7): 583–589.

65 Krizhevsky, A., Sutskever, I., and Hinton, G.E. (2012). Imagenet classification with deep convolutional neural networks. In: Pereira, F., Burges, C.J., Bottou, L., et al. (eds) *Advances in Neural Information Processing Systems.* NeurIPS Proceedings. vol. 25.

66 Chen, C., Seff, A., Kornhauser, A., and Xiao, J. (2015). Deepdriving: learning affordance for direct perception in autonomous driving. In: *Proceedings of the IEEE International Conference on Computer Vision*, pp. 2722–2730.

67 Lei, Y.M., Yin, M., Yu, M.H. et al. (2021). Artificial intelligence in medical imaging of the breast. *Front. Oncol.* 2892. 11: 600557.

68 Hamet, P. and Tremblay, J. (2017). Artificial intelligence in medicine. *Metabolism* 69: S36–S40.

69 Viceconti, M., Hunter, P., and Hose, R. (2015). Big data, big knowledge: big data for personalized healthcare. *IEEE J. Biomed. Health Inf.* 19 (4): 1209–1215.

70 Ziad Obermeyer, E.J. (2016). Predicting the future—big data. Machine learning, and clinical medicine. *N. Engl. J. Med.* 375 (13): 1216–1219.

71 Ting, D.S., Cheung, C.Y., Lim, G. et al. (2017). Development and validation of a deep learning system for diabetic retinopathy and related eye diseases using retinal images from multiethnic populations with diabetes. *JAMA* 318 (22): 2211–2223.

72 Larson, D.B., Chen, M.C., Lungren, M.P. et al. (2018). Performance of a deep-learning neural network model in assessing skeletal maturity on pediatric hand radiographs. *Radiology* 287 (1): 313–322.

73 Rajpurkar, P., Irvin, J., Ball, R.L. et al. (2018). Deep learning for chest radiograph diagnosis: a retrospective comparison of the Che XNeXt algorithm to practicing radiologists. *PLoS Med.* 15 (11): e1002686.

74 Wang, S. and Summers, R.M. (2012). Machine learning and radiology. *Med. Image Anal.* 16 (5): 933–951.

75 Adimule, V., Nandi, S.S., Yallur, B.C. et al. (2021). Enhanced photoluminescence properties of $Gd_{(x-1)}$ Sr_xO: CdO nanocores and their study of optical structural and morphological characteristics. *Mater. Today Chem.* 20: 00438.

76 Al-Antari, M.A., Al-Masni, M.A., Choi, M.T. et al. (2018). A fully integrated computer-aided diagnosis system for digital X-ray mammograms via deep learning detection, segmentation, and classification. *Int. J. Med. Inform.* 117: 44–54. https://doi.org/10.1016/j.ijmedinf.2018.06.003.

77 Rodriguez-Ruiz, A., Lång, K., Gubern-Merida, A. et al. (2019). Can we reduce the workload of mammographic screening by automatic identification of normal exams with artificial intelligence? A feasibility study. *Eur. Radiol.* 29 (9): 4825–4832.

78 Le, E.P., Wang, Y., Huang, Y. et al. (2019). Artificial intelligence in breast imaging. *Clin. Radiol.* 74 (5): 357–366.

79 Bengio, Y. (2012). Deep learning of representations for unsupervised and transfer learning. In: *Proceedings of ICML Workshop on Unsupervised and Transfer Learning* (27 June 2012), pp. 17–36. JMLR Workshop and Conference Proceedings.

80 Chartrand, G., Cheng, P.M., Vorontsov, E. et al. (2017). Deep learning: a primer for radiologists. *Radiographics* 37 (7): 2113–2131.

81 LeCun, Y., Bengio, Y., and Hinton, G. (2015). Deep learning. *Nature* 521: 436–444.

82 Dubrovina, A., Kisilev, P., Ginsburg, B. et al. (2018). Computational mammography using deep neural networks. *Comput. Meth. Biomech. Biomed. Eng.* 6 (3): 243–247.

83 Yamashita, N., Tokunaga, E., Iimori, M. et al. (2018). Epithelial paradox: clinical significance of coexpression of E-cadherin and Vimentin with regard to invasion and metastasis of breast cancer. *Clin. Breast Cancer* 18 (5): e1003–e1009. https://doi.org/10.1016/j.clbc.2018.02.002.

84 Mendel, K., Li, H., Sheth, D. et al. (2019). Transfer learning from convolutional neural networks for computer-aided diagnosis: a comparison of digital breast tomosynthesis and full-field digital mammography. *Acad. Radiol.* 26 (6): 735–743.

85 Adimule, V., Nandi, S.S., Yallur, B.C. et al. (2021). Optical structural and photoluminescence properties of Gd_xSrO: CdO nanostructures synthesized by co precipitation method. *J. Fluoresc.* 31 (2): 487–499.

86 Adimule, V., Yallur, B.C., Bhowmik, D., and Gowda, A.H.J. (2021). Morphology structural and photoluminescence properties of shaping triple semiconductor Y_xCoO: ZrO_2 nanostructures. *J. Mater. Sci. - Mater. Electron.* 32 (9): 2164–12181.

87 Adimule, V., Revaiah, R.G., Nandi, S.S., and Jagadeesha, A.H. (2021). Synthesis characterization of Cr doped TeO_2 nanostructures and its application as EGFET pH sensor. *Electroanalysis* 33 (3): 579–590.

88 Charniak, E. and McDermott, D. (1985). *Introduction to Artificial Intelligence*. Reading, MA: Addison-Wesley.

89 Buchanan, B.G. and Shortliffe, E.H. (1984). *Rule Based Expert Systems: The Mycin Experiments of the Stanford Heuristic Programming Project (the Addison-Wesley Series in Artificial Intelligence)*. Addison-Wesley Longman Publishing Co., Inc.

90 Rowlands, C.F., Baralle, D., and Ellingford, J.M. (2019). Machine learning approaches for the prioritization of genomic variants impacting pre-mRNA splicing. *Cells* 8 (12): 1513.

91 Watanabe, A.T., Lim, V., Vu, H.X. et al. (2019). Improved cancer detection using artificial intelligence: a retrospective evaluation of missed cancers on mammography. *J. Digital Imaging* 32 (4): 625–637.

92 Lehman, C.D., Arao, R.F., Sprague, B.L. et al. (2017). National performance benchmarks for modern screening digital mammography: update from the breast cancer surveillance consortium. *Radiology* 283 (1): 49.

93 Oeffinger, K.C., Fontham, E.T., and Wender, R.C. (2016). Clinical breast examination and breast cancer screening guideline – reply. *JAMA* 315 (13): 1404.

94 Fenton, J.J., Taplin, S.H., Carney, P.A. et al. (2007). Influence of computer-aided detection on performance of screening mammography. *N. Engl. J. Med.* 356 (14): 1399–1409.

95 Aboutalib, S.S., Mohamed, A.A., Berg, W.A. et al. (2018). Deep learning to distinguish recalled but benign mammography images in breast cancer screening. *Clin. Cancer Res.* 24 (23): 5902–5909.

96 Kim, C., Gao, R., Sei, E. et al. (2018). Chemoresistance evolution in triple-negative breast cancer delineated by single-cell sequencing. *Cell* 173 (4): 879–893.e13. https://doi.org/10.1016/j.cell.2018.03.041.

97 Hamidinekoo, A., Denton, E., Rampun, A. et al. (2018). Deep learning in mammography and breast histology, an overview and future trends. *Med. Image Anal.* 47: 45–67.

98 Kooi, T., Litjens, G., Van Ginneken, B. et al. (2017). Large scale deep learning for computer aided detection of mammographic lesions. *Med. Image Anal.* 1 (35): 303–312.

99 Agarwal, R., Diaz, O., Lladó, X. et al. (2019). Automatic mass detection in mammograms using deep convolutional neural networks. *J. Med. Imaging* 6 (3): 031409.

100 Rajkomar, A., Hardt, M., Howell, M.D. et al. (2018). Ensuring fairness in machine learning to advance health equity. *Ann. Intern. Med.* 169 (12): 866–872.

101 McKinney, S.M., Sieniek, M., Godbole, V. et al. (2020). International evaluation of an AI system for breast cancer screening. *Nature* **577**: 89–94. https://doi .org/10.1038/s41586-019-1799-6.

102 Redberg, R.F. and Dhruva, S.S. (2019). Moving from substantial equivalence to substantial improvement for 510 (k) devices. *JAMA* 10, 322 (10): 927–928.

103 Patel, V. (2019). A framework for secure and decentralized sharing of medical imaging data via blockchain consensus. *Health Inform. J.* 25 (4): 1398–1411.

104 Becker, A.S., Jendele, L., Skopek, O. et al. (2019). Injecting and removing suspicious features in breast imaging with Cycle GAN: a pilot study of automated adversarial attacks using neural networks on small images. *Eur. J. Radiol.* 1 (120): 108649.

105 Mendelson, E.B. (2019). Artificial intelligence in breast imaging: potentials and limitations. *Am. J. Roentgenol.* 212 (2): 293–299.

106 Park, S.H. and Kressel, H.Y. (2018). Connecting technological innovation in artificial intelligence to real-world medical practice through rigorous clinical validation: what peer-reviewed medical journals could do. *J. Korean Med. Sci.* 33 (22): e152.

107 Yassin, N.I., Omran, S., El Houby, E.M., and Allam, H. (2018). Machine learning techniques for breast cancer computer aided diagnosis using different image modalities: a systematic review. *Comput. Methods Progr. Biomed.* 156: 25–45.

108 Kim, J.H., Kim, B.G., Roy, P.P., and Jeong, D.M. (2019). Efficient facial expression recognition algorithm based on hierarchical deep neural network structure. *IEEE Access,* 7: 41273–41285.

9

Establishing Nanotechnology-Based Drug Development for Triple-Negative Breast Cancer Treatment

Ravinder Verma[1], Shailendra Bhatt[2], Rohit Dutt[3], Manish Kumar[4], Deepak Kaushik[5] and Rupesh K. Gautam[6]

[1] Chaudhary Bansi Lal University, Department of Pharmaceutical Sciences, Bhiwani 127021, India
[2] G. D. Goenka University, Department of Pharmacy, Sohna Road, Gurugram 122103, India
[3] Gandhi Memorial National College, Ambala Cantt 133001, India
[4] M.M. College of Pharmacy, Maharishi Markandeshwar (Deemed to be University), Mullana-Ambala, Haryana 133207, India
[5] Maharshi Dayanand University, Department of Pharmaceutical Sciences, Rohtak, Haryana 124001, India
[6] Indore Institute of Pharmacy, Department of Pharmacology, IIST Campus, Opposite IIM Indore, Rau-Pithampur Road, Indore, Madhya Pradesh 453331, India

9.1 Introduction

Worldwide, breast cancer (BC) is the second foremost reason for cancer-related expiries among females. It is accountable for 1700 deceases per day and one in three females have a threat of evolving BC during their life [1]. More than 3.5 million females were detected with this in the US only [2, 3]. According to GLOBOCAN 2020, globally 3 465 951 incidences and 1 121 413 mortalities were reported; and 1 204 532 incidences and 436 417 mortalities were reported in India [4]. It has inter- and intratumoral heterogeneities that are the foremost issues in its effective treatment [5].

Immunohistochemical (ICH) examination of BC is the gold-standard approach that is utilized in clinics for their classification based on hormone receptor expression intended for enhanced therapeutic results as shown in Table 9.1. On this basis, it is broadly classified into five types:

(i)   Luminal A
(ii) Luminal B
(iii) HER2 overexpressing
(iv) Basal like or triple negative
(v)   Normal breast-like cancer (unclassified) [7–10].

According to a literature survey, triple-negative breast cancer (TNBC) accounts for 10–20% of all BC that have high rates of degeneration, recurrence, and expiries [11].

Table 9.1 Characteristics of BC subtypes.

Molecular subtypes	Estrogen receptor (ER)	Progesterone receptor (PR)	Human epidermal growth factor-2 (HER-2)	Cytokeratin 5 and 6	Prevalence (%)
Luminal A	+	+	−	−	40
Luminal B	+	+	+/−	−	20
HER-2 enriched	−	−	+	−	10–15
Basal-like	−	−	−	+	15–20
Unclassified	−	−	−	−	2–8

Source: Kanwal [6]/Public Domain CC BY 4.0.

It is more dominant in Hispanic, African, and American women, and generally, young women are more vulnerable to it [12]. Its heterogeneous nature is responsible for its increased aggressiveness and poorer quality clinical results. Neoadjuvant chemotherapy (NACT) is the best available cure option for TNBC patients [13].

With this chapter, we aimed to discuss the existing and imperative results on TNBC and various signaling pathways. Novel approaches based on nanotechnology that can be clinically applied in the future are also emphasized. This chapter presents recent information associated with different nanoformulations including nanoparticles (NPs), quantum dots (QDs), liposomes, nanostructured lipid carriers, dendrimers, nanoconjugates, micelles, and many others with their potential utility in medical practice in next future for TNBC treatment. Detailed information related to various patented formulations and nano vaccines is also provided in this chapter.

9.2 Triple-Negative Breast Cancer

Immune cells have vital importance for BC identification and early suppression, but they are also responsible for tumor progression [14]. TNBC is often diagnosed in young women. It is a particularly devastating subtype of BC that is recognized by transmutations in *Breast Cancer BRCA*1 or *BRCA*2 liable gene activity. It is not treatable with antiestrogen hormonal therapy because of the nonexistence of ER, PR, and, HER-2 receptor expression. Statistics of the hormonal status of the Indian populace are threatened because economic limitations are a major hurdle in the execution of immunohistochemistry (IHC) estimation [15].

TNBC is diagnosed through IHC [16]. Morphological features include a hyperdense mass that lacks calcification, which commonly occur in women who are older than 50 years. Histologically, it includes significant infiltration of lymphocytes, central necrosis, central necrosis, aggressive tumor borders, and fibrosis [17]. Basal-like TNBC is diagnosed by the manifestation of cytokeratins, fascin, EGFR, caveolin, and vimentin. Molecular overlapping between triple-negative and basal-like subtypes of BC is a major hurdle in their diagnosis. TNBC is quite complex due to poor growth of the cell, while molecular heterogeneity, chemoresistance, and recurrences of the disease are the major challenges in TNBC treatment [18]. Unquestionably, innovations in the field of BC exploration are responsible for an incredible rise in the rate of survival in BC patients when it is identified at the primary stage and without metastasis. It is the most challenging BC subtype to treat because it resembles different BC subtypes [17].

The usage of nanotechnology is developed as a weapon for the deliverance of therapeutic agents and biomolecules in a single nanosystem thereby surpassing the disadvantages of traditional drug deliverance approaches. Nanomedicine is a nanotechnology-based novel approach that is a novel therapeutic alternative for biomedical usage of nanocarrier with at least one aspect >100 nm. But nanosystems with 100–200 nm are also considered. Liposomes (Doxil®) and nanoparticles (Abraxane®) are currently used for the successful treatment of TNBC. These marketed products were developed as generic anticancer agents. For a better understanding of the molecular biology of TNBC, various alternative nano delivery approaches further designed for TNBC were actively explored during the last decade as shown in Figure 9.1 [19].

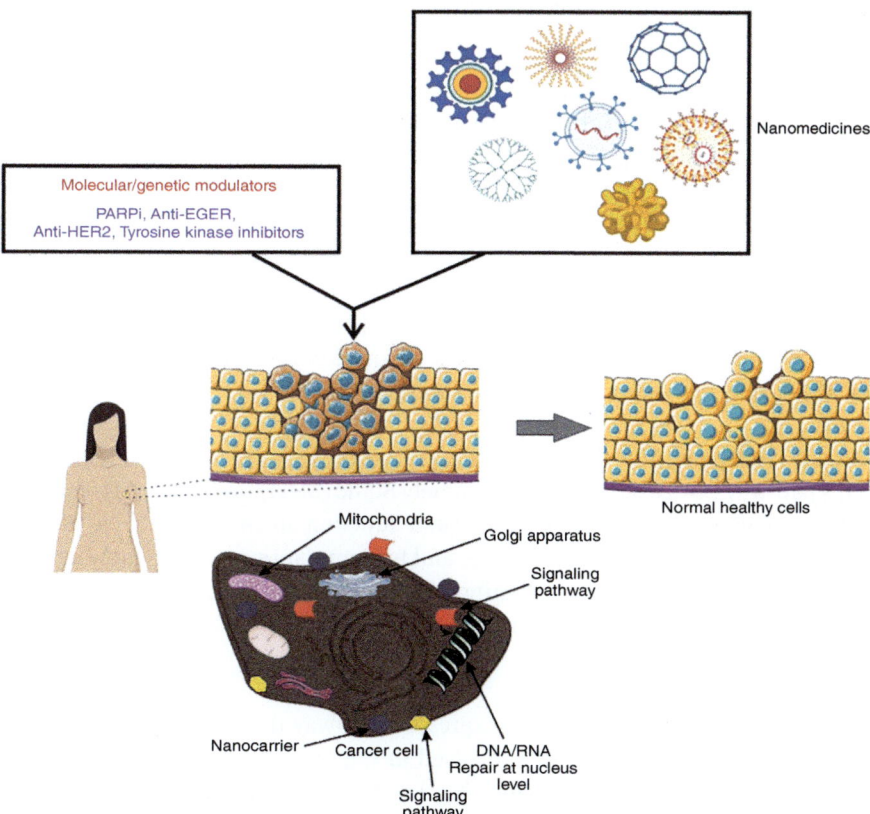

Figure 9.1 Role of nanomedicines and molecular pathways in TNBC treatment.

Nanomedicine has various benefits such as greater surface-area-to-volume ratio (helps to deploy their surface characteristics for better treatment), entrapment/binding onto nanocarriers (drug achieves enhanced stability, enhanced solubility, and CR kinetics), and delivery of therapeutic agents in combinations (co-delivered for augmented synergistic antitumor activity). Several nanotechnologies-based nanoformulations are being widely used such as nanoparticles, quantum dots, liposomes, nanostructured lipid carriers, dendrimers, nanoconjugates, micelles, and many others. These nanoformulations offer various unique features such as nanometric size range, effectiveness, enhancing the aqueous solubility of anticancer agents, enhancing drug delivery efficacy at targeted sites, and aiding site-targeted deliverance of anticancer agents. Although the benefits are alluring, these nanoformulations also hold certain limits such as toxicity, immunogenicity, and excretion mechanism [20]. The major advantage of nanotechnology-based methodologies is a significant decrease in blood toxicity with greater therapeutic efficiency in contrast to surgery, chemotherapy, etc. Other desirable advantages like quick identification, more circulation time, greater effectiveness, and greater potency can be achieved with these nanoformulations [21].

9.2.1 Molecular Mechanisms (Signaling Pathways) Involved in TNBC Therapeutics

The latest innovations in omics technologies help in understanding its molecular mechanisms [22]. The various mechanisms involved in TNBC development and diagnosis are summarized in Table 9.2 and discussed below.

9.2.1.1 Notch Signaling Pathway

This pathway is engaged in cell proliferation and differentiation [24]. This pathway comprises four receptors of Notch, namely Notch 1, 2, 3, and 4, and five ligands namely Delta-like 1, 3, and 4, Jagged 1 and 2 [25–27]. Notch 1 is identified as the cause of human mammary tumors [28], various kinds of hematological malignancies [29], pancreatic tumors [30], and several others.

9.2.1.2 Hedgehog Signaling Pathway

This pathway involves three ligands namely Sonic (SHH), Indian (IHH), and Desert (DHH). This pathway plays an important function in the invasion of cancerous cells, metastasis and chemoresistance, and tumor relapse cancer after therapy. SHH is involved in the erroneous beginning of malignancy in BC as it helps in maintaining atypical proliferation and stimulates metastasis.

9.2.1.3 Wnt/β-Catenin

This is the utmost frequently overexpressed pathway that leads activation of transcriptional factors. This factor is involved in the activation of epithelial to mesenchymal cells (EMTs) transitions in CSCs. In TNBC, this pathway is also erroneously regulated in canonical and no-canonical molecules [31]. To date, this pathway consists of 19 human Wnts and 10 Frizzled (FZD) receptors and co-receptors [32, 33].

Table 9.2 Various signaling pathways responsible for the development and diagnosis of TNBC.

Pathway	Mechanism of action	Drug	Phase
Notch	γ-secretase/aspartyl protease inhibitors	RO-4929097	Phase I/II
Hedgehog	SMO antagonist and Pathed1 antagonists	Cyclopamine	Phase II
Wnt/β-catenin	FZD7 receptor and Wnt co-receptor LDL receptor-related protein-6 inhibitors	Salinomycin	Phase I/II
PARP inhibitors	Inhibits repair DNA damage	Iniparib	Phase II/III
		Olaparib	Phase II
		Rucaparib	Phase II
		Veliparib	Phase I
EGFR inhibitors	Inhibits overexpressed epidermal growth factor	Gefltinib	Phase II
mTOR inhibitors	Inhibit FK506 binding protein 12-rapamycin linked protein 1	Everolimus	Phase II
		Temsirolimus	Phase II
TGF-β	TGF-β1 inhibitor	LY2157299	Phase I
CSPG4 proteins	CSPG4 targeted mAb	CSPG4-specific mAb	Tested in a few patients with BC

Source: Adapted from Jamdade et al. [23].

FZD6 receptor is vital descriptive in TNBC because it has the potential to yield metastasis by raising motility features of tumor cells [34].

9.2.1.4 Poly(ADP-Ribose) Polymerase (PARP) Inhibitors

It affects all molecular actions that are responsible for regaining cells from DNA impairment, gene transcription, apoptosis, and genomic stability. When poly(ADP-ribose) polymerase (PARP) action is suppressed, it results in inhibition of the development of ADP-ribose composite [35]. Trapped PARP–DNA complexes are highly cytotoxic showing high anticancer potential [36].

9.2.1.5 EGFR

Approximately 89% of TNBC cases involve EGFR expression. That's why, it is well thought out to be an effective therapeutic target, particularly for BL2-subtype cancers that are increased in EGFR gene expression [37]. Stimulation of the EGFR gene excites primary tumorigenesis and metastasis [38, 39].

9.2.1.6 Mammalian Target of Rapamycin (mTOR) Inhibitors

The dysregulation of this pathway has a direct relationship with malignancy. Alteration in this pathway is responsible for the poor diagnosis of TNBC. Phosphorylation reactions are initiated because PI3K/Akt/mammalian target of rapamycin (mTOR) is liable for the growth of tumor cells, propagation of cells, and angiogenesis. Furthermore, overexpression of Akt is also responsible for tumor metastasis and invasion [9].

9.2.1.7 TGF-β Signaling Pathway

It belongs to the cytokines superfamily that helps in encrypting of TGF-β1 gene. Platelets cells are involved in wound healing and regulation of the immune system. Thus, they inhibit the emission and functions of various cytokines including IFN-ɣ, TNF-α, and IL-2. It also plays a vital role in BCSCs as they express it and its receptor exponentially [15].

9.2.1.8 CSPG4 (Chondroitin Sulfate Proteoglycan) Protein Signaling Pathway

This is a cell-surface proteoglycan that is present in basal breast tumor cells. So, its inhibition is medicinally effective for BC therapy. This protein stabilizes cell-substratum interaction via the spreading of endothelial basement membrane protein [40] (Figure 9.2).

9.2.2 Conventional Therapeutics

Neoadjuvant therapy, adjuvant therapy, surgery, and radiotherapy are the available traditional treatments for TNBC that are summarized in Table 9.3.

Neoadjuvant therapy is used before the surgery. This therapy reduces tumor size and evades mastectomy. In neoadjuvant and adjuvant therapy, taxanes (paclitaxel

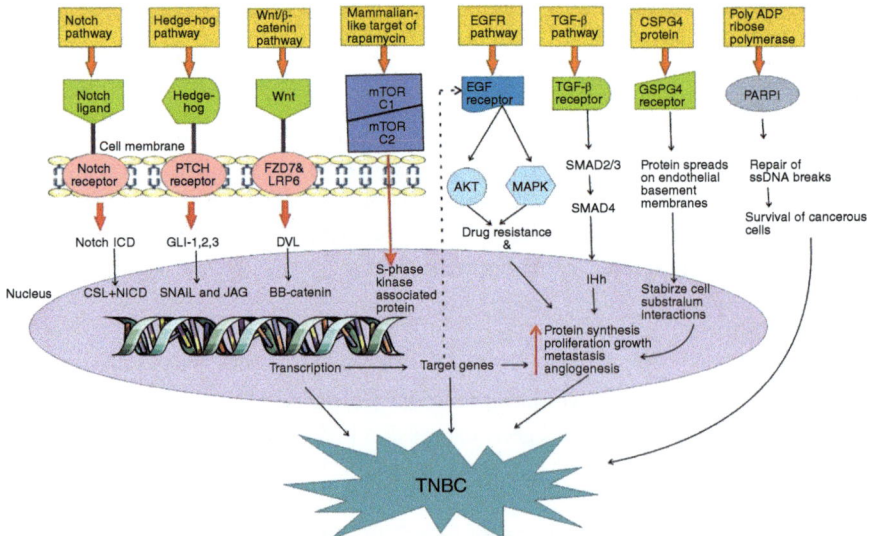

Figure 9.2 Illustration of various signaling pathways involved in the development of TNBC that can be targeted in TNBC treatment [15].

Table 9.3 Traditional treatment available for TNBC.

Conventional treatment	Therapeutic moiety used (Dose used)	Mechanism
Neoadjuvant treatment for early TNBC	Doxorubicin ($20 \, mg \, m^{-2}$) with cyclophosphamide ($600 \, mg \, m^{-2}$) followed by paclitaxel ($80 \, mg \, m^{-2}$) Capecitabine ($1250 \, mg \, m^{-2}$) for 14 days with docetaxel ($75 \, mg \, m^{-2}$) and ixabepilone ($40 \, mg \, m^{-2}$)	Cytotoxicity, Stabilization of microtubules
New neoadjuvant agents	Abraxane ($125 \, mg \, m^{-2}$), carboplatin AUC, bevacizumab ($10 \, mg \, kg^{-1}$)	Cytotoxicity and VEGF immunotherapy
Adjuvant agents	Cyclophosphamide ($600 \, mg \, m^{-2}$) + DOX ($20 \, mg \, m^{-2}$) + docetaxel ($75 \, mg \, m^{-2}$)	Cytotoxicity

Source: Adapted from Medina et al. [41].

and docetaxel) and anthracyclines (epirubicin and doxorubicin) are mostly used therapeutics.

Adjuvant therapy uses anthracycline-based chemotherapy (cyclophosphamide and 5-fluorouracil). But their acute toxicity is the foremost issue. Capecitabine, gemcitabine, and vinorelbine are effective for metastatic patients who show resistance to anthracycline drugs. Docetaxel + capecitabine showed better results with the OS of patients with metastatic TNBC [21].

New neoadjuvant therapy includes bevacizumab and nab-paclitaxel carboplatin [42–44]. Recently, NCT02441933 (PEARLY trial) discovered the combinational therapy of taxanes and carboplatin neoadjuvant therapy [45]. The combination of carboplatin with docetaxel/paclitaxel revealed auspicious efficiency in TNBC and brain metastasis [46].

Sensitivity to chemotherapy and early diagnosis are the major issues that are related to TNBC treatment. Therefore, there is a need of optimizing developed standard regimes to address chemotherapy challenges including toxicity and diagnosis.

9.2.3 Promising Nanotechnology Innovations for TNBC Therapy

Nanomedicine plays important role in solving complications of cytotoxicity and the absence of specificity of traditional chemotherapies. Nanotechnology-based drug delivery systems (DDSs) have a perspective to increase the therapeutic index of traditional anticancer agents [47]. During the last 20 years, advancements in nanotechnology explored their role biomedical domain for TNBC for augmented bioavailability, and targeted cellular uptake with the least toxicity for both therapeutic and diagnostic agents. Nanoformulations contain polymers, lipids, nucleic acid, proteins, micelles, dendrimers, liposomes, and many others. These smart nanoformulations encapsulate anticancer agents (arsenal) and the surface is coated with a particular ligand that ultimately binds with a particular type of receptor expressing on the target site. Thus, they help in destroying tumor cells as well as molecular imaging (tracer agents) [48–50].

These nanoformulations have the potential to carry all required arsenals (drug, tracking probe, and ligand) and target specifically TNBC cells. Doxil, Abraxane, Feridex®, and Oncaspar® are some available marketed products based on nanomedicine [48]. Some examples of nanomedicines used for the treatment of TNBC include nanoparticles, nanoconjugates, nano-diamonds (NDs), nanocomposites, polymeric micelles (PM), virus-like particles (VLPs), liposomes, Au-nano matryoshkas, and many others as shown in Figure 9.3 [51]. These nanomedicines are described briefly below.

9.2.3.1 Nanoparticles (NPs)

Paul Ehrlich invented the word "magic bullet" for NPs that are used in drug targeting. They have a small size of particles, greater surface area, and good biocompatibility and degradability. They have the potential to improve pharmacodynamics and lessen the cytotoxicity of anticancer agents via active targeting that can be attained by modifying their surface [52]. Paclitaxel-DART NPs are FDA-approved nanoformulations for TNBC and intracranial models [53].

Xu et al. designed hyaluronic acid-coated pH-sensitive poly-β-amino ester NPs for embelin delivery and pTRAIL (TNF-related apoptosis-inducing ligand plasmid) that have anti-TNBC activity [54].

Silver NPs (AgNPs) are effective contrary to tumor cells in TNBC because they can induce DNA impairment. These NPs aid in lessening TNBC growth and enhance radiation therapy [55].

Sears et al. evaluated the potential for selective treatment of MDA-MB-231 TNBC cells without influencing noncancerous MCF-10A breast cells by employing

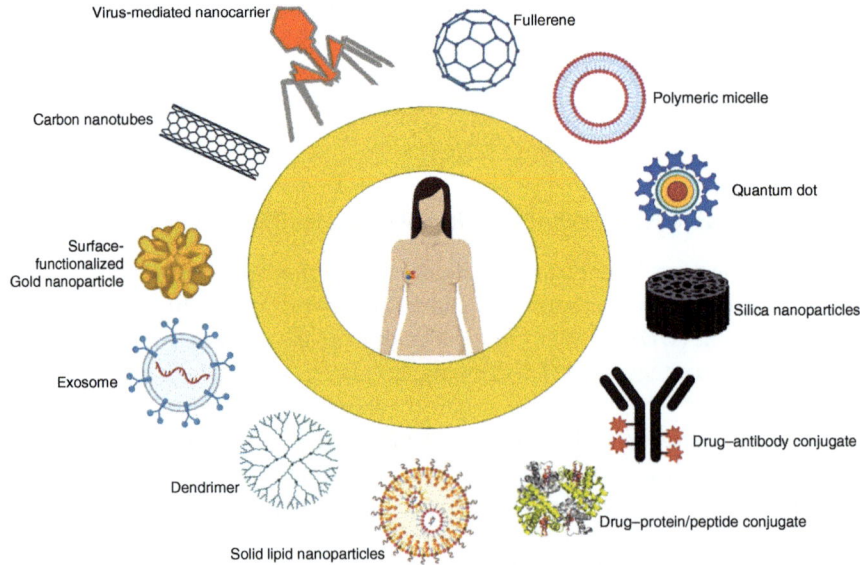

Various nanomedicines utilized for TNBC treatment

Figure 9.3 Various nanoformulations utilized for TNBC treatment.

triangular AgNPs using a multimodal approach. Their results revealed that developed NPs alone could be adequate for the treatment of tumors. It was also reported that increasing the dose of one or all of these modalities increases off-target effects [56].

Mittal et al. aimed to tailor a method of the electrical pulse (EP)-mediated turmeric AgNPs therapy for effective targeting of TNBC cells. This technique efficiently delivered natural bioactive compounds (turmeric) with anticancer activity via biophysical means [57].

Swanner et al. found that silver nanoparticles are extremely cytotoxic for TNBC cells at doses that have a slight influence on non-tumorigenic BC. These developed NPs induced more DNA and oxidative impairment in TNBC cells than in other breast cells [58].

Gold NPs are photothermally tunable as they demonstrate plasmonic actions when exposed to light, an inimitable feature of substance at the nanoscale. These NPs are beneficial for generating heat and carrying apoptosis via hyperthermia. Hyperthermia/photothermal therapy endures being the subject of exploration in nanomedicine because at >100 nm electromagnetic characteristics of materials permit heat production, stimulating advanced diagnostics and treatments [59].

Nirmala et al. synthesized chemically modified tryptone-stabilized gold NPs with tryptone and $HAuCl_4 \cdot 3H_2O$. They reported that these NPs have strong antiproliferative and anti-clonogenic activity against MDA-MB-321 cells [60].

Iron oxide and superparamagnetic iron oxide nanoparticles (IONPs) are the types of magnetic nanoparticles [61, 62]. IONPs have tunable-augmented optoelectronics and magnetic properties. They have the potential to yield strong contrast images in MRI, which has provided them a position in theranostic of cancer. IONP is a novel imaging technique that has been employed in various xenograft models and for MRI diagnostic [63].

Medina et al. synthesized cobalt ferrite nanoparticles (CFNPs) with 10 nm size by thermal decomposition technique with Triton-X100. They found that developed CFNPs enhanced cell cytotoxicity in presence of microwave as compared to CFNPs alone [41].

Oei et al. determined the efficiency of magnetic IONPs in a metastatic TNBC model for improving the abscopal effect of immune checkpoint inhibitors and radiation treatments [64].

Superparamagnetic iron oxide nanoparticles (SPIONs) have greater magnetic characteristics than paramagnetic compounds. They can produce heat inside the cancerous cells resulting in apoptosis by hyperthermia [65–67]. It has been reported that they are useful in the deliverance of anticancer agents like gambogic acid (GA) in TNBC treatment [63].

Vyas et al. investigated doxorubicin-hyaluronan (DOX-HA) SPIONPs therapeutic efficacy in TNBC cell lines (MDA-MB-231). They revealed that developed nanoformulation improved cytoplasmic retention of DOX and lessened apoptosis, Il-6 release, and NF-kappa-b action [68].

9.2.3.2 Nanoconjugates

In cancer therapeutics, the main disadvantage of various nanoformulations is non-specificity. Various nanoformulations can bind to several cellular and extracellular components; thus, it becomes challenging for effective delivery of therapeutic agents. So, nanoconjugates are the best alternative for avoiding this. They comprise active functional groups that are covalently bound to drugs. At present, various mAbs and polymer–drug conjugates are employed in targeted anticancer therapy.

Recently, EC1456 (drug–folate conjugates) and IMMU-132 (drug-antibody conjugates) passed clinical trials (CTs) for the TNBC subtype. EC1456 consists of folate particles that are associated with tubulysin B hydrazine, whereas IMMU-132 contains RS7, an anti-TROP-2 antibody, conjugated to the active metabolite of irinotecan [69].

9.2.3.3 Quantum Dots (QDs)

QD-based nanotechnology has broader utility in the imaging and quantitative assessment of cancer molecules [70]. Many kinds of research revealed that the QD approach could substitute for IHC because they have superior fluorescent signaling and provide more precise quantitative analysis for the assessment and diagnosis of TNBC [71, 72].

Adimule et al. engineered a combinational photothermal immunotherapy approach based on tumor cell membrane-coated biomimetic black phosphorous (BBP) QDs for PTT and anti-PD-L1 mediated immunotherapy. The results of *in vivo* investigation revealed that this combinational approach has a stronger antitumor action than monotherapy of anti-PD-L1 [73].

Kim et al. fabricated aptamer-coupled lipid nanosystem that encapsulated QD and siRNA for theragnosis of TNBC. Engineered quantum dots-based micelle conjugated with an EGFR nanobody and loaded with aminoflavone. It was observed that the developed formulation has no systemic toxicity with this treatment [74].

9.2.3.4 Nano-Diamonds (NDs)

NDs have several inimitable chemical–mechanical characteristics on their surfaces that make them suitable nanomaterials for biomedical utility. Their surface can be altered with numerous functional groups to regulate interaction with biologically relevant conjugates as well as water molecules.

Cui et al. integrated multicomponent-assembled NDs hybrids using hyaluronic acid, protamine, NDs, curcumin, and IR 780 into single nanoplatforms for targeted and imaging-guided TNBC [75].

Liao et al. reported that ND-paclitaxel conjugated with cetuximab and an anti-EGFR chimeric antibody. They found that developed conjugate had favorable results for ND-paclitaxel because of boosted cellular uptake via receptor-mediated endocytosis [76].

Zhang et al. also got similar results in multifunctional NDs conjugated with PTX and anti-EGFR mAbs that revealed greater intracellular uptake and cytotoxic potential against cell lines contrasted to unmodified PTX [77].

9.2.3.5 Nanocomposites

Nanocomposites comprise biphasic/multiphasic constituents concerning the condition that at least one aspect of the compound has >100 nm of dimension. These are the creation of alternatives in nanoengineering [78, 79]. The upgraded optoelectronic features permit nanocomposites to be suitable aspirants for the deliverence of drugs, food packaging, and sensing devices [80].

Li et al. developed biodegradable mesoporous manganese carbonate nanocomposites that were added with riboflavin and surviving shRNA-expressing plasmid DNA (MRp NCs), for a synergistic effect in TNBC treatment [81].

9.2.3.6 Nano-Matryoshkas

It is a singular design that has been established as diagnosis and therapeutics. It is defined as a multilayer NPs repeat of a Russian doll that comprises several other dolls and is esoteric. They are used to deliver multiple drug loads developed in multilayers, which can be fabricated with altered constituents as recommended by an MDA-MB-231 murine xenograft investigation [82].

9.2.3.7 Polymeric Micelles (PM): A Miracle Ball in Cancer Therapy

These possessed the most advantageous nanotechnology in the previous few decades. These are spherically shaped with core/shell, developed by self-assembly of core and shell components. These have great prospects for designing target DDSs because of their high flexibility in altering their molecular weight, HLB value, lesser size, surface shape, and chemistry [83].

Biancacci et al. evaluated core-crosslinked polymeric micelless (CCPMs) for TNBC treatment. The developed nanocarrier formulation of PMs demonstrated effective tumor targeting in 4T1 TNBC mouse model [84].

Kutlehria et al. designed DOX-loaded cholecalciferol-PEG conjugate-based nano micelles TNBC. The developed nanomicelles are good alternative DDSs for enhancing the anti-tumor potential of DOX in TNBC. These nanosystems may be a good carrier for other anticancer drugs because they minimize the side effects [85].

Paulmurugan et al. designed orlistat-loaded folate receptor-targeted PM for TNBC treatment. *In vivo* data of mice showed significant antitumor potential for tumor-targeted treatment. Moreover, mice treated with this nanoformulation revealed a significantly greater decline in tumor volume [86].

9.2.3.8 Dendrimers

These are synthetic macromolecules having 10–100 nm dimensions that are produced by repetitive entities of branched monomers originating outwardly from a central core. Divergent and convergent approaches are used for the preparation of synthesis [87]. These are the alternate prototype for the development of nanoparticulate DDSs. Their smaller particle size helps in improving the potential for cell penetration while still evading toxicity.

Zhang et al. designed dendrimer-based NPs having GdDOTA (MRI contrast agent) and DL680 (fluorescent dye). The systemic deliverance of the developed formulation accumulated in a flank mouse model of TNBC that was identified by optical and MRI. The outcomes of *in vivo* and *ex vivo* investigations revealed that nanoparticles accumulated into tumors selectively [88].

Finlay et al. developed a siRNA-based TWIST1 silencing approach for modified PAMAM-based dendrimer. They reported that TNBC cells (SUM1315) proficiently retained PAMAM–siRNA complexes. This revealed that it reduced TWIST1 and EMT-related target genes [89].

9.2.3.9 Folded Graphene: Carbon nanotubes (CNTs)

Carbon nanotubes (CNTs) are nanostructures of carbon allotropes that have achieved great importance in the biomedical field because they provide many benefits including greater surface area, thermal characteristics, cell penetration potential, electronic properties, size stability on the nanoscale, and many others. These are mainly used as mediators and carriers for cancer treatment. These are used to deliver various bioactive therapeutic moieties including peptides, genes, proteins, and drugs, and in the regeneration of bone and neural tissue [90, 91].

Luo et al. reported the inhibition of cell viability of TNBC cells with newly developed Rg3-CNTs. They investigated the effect of Rg3-CNTs on PD-1/PD-L1 signaling and the progression of TNBC [92]. Badea et al. reported a technique for multi-walled CNTs for their utility as carriers for cisplatin in BC therapy [93].

9.2.3.10 Virus-Like Particles (VLPs) as Novel Nanovesicles

These nanoformulations have a great potential to produce strong immune responses because these systems have optimum size range, particulate nature, and potent intrinsic adjuvant potential. AX09 is a VLP-based vaccine that is currently authorized for communicable infections and cancer-related pathogens. But, to date, no VLPs to target cancer antigens have been licensed for usage in the clinic so far [94].

Rolih et al. designed a VLP-based vaccine (AX09) for inhibiting the formation of de novo metastasis and finally prolonging the persistence of patients with cancer. They developed bacteriophage MS2 VLP to demonstrate an extracellular loop of xCT. It was reported that the rising level of this protein was responsible for TNBC, which correlates with a poor survival rate for the patient. They observed that antibody-based response showed inhibition of xCT's role and decline in the formation of metastasis [95].

Bolli et al. developed VLP (AX09-0M6)-based immunotherapy to target xCT. This nanoformulation produced a strong antibody reaction against xCT including greater levels of IgG2a. IgG antibodies were sequestered from VLPs treated mice associated with tumorspheres, inhibited xCT activity, and lessened BCSC growth and regeneration. The administration of developed nanoformulation significantly hampered the growth of the tumor and pulmonary metastases [96].

9.2.3.11 Liposomes

These are mostly employed as ideal drug-carrier systems that are micro- and nano-sized colloidal multilayer vesicles. They contain an aqueous portion enclosed by lipid content. Liposomes have a plethora of benefits because of their resemblance with cellular membranes and potential to add several substances and their biocompatibility, improved bioavailability of drugs, and excellent biodegradability.

George et al. developed anthraquinone-loaded liposomes for improving efficiency and safety in TNBC. They developed PEG-modified liposomal preparation (LipoRV) of novel anthraquinone derivatives, which showed an effective impact on multiple TNBC cell lines. Liposomal formulation cleared cancer and resulted in good safety and efficacy profile without detrimental effects on biochemical markers in xenograft animal model [97].

Albakr et al. developed a cationic nanoliposome and investigated its efficacy in TNBC treatment. The developed nanoliposomes can act as a miR-1296 carrier [98].

Ding et al. designed chitosan oligosaccharide-based liposomes with Photochlor (HPPH) and evofosfamide as hypoxia-activated prodrugs. They reported that the developed liposomal formulation showed considerably better anticancer activity than other monotherapy groups during studies [99].

De Vita et al. engineered an innovative extracellular matrix targeting with a lipid-based NP chemically associated with lysyl oxidase 1 (LOX). The results of *in vivo* study revealed extended existence, least cytotoxicity, and higher biocompatibility in contrast to epirubicin-loaded NPs and pure epirubicin [100].

9.2.4 Vaccines Under the Clinical Trial (CT) for TNBC Treatment

During the last two decades, most cancer vaccines (CVs) gave poor results that resulted in the evolution of active immunotherapy due to certain technological advancements, eventually augmented by the coronavirus infection 2019 pandemic. Especially, gene and viral vector-based platforms are evolving as favored targets for CVs as shown in Figure 9.4 [101].

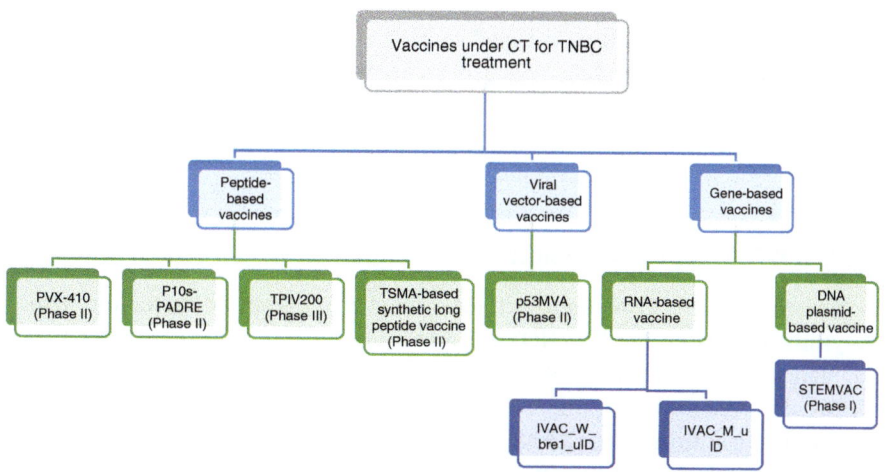

Figure 9.4 Schematic representation of various vaccines under CT for TNBC treatment.

9.2.4.1 Peptide-Based Vaccines

PVX-410 (PVX, OncoPep) is an innovative tetrapeptide HLA-A2-inhibition vaccine that consists of X-box-binding protein 1, with 2 splice variants, syndecan-1 (CD138) and cell surface gP SLAM7 (CD319) as an antigen, which is most commonly overexpressed in TNBC. This vaccine is under CT (phase II) [94, 102]. P10s-PADRE is another vaccine that consists of a carbohydrate mimetic peptide P10s, fused to pan-HLA-DR-binding epitope (PADRE) peptide, with immunoadjuvant and anticancer potential. Montanide ISA 51 VG is prescribed with this vaccine [103]. This vaccine is under CT (phase II) in the early TNBC treatment.

TPIV 200 is a penta-epitope vaccine that consists of five fragments of FR-α, capable to stimulate CD4$^+$ and CD8$^+$ immune responses [104]. This vaccine is under CT (phase III) in patients with early TNBC. The vaccine demonstrated a nontoxicity profile with lymphopenia, neutropenia, and injection site reactions as the most severe AEs (grade II) during the Phase I study [105].

TSMA-based synthetic long peptide vaccines are under Phase II CT to target TNBC neoantigens in metastatic treatment. Galinpepimut-S is another new vaccine that consists of four peptide chains of Wilms' tumor gene protein [106]. This vaccine is under CT Phase II in combination with pembrolizumab.

9.2.4.2 Viral Vector-Based Vaccines

p53MVA is a viral vector-based vaccine that comprises MVA virus that has been designed to express wild-type TP53 transgene (p53MVA) as an immunotherapeutic approach. For evaluation of the safety and tolerability profiles of this vaccine with pembrolizumab, Phase I CT was designed [107].

9.2.4.3 Gene-Based Vaccines

IVAC_W_bre1_uID and IVAC_M_uID vaccines are RNA-based vaccines that target to TAAs (IVAC_W_-bre1_uID) gene. This vaccine consists of on-demand RNA manufacturing for targeting up to 20 individual cancer neo-antigens obtained from mutated epitopes (IVAC_M_uID) [101].

STEMVAC is a DNA plasmid-based vaccine that is useful for the III-IV stage of HER2-negative BC. The data revealed an improved humoral immune response against eight BCSCs proteins in progressive BC and recommended an association between autoantibodies and more violent disease. This vaccine is under Phase I CT [108].

9.2.5 USFDA-Approved Clinical Trials

In November 2020, FDA approved accelerated authorization for pembrolizumab (Keyruda®) based on the KEYNOTE-355 trial, as an adjuvant to chemotherapy for the management of TNBC [109].

FDA approved the use of GeparNUEVO, which is the combination of durvalumab (Imfinzi) with anthracycline and taxane-based chemotherapy. The results of this CT demonstrated that this product significantly enhanced rates of survival in early TNBC patients [110].

In April 2021, FDA also granted the usage of sacituzumab govitecan (Trodelvy®) for metastatic TNBC patients. This product is based on antibody–drug conjugate with topoisomerase-I inhibitor SN-38 that interacts with antibodies of anti-Trop-2 [111].

In another trial of CREATE-X, it was reported that adjuvant capecitabine after NACT and surgery with HER2 negative is also effective and safe for TNBC patients [112].

Moreover these novel platforms, liposomes can be employed as a vehicle for various vaccines.

9.2.6 Current Status of TNBC Treatment

At present, patients suffering from TNBC have inadequate choices of treatment due to the lack of availability of effective treatment therapy for this. For early-stage TNBC, anthracycline and taxane-based chemotherapy are available as systemic treatment regimens. But drug resistance is the most problematic in TNBC as well as other types of tumors. Opportunely, various targeted-based therapies are ongoing for the treatment of TNBC patients. For BReast CAncergene (BRCA) 1 and 2 mutagens, PARP inhibitors got approval from the FDA. Various other inhibitors, including PI3K, AKT, EGFR, and angiogenesis inhibitors, and many others may be highly effective as targeted therapies for TNBC treatment. Approximately 90% of chemoresistant TNBC have variations in pathways in which numerous targeted agents are presently under CT.

Recently, immunotherapy with immune check inhibitors combined with chemotherapy got approval from FDA and European Medicines Agency (EMA) for its treatment [113].

9.2.7 Recent Patents Based on Nanoformulations for TNBC Treatment

The Google patent, United States Patent and Trademark Office (USPTO), Escapenet, and many others were used as the search engine for patent search, and data were collected and summarized below.

Lieberman et al. patented method and composition related to BC treatment, using aptamer-siRNA chimera molecules. It contained an Epithelial Cell Adhesion Molecule (EpCAM) binding aptamer sphere and inhibitory nucleic acid domain that inhibits Polo-like Kinase 1 (Plk1). It precisely prevents cell proliferation in basal BC cells [114].

Adimule et al. patented unimolecular NPs for proficient deliverance of RNA that constrains manifestation of a gene required for survival/growth of cancer cells. The core of unimolecular NPs may be a dendrimer including PAMAM dendrimer having three to seven generations of polyester hyperbranched polymer. The patented compositions can be developed for parenteral, rectal, nasal, vaginal, and implants route [115].

Kim et al. invented a nanoparticle delivery system suitable for delivery of treatment, imaging, and other agents to Fn14-(fibroblast growth factor-inducible 14) positive tumors. They used PLGA-containing nanoparticles with a low-molecular-weight PEG coating to encapsulate PTX. Surface modifications have been made to enable nanoparticle targeting to Fn14-positive TNBC tumors and to prevent non-specific binding to tumor ECM. They assessed TNBC tumor growth inhibition of these PTX-loaded nanoparticles.

The invented nanoparticles were designed to overcome the specific barriers to effective treatment of primary and metastatic brain and BCs, tumor heterogeneity, lack of efficacious primary and metastatic tumor-specific drug targets, off-target delivery to healthy tissues, and inability to deliver sufficient quantities of the treatment or other agent into the tumor microenvironment [116].

Stafford et al. invented compositions and methods of antibody conjugates with binding specificity for folate receptor alpha (FOLR1) to treat TNBC, ovarian cancer, and endometrial cancer [117].

Adimule et al. invented compositions for polymalic acid-based nano biopolymers. The invented formulation was suitable for intravenous administration. In this invention, nano biopolymers conjugate of poly (β-L-malic acid) PMLA was covalently associated with molecular modules that comprise morpholino antisense oligonucleotides, an siRNA/Ab specific for a cancer protein in tumor cell and Ab specific for transferrin receptor protein, were used [115].

Glackin et al. invented a siRNA-based TWIST1 silencing method with a modified PAMAM dendrimer complex for reducing metastasis of BC cells. Their results showed that TNBC cells have the potential to take up developed dendrimer complexes that result in targeting TWIST1 and EMT-related genes. They concluded that developed dendrimers are effective as adjunctive therapy for TNBC patients [118].

Nandi et al. polymalic acid-based nanoconjugates for imaging of TNBC cells. It contained polymalic acid-based molecular framework, Gd-DOTA (imaging molecule), and cetuximab (targeting module). These nanoconjugates had the potential for targeting and differentiation between different kinds of cancers [119].

9.3 Challenges

Key challenges in nanoformulations are preclinical-to-clinical translation, pharmaceutical barriers (scale-up and production, safety/toxicity, and biocompatibility), biological barriers, and regulatory and industry barriers.

9.4 Future Perspectives on TNBC Metastasis Therapy

Novel innovations in various domains particularly nanotechnology, immunology, molecular biology, and computer systems will help the clinician in early and more accurate diagnosis that will help in personalized treatment. Presently, various researchers are focused on micro- and nanofluidics for understanding about dynamic processes of cancer and investigating the signaling phenomena in the tumor microenvironment. Some of the future perspectives on TNBC therapy are discussed below.

9.4.1 NEO Adjuvant Modeling

The genomic data of the cancer patients resulted in the development of new targeting specific molecular drivers and treatment management. Considering this, (NEO) adjuvant modeling is a novel platform for exploration and inventive strategies for research in TNBC. It may be a tremendous platform for drug development, innovation, validation of biomarkers, and identification of mechanisms of drug resistance [120].

9.4.2 Execution of *In Vivo* Genetic Screening

Based on a preclinical mouse modeling of TNBC data, about 150 genes (after screening the whole human genome that contains about 20 000 genes) have been identified that are responsible for inducing and inhibiting tumor formation. The gene editing approach CRISPR/Cas9 was implemented and genes were cut individually with a gene knockout process. To date, very limited investigations used *in vivo* genetic screening at the genome-wide scale so far.

9.4.3 Identification of Effective Drugs for TNBC

To date, two major pathways are recognized that are involved in the regulation of tumor formation. It is reported that in TNBC an oncogenic signaling pathway (mTOR) is stimulated while the tumor suppressor pathway (HIPPO) is repressed. That's why TNBC is so deadly and aggressive. It is hypothesized that the identification of the existing drug that can inhibit the mTOR pathway and aid HIPPO pathway will help in TNBC treatment.

9.4.4 Synergistic Effect of Drugs that Almost Eliminate Tumor

In an investigation, scientists observed that verteporfin induced cell death via apoptosis, while Torin1 induced cell death via macropinocytosis. It was observed that when these two drugs were administered together, Torin1 exploits this mechanism to favor the entry of verteporfin in tumor cells. This resulted in increased later apoptotic cell death impact. It is this synergistic mechanism that permits two medicaments to efficiently prevent the development of tumors [1].

9.5 Conclusion

The development of advanced treatment of TNBC is the extreme prerequisite of an era. The objectives of current investigations are developing and strategies for augmented bioavailability, target specificity, lessening systemic toxicity, and increasing the therapeutic effect of FDA-approved treatment regimens are being implemented. A large number of chemotherapeutics have inadequate water solubility. Therefore, there is a need for specialized nanosized delivery vehicles (nanoparticles, quantum dots, liposomes, nanostructured lipid carriers, dendrimers, nanoconjugates, micelles, and many others) for parenteral administration. Abundant efforts have been devoted to drug development of nanoformulations to targeted specific sites, but these efforts have been encountered with inadequate success. Though there are many domains where amendments are still required for improving these therapeutic strategies and further authenticating their effectiveness, these therapies have great prospects for TNBC treatment. In brief, nanotechnology-based molecular therapy of TNBC may be accomplished through several modalities.

Nanoformulations are effective against TNBC because they overcome the limitations of current and cytotoxic therapies. But these developments are still in the early stages. Apart from surgery and radiotherapy, nonspecifically developed anticancer agents including anthracycline and taxane are currently available treatment options for TNBC treatment. The major problem with TNBC treatment is the lack of effective and safe treatment availability. Various researches and CTs are ongoing in this domain. Apart from this, several targeted DDSs are being developed for the deliverance of medicaments to specific tissue.

At present, there are only a few FDA-approved targeted therapies for TNBC that are under CT. These ongoing CTs are investigating the efficiency of single/combinatorial approaches that tackle diverse TNBC molecular variations. More efforts are required for the discovery of novel/new targets of TNBC and the identification of the molecular profile of each patient that will help in the recognition of mutation at genomic levels. It will be beneficial for more individualized treatment of TNBC. At present, only PARP inhibitors are approved by FDA for TNBC patients with *BRCA*1/2 mutations and atezolizumab for PD-L1$^+$ cancer. Novel targeted therapies provide anticipation for TNBC treatment [121]. However, based on current CTs data, combination therapies will be included in the treatment of TNBC prototypes sooner in the future [122]. High production costs, variation in biodistribution, and high liabilities are the major problems for translation to clinical practice.

Despite the increasing challenges in TNBC treatment, it is assumed that in the upcoming future, several efforts for the identification of novel precise inhibitors of signaling pathways will be identified and new CTs initiated for analysis of these novel, potential anticancer drugs. The recognition of particular biomarkers that would help treatment decision-making in TNBC has not yet been attained.

References

1 New study paves the way to novel treatment for Triple Negative Breast Cancer (2021). https://muhc.ca/news-and-patient-stories/releases/ new-study-paves-way-novel-treatment-triple-negative-breast-cancer.

2 Cancer facts and figures (2021). https://www.cancer.org/content/dam/cancer-org/ research/cancer-facts-and-statistics/annual-cancer-facts-and-figures/2021/ cancer-facts-and-figures-2021.pdf.

3 Henriques, B., Mendes, F., and Martins, D. (2021). Immunotherapy in breast cancer: when, how, and what challenges? *Biomedicines* 9 (11): 1687. https://doi .org/10.3390/biomedicines9111687.

4 Sung, H., Ferlay, J., Siegel, R.L. et al. (2021). Global cancer statistics 2020: GLOBOCAN estimates of incidence and mortality worldwide for 36 cancers in 185 countries. *CA Cancer J. Clin.* https://doi.org/10.3322/caac.21660.

5 Mohan Gift, M.D., Pattnaik, B., Nandi, S.S. et al. (2022). Determination of prohibition mechanism of cationic polymer/SiO_2 composite as inhibitor in water using drilling fluid. *Mater. Today Proc.* https://doi.org/10.1016/j .matpr.2022.08.171.

6 Kanwal, B. (2021). Untangling triple-negative breast cancer molecular peculiarity and chemo-resistance: trailing towards marker-based targeted therapies. *Cureus* 13 (7): e16636. https://doi.org/10.7759/cureus.16636.

7 Perue, C.M., Sorlie, T., Elsen, M.B. et al. (2000). Molecular portraits of human breast tumors. *Nature* 406: 747–752.

8 Kadapure, S.A., Kadapure, P., Nandi, S.S., and Shet, A. (2022). Overview on catalyst and co-solvents for sustainable biodiesel production. *Proc. Inst. Civ. Eng. Energy* 1–9. https://doi.org/10.1680/jener.21.00092.

9 Adimule, V., Nandi, S.S., and Yallur, B.C. (2022). Devices and sensors based on additively manufactured shape-memory of hybrid nanocomposites. In: *Shape Memory Composites Based on Polymers and Metals for 4D Printing* (ed. M.R. Maurya, K.K. Sadasivuni, J.J. Cabibihan, et al.). Cham: Springer https://doi .org/10.1007/978-3-030-94114-7_15.

10 Eliyatkın, N., Yalçın, E., Zengel, B. et al. (2015). Molecular classification of breast carcinoma: from traditional, old-fashioned way to a new age, and a new way. *J. Breast Health* 11 (2): 59–66. https://doi.org/10.5152/tjbh.2015.1669.

11 Buller, C.W., Mathew, P.A., and Mathew, S.O. (2020). Roles of NK cell receptors 2B4 (CD244), CS1 (CD319), and LLT1 (CLEC2D) in cancer. *Cancers (Basel)* 12 (7): 1755.

12 Adimule, V., Yallur, B.C., Pai, M.M. et al. (2022). Biogenic synthesis of magnetic palladium nanoparticles decorated over reduced graphene oxide using piper betle petiole extract (Pd-rGO@Fe_3O_4 NPs) as heterogeneous hybrid nanocatalyst for applications in Suzuki-Miyaura coupling reactions of biphenyl compounds. *Top. Catal.*

13 Adimule, V., Jagadeesha Gowda, A.H., Nandi, S.S., and Bowmik, D. (2022). Antimalarial activity of novel class of 1,3-benzoxaborole derivatives containing 1,3,4-oxadiazole moiety. In: *Drug Development for Malaria* (ed. P. Kendrekar). https://doi.org/10.1002/9783527830589.ch12.

14 Mendes, F., Domingues, C., Rodrigues-Santos, P. et al. (2016). The role of immune system exhaustion on cancer cell escape and anti-tumor immune induction after irradiation. *Biochim. Biophys. Acta Rev. Cancer* 1865: 168–175.

15 Akhtar, M., Dasgupta, S., and Rangwala, M. (2015). Triple negative breast cancer: an Indian perspective. *Breast Cancer (Dove Med Press)* 7: 239–243. https://doi .org/10.2147/BCTT.S85442.

16 Shaikh, N.M., Adimule, V., Bagihalli, G.B. et al. (2022). A novel mixed Ag–Pd nanoparticles supported on SBA silica through [DMAP-TMSP-DABCO]OH basic ionic liquid for suzuki coupling reaction. *Top. Catal.*

17 Chang-Qing, Y., Jie, L., Shi-Qi, Z. et al. (2020). Recent treatment progress of triple negative breast cancer. *Prog. Biophys. Mol. Biol.* 151: 40–53.

18 Manjunath, M. and Choudhary, B. (2021). Triple-negative breast cancer: a run-through of features, classification and current therapies. *Oncol. Lett.* 22 (1): 512. https://doi.org/10.3892/ol.2021.12773.

19 Wu, D., Si, M., Xue, H.Y., and Wong, H.L. (2017). Nanomedicine applications in the treatment of breast cancer: current state of the art. *Int. J. Nanomed.* 12: 5879–5892. https://doi.org/10.2147/IJN.S123437.

20 Karthika, C., Hari, B., Mano, V. et al. (2021). Curcumin as a great contributor for the treatment and mitigation of colorectal cancer. *Exp. Gerontol.* 152: 111438. https://doi.org/10.1016/j.exger.2021.111438.

21 Adimule, V., Yallur, B.C., Batakurki, S., and Nandi, S.S. (2022). Synthesis, morphology and enhanced optical properties of novel $Gd_xCo_3O_4$ nanostructures. *Adv. Mater. Res.* 1173: 71–82. Trans Tech Publications, Ltd.

22 Adimule, V., Batakurki, S., Yallur, B.C. et al. (2022). Enhanced photoluminescence, optical, structural properties of ZrO_2-incorporated Sm_2O_3:Co_3O_4 nanocomposite and their applications in photocatalytic degradation of methylene blue. *J. Mater. Res.* 37: 2396–2405. https://doi.org/10.1557/s43578-022-00641-y.

23 Jamdade, V.S., Sethi, N., Mundhe, N.A. et al. (2015). Therapeutic targets of triple-negative breast cancer: a review. *Br. J. Pharmacol.* 172 (17): 4228–4237. https://doi.org/10.1111/bph.13211.

24 Palomero, T., Barnes, K.C., Real, P.J. et al. (2006). CUTLL1, a novel human T-cell lymphoma cell line with t(7;9) rearrangement, aberrant NOTCH1 activation and high sensitivity to gamma-secretase inhibitors. *Leukemia* 20: 1279–1287.

25 Speiser, J.J., Ersahin, C., and Osipo, C. (2013). The functional role of Notch signaling in triple-negative breast cancer. *Vitam. Horm.* 93: 277–306.

26 Brennan, K. and Clarke, R.B. (2013). Combining Notch inhibition with current therapies for breast cancer treatment. *Ther. Adv. Med. Oncol.* 5: 17–24.

27 Soares, R., Balogh, G., Guo, S. et al. (2004). Evidence for the notch signaling pathway on the role of estrogen in angiogenesis. *Mol. Endocrinol.* 18: 2333–2343.

28 Weijzen, S., Rizzo, P., Braid, M. et al. (2002). Activation of Notch-1 signaling maintains the neoplastic phenotype in human Ras-transformed cells. *Nat. Med.* 8: 979–986.

29 Weng, A.P., Ferrando, A.A., Lee, W. et al. (2004). Activating mutations of NOTCH1 in human T cell acute lymphoblastic leukemia. *Science* 306: 269–271.

30 Adimule, V.M., Nandi, S.S., Kerur, S.S. et al. (2022). Recent advances in the one-pot synthesis of coumarin derivatives from different starting materials using nanoparticles: a review. *Top. Catal.* https://doi.org/10.1007/s11244-022-01571-z.

31 Nandi, S.S., Adimule, V., and Yallur, B.C. (2022). Synthesis, structural and optical properties of co doped Sm_2O_3 nanostructures. *Adv. Mater. Res.* 1173: 59–69. Trans Tech Publications, Ltd.

32 Gurney, A., Axelrod, F., Bond, C.J. et al. (2012). Wnt pathway inhibition via the targeting of Frizzled receptors results in decreased growth and tumorigenicity of human tumors. *Proc. Natl. Acad. Sci. U.S.A.* 109: 11717–11722.

33 Zhu, Y., Tian, Y., Du, J. et al. (2012). Dvl2-dependent activation of Daam1 and RhoA regulates Wnt5a-induced breast cancer cell migration. *PLoS One* 7: e37823.

34 Corda, G., Sala, G., Lattanzio, R. et al. (2017). Functional and prognostic significance of the genomic amplification of frizzled 6 (FZD6) in breast cancer. *J. Pathol.* 241: 350–361.

35 Adimule, V., Bhat, V.S., Yallur, B.C. et al. (2022). Facile synthesis of novel $SrO_{0.5}{:}MnO_{0.5}$ bimetallic oxide nanostructure as a high-performance electrode material for supercapacitors. *Nanomater. Nanotechnol.* 12: 1–14. https://doi.org/10.1177/18479804211064028.

36 Murai, J., Huang, S.Y., Das, B.B. et al. (2012). Trapping of PARP1 and PARP2 by clinical PARP inhibitors. *Cancer Res.* 72: 5588–5599.

37 Sobande, F., Dusek, L., Matejkova, A. et al. (2015). EGFR in triple negative breast carcinoma: significance of protein expression and high gene copy number. *Ceskoslovenska Patol.* 51: 80–86.

38 Eccles, S.A. (2011). The epidermal growth factor receptor/Erb-B/HER family in normal and malignant breast biology. *Int. J. Dev. Biol.* 55: 685–696.

39 Hsiao, Y., Yeh, M., Chen, Y. et al. (2015). Lapatinib increases motility of triple-negative breast cancer cells by decreasing miRNA-7 and inducing Raf-1/MAPK-dependent interleukin-6. *Oncotarget* 6: 37965–37978.

40 Bhola, N.E., Balko, J.M., Dugger, T.C. et al. (2013). TGF-beta inhibition enhances chemotherapy action against triple-negative breast cancer. *J. Clin. Invest.* 123: 1348–1358.

41 Medina, M.A., Oza, G., Ángeles-Pascual, A. et al. (2020). Synthesis, characterization and magnetic hyperthermia of monodispersed cobalt ferrite nanoparticles for cancer therapeutics. *Molecules* 25 (19): 4428. https://doi.org/10.3390/molecules25194428.

42 Adimule, V., Yallur, B.C., Challa, M., and Joshi, R.S. (2021). Synthesis of hierarchical structured Gd doped α-Sb_2O_4 as an advanced nanomaterial for high performance energy storage devices. *Heliyon* 7 (12): e08541. https://doi.org/10.1016/j.heliyon.2021.e08541.

43 Isakoff, S.J., Mayer, E.L., He, L. et al. (2015). TBCRC009: a multicenter phase II clinical trial of platinum monotherapy with biomarker assessment in metastatic triple-negative breast cancer. *J. Clin. Oncol.* 33: 1902–1909.

44 Byrski, T., Dent, R., Blecharz, P. et al. (2012). Results of a phase II open-label, non-randomized trial of cisplatin chemotherapy in patients with BRCA1-positive metastatic breast cancer. *Breast Cancer Res.* 14: R110.

45 Kim, G.M., Jeung, H.C., Jung, K.H. et al. (2017). PEARLY: a randomized, multicenter, open-label, phase III trial comparing anthracyclines followed by taxane versus anthracyclines followed by taxane plus carboplatin as (neo) adjuvant therapy in patients with early triple-negative breast cancer. *J. Clin. Oncol.* 35 (15): TPS587–TPS587.

46 Chen, X.S., Nie, X.Q., Chen, C.M. et al. (2010). Weekly paclitaxel plus carboplatin is an effective nonanthracycline-containing regimen as neoadjuvant chemotherapy for breast cancer. *Ann. Oncol.* 21: 961–967.

47 Adimule, V., Bowmik, D., and Adarsha, H.J. (2020). A facile synthesis of Cr doped WO_3 nanocomposites and its effect in enhanced current-voltage and impedance characteristics of thin films. *Lett. Mater.* 10 (4): 481–485. https://doi.org/10.22226/2410-3535-2020-4-481-485.

48 Thakur, V. and Kutty, R.V. (2019). Recent advances in nanotheranostics for triple negative breast cancer treatment. *J. Exp. Clin. Cancer Res.* 38 (1): 430. https://doi.org/10.1186/s13046-019-1443-1.

49 Montaseri, H., Kruger, C.A., and Abrahamse, H. (2021). Targeted photodynamic therapy using alloyed nanoparticle-conjugated 5-aminolevulinic acid for breast cancer. *Pharmaceutics* 13 (9): 1375. https://doi.org/10.3390/pharmaceutics13091375.

50 Nandi, S.S., Suryavanshi, A., Adimule, V., and Maradur, S.R. (2020). Semiconductor current-voltage characteristics of some novel perovskite ionic nanocomposites of Sr0.5 Cu0.4 Y0.1 and Sr0.5 Mn0.5 and their electronic sensor applications. In: *AIP Conference Proceedings*, vol. 2274, p. 020006. AIP Publishing LLC.

51 Adimule, V., Yallur, B.C., Kamat, V., and Krishna, P.M. (2021). Characterization studies of novel series of cobalt(II) nickel(II) and copper(II) complexes: DNA binding and antibacterial activity. *J. Pharm. Invest.* 51 (3): 347–359.

52 Wang, J., Zhou, T., Liu, Y. et al. (2022). Application of nanoparticles in the treatment of lung cancer with emphasis on receptors. *Front. Pharmacol.* 12: 781425. https://doi.org/10.3389/fphar.2021.781425.

53 Dancy, J.G., Wadajkar, A.S., Connolly, N.P. et al. (2020). Decreased nonspecific adhesivity, receptor-targeted therapeutic nanoparticles for primary and metastatic breast cancer. *Sci. Adv.* 6: 1–14.

54 Xu, Y., Liu, D., Hu, J. et al. (2020). Hyaluronic acid-coated pH sensitive poly(β-aminoester) nanoparticles for co-delivery of embelin and TRAIL plasmid for triple negative breast cancer treatment. *Int. J. Pharm.* 573: 118637.

55 Adimule, V., Nandi, S.S., and Jagadeesha Gowda, A.H. (2020). Enhanced power conversion efficiency of the P3BT (poly-3-butyl thiophene) doped nanocomposites of Gd-TiO_3 as working electrode. In: *Techno-Societal* (ed. P.M. Pawar, R. Balasubramaniam, B.P. Ronge, et al.). Cham: Springer https://doi.org/10.1007/978-3-030-69925-3_6.

56 Sears, J., Swanner, J., Fahrenholtz, C.D. et al. (2021). Combined photothermal and ionizing radiation sensitization of triple-negative breast cancer using triangular silver nanoparticles. *Int. J. Nanomed.* 16: 851–865. https://doi.org/10.2147/IJN.S296513.

57 Mittal, L., Camarillo, I.G., Varadarajan, G.S. et al. (2020). High-throughput, label-free quantitative proteomic studies of the anticancer effects of electrical pulses with turmeric silver nanoparticles: an in vitro model study. *Sci. Rep.* 10 (1): 7258. https://doi.org/10.1038/s41598-020-64128-8.

58 Swanner, J., Mims, J., Carroll, D.L. et al. (2015). Differential cytotoxic and radiosensitizing effects of silver nanoparticles on triple-negative breast cancer and non-triple-negative breast cells. *Int. J. Nanomed.* 10: 3937–3953. https://doi .org/10.2147/IJN.S80349.

59 Adimule, V., Suryavanshi, A., and BC Y, Nandi SS. (2020). A facile synthesis of poly(3-octyl thiophene):$Ni_{0.4}Sr_{0.6}TiO_3$ hybrid nanocomposites for solar cell applications. *Macromol. Symp.* 392: 2000001. https://doi.org/10.1002/ masy.202000001.

60 Nirmala, J.G. and Lopus, M. (2019). Tryptone-stabilized gold nanoparticles induce unipolar clustering of supernumerary centrosomes and G1 arrest in triple-negative breast cancer cells. *Sci. Rep.* 9 (1): 19126. https://doi.org/10.1038/ s41598-019-55555-3.

61 Surapaneni, S.K., Bashir, S., and Tikoo, K. (2018). Gold nanoparticles-induced cytotoxicity in triple negative breast cancer involves different epigenetic alterations depending upon the surface charge. *Sci. Rep.* 8 (1): 12295. https://doi.org/10.1038/ s41598-018-30541-3.

62 Haynes, B., Zhang, Y., Liu, F. et al. (2016). Gold nanoparticle conjugated Rad6 inhibitor induces cell death in triple negative breast cancer cells by inducing mitochondrial dysfunction and PARP-1 hyperactivation: synthesis and characterization. *Nanomedicine* 12 (3): 745–757. https://doi.org/10.1016/j .nano.2015.10.010.

63 Hayashi, K., Nakamura, M., Sakamoto, W. et al. (2013). Superparamagnetic nanoparticle clusters for cancer theranostics combining magnetic resonance imaging and hyperthermia treatment. *Theranostics* 3 (6): 366–376. https://doi .org/10.7150/thno.5860.

64 Oei, A.L., Korangath, P., Mulka, K. et al. (2019). Enhancing the abscopal effect of radiation and immune checkpoint inhibitor therapies with magnetic nanoparticle hyperthermia in a model of metastatic breast cancer. *Int. J. Hyperthermia* 36 (1): 47–63. https://doi.org/10.1080/02656736.2019.1685686.

65 Xu, C., Feng, Q., Yang, H. et al. (2018). A light-triggered mesenchymal stem cell delivery system for photoacoustic imaging and chemo-photothermal therapy of triple negative breast cancer. *Adv. Sci. (Weinh)* 5 (10): 1800382. https://doi .org/10.1002/advs.201800382.

66 Adimule, V., Revaiah, R.G., Nandi, S.S., and Jagadeesha, A.H. (2021). Synthesis characterization of Cr doped TeO_2 nanostructures and its application as EGFET pH sensor. *Electroanalysis* 33 (3): 579–590.

67 Adimule, V., Nandi, S.S., Yallur, B.C., and Shaikh, N. (2021). CNT/graphene-assisted flexible thin-film preparation for stretchable electronics and superconductors. In: *Sensors for Stretchable Electronics in Nanotechnology* (ed. P. Kaushik), 89–103. CRC Press.

68 Vyas, D., Lopez-Hisijos, N., Gandhi, S. et al. (2015). Doxorubicin-hyaluronan conjugated super-paramagnetic iron oxide nanoparticles (DOX-HA-SPION) enhanced cytoplasmic uptake of doxorubicin and modulated apoptosis, Il-6 release and NF-kappab activity in human MDA-MB-231 breast cancer cells. *J. Nanosci. Nanotechnol.* 15 (9): 6413–6422. https://doi.org/10.1166/jnn.2015.10834.

69 Adimule, V., Yallur, B.C., and Sharma, K. (2022). Studies on crystal structure, morphology, optical and photoluminescence properties of flake-like Sb doped Y_2O_3 nanostructures. *J. Opt.* 51: 173–183. https://doi.org/10.1007/s12596-021-00746-3.

70 Devi, S., Kumar, M., Tiwari, A. et al. (2022). Quantum dots: an emerging approach for cancer therapy. *Front. Mater.* 8: 798440. https://doi.org/10.3389/fmats.2021.798440.

71 Zheng, H.M., Chen, C., Wu, X.H. et al. (2016). Quantum dot-based in situ simultaneous molecular imaging and quantitative analysis of EGFR and collagen IV and identification of their prognostic value in triple-negative breast cancer. *Tumour Biol.* 37 (2): 2509–2518. https://doi.org/10.1007/s13277-015-4079-6.

72 Adimule, V., Nandi, S.S., Yallur, B.C. et al. (2021). Optical structural and photoluminescence properties of Gd_xSrO: CdO nanostructures synthesized by co precipitation method. *J. Fluoresc.* 31 (2): 487–499.

73 Adimule, V., Yallur, B.C., Bhowmik, D., and Gowda, A.H.J. (2021). Morphology structural and photoluminescence properties of shaping triple semiconductor Y_xCoO: ZrO_2 nanostructures. *J. Mater. Sci. Mater. Electron.* 32 (9): 2164–12181.

74 Kim, M.W., Jeong, H.Y., Kang, S.J. et al. (2019). Anti-EGF receptor aptamer-guided co-delivery of anti-cancer siRNAs and quantum dots for theranostics of triple-negative breast cancer. *Theranostics* 9 (3): 837–852. https://doi.org/10.7150/thno.30228.

75 Cui, X., Deng, X., Liang, Z. et al. (2021). Multicomponent-assembled nanodiamond hybrids for targeted and imaging guided triple-negative breast cancer therapy via a ternary collaborative strategy. *Biomater. Sci.* 9 (10): 3838–3850. https://doi.org/10.1039/d1bm00283j.

76 Liao, W.S., Ho, Y., Lin, Y.W. et al. (2019). Targeting EGFR of triple-negative breast cancer enhances the therapeutic efficacy of paclitaxel- and cetuximab-conjugated nanodiamond nanocomposite. *Acta Biomater.* 86: 395–405. https://doi.org/10.1016/j.actbio.2019.01.025.

77 Zhang, X.Q., Lam, R., Xu, X. et al. (2011). Multimodal nanodiamond drug delivery carriers for selective targeting, imaging, and enhanced chemotherapeutic efficacy. *Adv. Mater.* 23 (41): 4770–4775. https://doi.org/10.1002/adma.201102263.

78 Fudala, R., Raut, S., Maliwal, B.P. et al. (2014). FRET enhanced fluorescent nanodiamonds. *Curr. Pharm. Biotechnol.* 14 (13): 1127–1133. https://doi.org/10.2174/1389201014131406051107111.

79 Arias, L.S., Pessan, J.P., Vieira, A.P.M. et al. (2018). Iron oxide nanoparticles for biomedical applications: a perspective on synthesis, drugs, antimicrobial activity, and toxicity. *Antibiotics (Basel)* 7 (2): 46. https://doi.org/10.3390/antibiotics7020046.

80 Jeevanandam, J., Barhoum, A., Chan, Y.S. et al. (2018). Review on nanoparticles and nanostructured materials: history, sources, toxicity and regulations. *Beilstein J. Nanotechnol.* 9: 1050–1074. https://doi.org/10.3762/bjnano.9.98.

81 Li, L., Chen, L., Huang, L. et al. (2021). Biodegradable mesoporous manganese carbonate nanocomposites for LED light-driven cancer therapy via enhancing photodynamic therapy and attenuating survivin expression. *J. Nanobiotechnol.* 19 (1): 310. https://doi.org/10.1186/s12951-021-01057-2.

82 Ayala-Orozco, C., Urban, C., Bishnoi, S. et al. (2014). Sub-100 nm gold nanomatryoshkas improve photo-thermal therapy efficacy in large and highly aggressive triple negative breast tumors. *J. Controlled Release* 191: 90–97. https://doi.org/10.1016/j.jconrel.2014.07.038.

83 Castelli, R., Ibarra, M., Faccio, R. et al. (2021). T908 polymeric micelles improved the uptake of sgc8-c aptamer probe in tumor-bearing mice: a co-association study between the probe and preformed nanostructures. *Pharmaceuticals (Basel)* 15 (1): 15. https://doi.org/10.3390/ph15010015.

84 Biancacci, I., Sun, Q., Möckel, D. et al. (2020). Optical imaging of the whole-body to cellular biodistribution of clinical-stage PEG-b-pHPMA-based core-crosslinked polymeric micelles. *J. Controlled Release* 328: 805–816. https://doi.org/10.1016/j.jconrel.2020.09.046.

85 Kutlehria, S., Behl, G., Patel, K. et al. (2018). Cholecalciferol-PEG conjugate based nanomicelles of doxorubicin for treatment of triple-negative breast cancer. *AAPS PharmSciTech.* 19 (2): 792–802. https://doi.org/10.1208/s12249-017-0885-z.

86 Paulmurugan, R., Bhethanabotla, R., Mishra, K. et al. (2016). Folate receptor-targeted polymeric micellar nanocarriers for delivery of orlistat as a repurposed drug against triple-negative breast cancer. *Mol. Cancer Ther.* 15 (2): 221–231. https://doi.org/10.1158/1535-7163.MCT-15-0579.

87 Adimule, V., Yallur, B.C., and Bhowmik, D. (2021). Morphology, structural and photoluminescence properties of shaping triple semiconductor $Y_xCoO:ZrO_2$ nanostructures. *J. Mater. Sci. Mater. Electron.* 32: 12164–12181. https://doi.org/10.1007/s10854-021-05845-2.

88 Zhang, L., Varma, N.R., Gang, Z.Z. et al. (2016). Targeting triple negative breast cancer with a small-sized paramagnetic nanoparticle. *J. Nanomed. Nanotechnol.* 7 (5): 404. https://doi.org/10.4172/2157-7439.1000404.

89 Finlay, J., Roberts, C.M., Lowe, G. et al. (2015). RNA-based TWIST1 inhibition via dendrimer complex to reduce breast cancer cell metastasis. *Biomed. Res. Int.* 2015: 382745. https://doi.org/10.1155/2015/382745.

90 Cao, Y. and Luo, Y. (2019). Pharmacological and toxicological aspects of carbon nanotubes (CNTs) to vascular system: a review. *Toxicol. Appl. Pharmacol.* 385: 114801. https://doi.org/10.1016/j.taap.2019.114801.

91 Adimule, V., Revaigh, M.G., and Adarsha, H.J. (2020). Synthesis and fabrication of Y-doped ZnO nanoparticles and their application as a gas sensor for the detection of ammonia. *J. Mater. Eng. Perform.* 29: 4586–4596. https://doi.org/10.1007/s11665-020-04979-4.

92 Luo, X., Wang, H., and Ji, D. (2021). Carbon nanotubes (CNT)-loaded ginsenosides Rb3 suppresses the PD-1/PD-L1 pathway in triple-negative breast cancer. *Aging (Albany NY)* 13 (13): 17177–17189. https://doi.org/10.18632/aging.203131.

93 Badea, M.A., Prodana, M., Dinischiotu, A. et al. (2018). Cisplatin loaded multiwalled carbon nanotubes induce resistance in triple negative breast cancer cells. *Pharmaceutics* 10 (4): 228. https://doi.org/10.3390/pharmaceutics10040228.

94 Adimule, V., Kerur, S.S., and Chinnam, S. (2022). Guar Gum and its nanocomposites as prospective materials for miscellaneous applications: a short review. *Top. Catal.* https://doi.org/10.1007/s11244-022-01587-5.

95 Rolih, V., Caldeira, J., Bolli, E. et al. (2020). Development of a VLP-based vaccine displaying an xCT extracellular domain for the treatment of metastatic breast cancer. *Cancers (Basel)* 12 (6): 1492. https://doi.org/10.3390/cancers12061492.

96 Bolli, E., O'Rourke, J.P., Conti, L. et al. (2017). A virus-like-particle immunotherapy targeting epitope-specific anti-xCT expressed on cancer stem cell inhibits the progression of metastatic cancer in vivo. *Oncoimmunology* 7 (3): e1408746. https://doi.org/10.1080/2162402X.2017.1408746.

97 George, T.A., Chen, M.M., Czosseck, A. et al. (2021). Liposome-encapsulated anthraquinone improves efficacy and safety in triple negative breast cancer. *J. Controlled Release* 342: 31–43. https://doi.org/10.1016/j.jconrel.2021.12.001.

98 Albakr, L., Alqahtani, F.Y., Aleanizy, F.S. et al. (2021). Improved delivery of miR-1296 loaded cationic nanoliposomes for effective suppression of triple negative breast cancer. *Saudi Pharm J.* 29 (5): 446–455. https://doi.org/10.1016/j.jsps.2021.04.007.

99 Ding, Y., Yang, R., Yu, W. et al. (2021). Chitosan oligosaccharide decorated liposomes combined with TH302 for photodynamic therapy in triple negative breast cancer. *J. Nanobiotechnol.* 19 (1): 147. https://doi.org/10.1186/s12951-021-00891-8.

100 De Vita, A., Liverani, C., Molinaro, R. et al. (2021). Lysyl oxidase engineered lipid nanovesicles for the treatment of triple negative breast cancer. *Sci. Rep.* 11 (1): 5107. https://doi.org/10.1038/s41598-021-84492-3.

101 Corti, C., Giachetti, P.P.M.B., Eggermont, A.M.M. et al. (2022). Therapeutic vaccines for breast cancer: has the time finally come? *Eur. J. Cancer* 160: 150–174. https://doi.org/10.1016/j.ejca.2021.10.027.

102 Criscitiello, C., Corti, C., Pravettoni, G., and Curigliano, G. (2021). Managing side effects of immune checkpoint inhibitors in breast cancer. *Crit. Rev. Oncol Hematol.* 162: 103354.

103 Hutchins, L.F., Makhoul, I., Emanuel, P.D. et al. (2017). Targeting tumor associated carbohydrate antigens: a phase I study of a carbohydrate mimetic-peptide vaccine in stage IV breast cancer subjects. *Oncotarget* 8 (58): 99161e78.

104 Norton, N., Youssef, B., and Hillman, D.W. (2020). Folate receptor alpha expression associates with improved disease-free survival in triple negative breast cancer patients. *NPJ Breast Cancer* 6: 4.

105 Murtaza Kasi, P., Kalli, K., and Block, M.S. (2015). A phase I trial of the safety and immunogenicity of a multi-epitope folate receptor alpha peptide vaccine used in combination with cyclophosphamide in subjects previously treated for breast or ovarian cancer. *J. Clin. Oncol.* 33 (15): e14028–e14028.

106 Jain, A.G., Talati, C., and Pinilla-Ibarz, J. (2021). Galinpepimut-S (GPS): an investigational agent for the treatment of acute myeloid leukemia. *Expert Opin. Invest. Drugs* 30 (6): 595–601.

107 Chung, V.M., Kos, F., Hardwick, N. et al. (2018). A phase 1 study of p53MVA vaccine in combination with pembrolizumab. *J. Clin. Oncol.* 36 (5): 206.

108 Nandi, S.S., Suryavanshi, A., Adimule, V., and Maradur, S.R. (2020). Semiconductor current-voltage characteristics of some novel perovskite ionic nanocomposites of $Sr_{0.5}$, $Cu_{0.4}$, $Y_{0.1}$ and $Sr_{0.5}$, $Mn_{0.5}$ and their electronic sensor applications. *AIP Conf. Proc.* 2274: 020006. https://doi.org/10.1063/5.0022453.

109 (2020). FDA grants accelerated approval to pembrolizumab for locally recurrent unresectable or metastatic triple-negative breast cancer. https://www.fda.gov/drugs/drug-approvals-and-databases/fda-grants-accelerated-approval-pembrolizumab-locally-recurrent-unresectable-or-metastatic-triple.

110 Loibl, S., Schneeweiss, A., and Huober, J. (2021). Durvalumab improves long-term outcome in TNBC: Results from the phase II randomized GeparNUEVO study investigating neodjuvant durvalumab in addition to an anthracycline/taxane based neoadjuvant chemotherapy in early triple-negative breast cancer (TNBC). *J. Clin. Oncol.* 39: 506.

111 (2021). FDA grants regular approval to sacituzumab govitecan for triple-negative breast cancer. https://www.fda.gov/drugs/resources-information-approved-drugs/fda-grants-regular-approval-sacituzumab-govitecan-triple-negative-breast-cancer.

112 Maya Pai, M., Yallur, B.C., and Batakurki, S.R. (2022). Synthesis and catalytic activity of heterogenous hybrid nanocatalyst of copper/palladium MOF, RIT 62-Cu/Pd for stille polycondensation of thieno[2,3-b]pyrrol-5-one derivatives. *Top. Catal.* https://doi.org/10.1007/s11244-022-01618-1.

113 Ehmsen, S. and Ditzel, H.J. (2021). Signaling pathways essential for triple-negative breast cancer stem-like cells. *Stem Cells* 39 (2): 133–143. https://doi.org/10.1002/stem.3301.

114 Lieberman, J., Gilboa-Geffen, A., and Wheeler, L.A. (2021). Methods and compositions for the treatment of cancer. US Patent 11,180,762.

115 Adimule, V., Medapa, S., Rao, P.K., and Kumar, L.S. (2014). Synthesis of Schiff bases of 5-[5-(4-fluorophenyl) thiophen-2-yl]-1, 3, 4-thiadiazol-2-amine and its anticancer activity. *Int. J. Adv. Pharm. Sci.* 5 (1): 1761–1768.

116 Adimule, V., Nandi, S.S., and Adarsha, H.J. (2021). A facile synthesis of Cr doped WO_3 nanostructures study of their current-voltage power dissipation and impedance properties of thin films. *J. Nano Res.* 67: 33–42.

117 Stafford, R., Yam, A., Li, X. et al. (2020). Anti-folate receptor antibody conjugates, compositions comprising anti-folate receptor antibody conjugates, and methods of making and using anti-folate receptor antibody conjugates. US Patent 10,596,270.

118 Glackin, C.A., Rossi, J., Zink, J.I. et al. (2020). Twist signaling inhibitor compositions and methods of using the same. US Patent 10,519,442.

119 Nandi, S.S., Suryavanshi, A., Adimule, V., and Yallur, B.C. (2020). Super capacitor characteristics of novel rare earth perovskite nanomaterials of $Sr_{0.5}$, $Cu_{0.4}$, $Y_{0.1}$. *AIP Conf. Proc.* 2274: 020007. https://doi.org/10.1063/5.0022454.

120 Adimule, V., Nandi, S.S., and Gowda, A.H.J. (2021). Enhanced power conversion efficiency of the P3BT (poly-3-butyl thiophene) doped nanocomposites of $GdTiO_3$ as working electrode. In: *Techno-Societal 2020* (ed. P.M. Pawar, R. Balasubramaniam, B.P. Ronge, et al.), 55–68. Cham: Springer.

121 Adimule, V., Yallur, B.C., Batakurki, S.R., and Gowda, A.H.J. (2021). Microwave assisted synthesis of Cr doped Gd_2O_3 nanostructures and investigation on morphology optical photoluminescence properties. *Nanosci. Technol. An Int. J.* https://doi.org/10.1615/NanoSciTechnolIntJ.2021039643.

122 Adimule, V., Nandi, S.S., Yallur, B.C. et al. (2021). Enhanced photoluminescence properties of $Gd_{(x-1)} Sr_xO$: CdO nanocores and their study of optical structural and morphological characteristics. *Mater. Today Chem.* 20: 00438.

10

Etiology and Therapy of Hormone Receptor-Positive Breast Cancer

Nalini Kurup and Darshana Warekar

University of Mumbai, Principal K. M. Kundnani College of Pharmacy, Department of Pharmaceutics, Mumbai, Maharashtra 400005, India

10.1 Introduction

Breast cancer is cancer that commences in the breast. It is the uncontrollable growth of healthy breast cells forming a mass of cells called tumors, which maybe malignant or benign. Breast cancer is second to skin cancer among women. According to WHO Fact Sheets, in the year 2020 breast cancer was diagnosed in approximately 2.3 million women, with 685 000 deaths worldwide. In the previous five years as of the end of 2020, breast cancer was in 7.8 million women making it the most common cancer in the world. The breast of a human starts to develop from six weeks of fetal development through 10 progressive stages. The breast is made up of ducts and lobules. The glands that can make milk are lobules and the role of ducts is to carry milk from lobules to nipples. Adult breast development starts at the onset of puberty. Hormones play a vital role in the development of the breast. The hormone which has a major influence on the breast during its development is estrogen. The steroid hormone progesterone and the growth hormone also play a role as regulators of breast development [1].

The development of breast cancer starts when the breast cells start growing abnormally. This is because of the mutation which takes place in genes whose function is to regulate the cell growth and as a result, the cells start developing more rapidly than the normal healthy cells, and their accumulation results in the formation of a lump or mass. This leads to the development of breast cancer [2].

The emergence of breast cancer can be in distinct areas of the breast such as the ducts, the lobules, and in some cases in the tissue in between. About 20 morphological subtypes of breast cancer have been conceded by WHO. Breast cancer can be invasive, noninvasive, and metastatic types. The invasive type is the one that starts in breast glands and ducts and further grows into the breast tissue. Then it starts spreading into nearby lymph nodes [3]. The further classifications are invasive ductal carcinoma, invasive lobular carcinoma, ductal carcinoma in situ, etc. Invasive ductal carcinoma is the most prevalent type which is further classified as tubular, medullary, mucinous, papillary, etc. Genetic profiling classifies the types of breast cancer as Luminal A breast cancer, Luminal B breast cancer, triple-negative/ basal type breast cancer, HER-2-enriched breast cancer, and normal-like breast cancer.

Noninvasive carcinoma sometimes also called precancer or carcinoma in situ (CIS), meaning "in the same place." They stop off within the lobules or milk duct in the breast and do not grow beyond the breast [4]. Noninvasive carcinoma is further classified as ductal CIS and lobular CIS [1]. In metastatic breast cancer, the cancerous cells start spreading from the breast to other parts of the body. Breast cancer cells may spread to the lungs, liver, brain, and bones. The symptoms depend on which part of the body is affected. Breast cancer at Stage 4 is also known as metastatic breast cancer [5]. There have been notable advancements in the treatment of breast cancer with more research on targeted therapies and hormonal therapies [5].

10.2 Etiology

Hormone receptor (HR)-positive breast cancer is a type that causes excessive growth of cells in the breast as a response to the hormones, especially estrogen and progesterone.

10.2.1 Role of Estrogen Hormone

Estrogen is a group of hormones which includes estriol, estrone, and estradiol. The hormone estrogen is produced by the ovary and in small amounts, is also produced by the adrenal gland and fatty tissue which are located at the top of the kidney [6]. Estrogen helps to stimulate ductal growth and also increases fat deposition. Receptors are the key mediators of the estrogen effect. ERαreceptor helps to mediate the estrogen effect [7].

10.2.2 Role of Progesterone Hormone

Progesterone is one of the female sex hormones. It is produced in the corpus luteum of the ovaries. In the small amount, it is also produced by the placenta and adrenal gland. The main role of the progesterone hormone is to regulate menstruation and support pregnancy in the female body [8]. It was observed that high progesterone concentration is associated with a heightened risk of breast cancer. The hazard ratio was calculated as 1.16 (95% CL, 1.00–1.35) per standard deviation of progesterone [9].

Breast cancer cells express some proteins, which are called estrogen and progesterone receptors (PRs). When there is an interaction of estrogen and progesterone hormone with estrogen and PRs respectively it leads to stimulating the growth of cancer cells. Based on this breast cancer is also termed HR-positive or HR-negative breast cancer. Progesterone is a key component of physiological functions like reproduction, menstrual cycle, central nervous system, cardiovascular system, immune system, mammary gland development, and biosynthesis of steroid hormone. Progesterone is a gestational agent but it has more wide applications than only being a gestational agent. Progesterone is the precursor of aldosterone, testosterone, cortisol, and oestradiol, which are gonadal and non-gonadal hormones mainly responsible for various functions such as blood pressure regulation, conversion of sodium in the kidney, development of secondary sexual characteristics of males and females, and regulation of blood glucose level [10]. Estrogen is a steroidal sex hormone. Estrogen is a key component of various physiological and biological functions such as metabolization of cholesterol, cardiovascular system, electrolyte balance, bone density regulation, skin physiology, female reproductive system, and central nervous system. Estrogen also plays important role in a gastrointestinal tract and comes up with the development of various GI diseases such as colon cancer, peptic ulcer, gastric acid syndrome, and inflammatory bowel diseases [11].

10.2.3 Estrogen Receptor (ER)

An essential biomarker of breast cancer has been the presence of estrogen receptors (ERs). The ER comes under the nuclear receptor superfamily, which consists of 553 amino acids. The ER is divided into six functional distinct domains. It comprises Domain A, Domain B, Domain C, Domain D, Domain E, and Domain F. Domain A and Domain B contain 17% of amino acids, Domain C is the central region, which is a DNA binding domain. Domain D contains nuclear localization signal and it links domain C with Domain E, a multifunctional carboxyl terminal. This is also called the ligand-binding domain (LBD). Domain E consists of 56% of amino acids. Domain F consists of 18% of amino acids and it is located at the carboxyl terminals of the receptor [12]. ER α and ER β are the two isoforms of the ERs. These are encrypted by two nonidentical genes. Female sex hormone 17 β-estradiol (E2) is the ligand for ER which in the physiological state mediate differentiation and growth of breast ducts.

When hormones diffuse through the cell membrane estrogen signaling pathway activates and binds to the nuclear ER. These ligand–receptor complex then promotes gene transcription [13].

Molecules which are agonists, antagonists, or mixed agonist–antagonists can bind to the ER. When the agonist molecule binds with this receptor it undergoes conformational changes. It binds as a dimer to the targeted genes, which in turn induces their expression. Antagonist molecules are nothing but agonist competitors for the binding site on receptors. There is a number of co-regulatory factors, which have been discovered recently, can bind to ER, and can help to stimulate or inhibit the activation of targeted genes. Based on the clinical and experimental data it was found that ERs play a remarkable part in breast cancer development which is why targeting the ER has become a useful treatment strategy for breast cancer. ERs are also targeted by the peptide ligands, which in turn disrupt ER/ERP mediated gene transcription [14].

10.2.4 Progesterone Receptor (PR)

Horwitz and Mcguire who have extensively studied ER proposed that PR might give out as a benchmark of the functional component of an estrogen signaling pathway in the breast. PRs come from the steroid receptor subgroup of ligand-activated transcription factor, which is part of the large nuclear receptor superfamily. PRs consist of 946 amino acids. There are two isoforms of PRs namely PR-A and PR-B, which are encoded by the same gene. PRs are disciplined by E2 and allied estrogen. In the case of breast cancer, PR is synthesized by tumor, cells that are energized by estrogen after ER interaction [15].

The PR is an effective and specific mediator of progestin action in breast and endometrium tissues. It is essential in the control of complex differentiation and proliferation at various stages of breast cancer development. PR complex of specific domain involved in DNA and hormone binding. Activation of PR in targeted tissue is initiated by phosphorylation and dimerization of the receptor. This in turn results in binding to progestin-responsive element on DNA. The PR is an important biomarker of responsiveness to breast cancer. The presence of PR expression in ER-positive breast cancer indicates better responsiveness to endocrine therapy. The ratio of PR A and PR B is of absolute importance to endocrine therapy [16].

10.3 Human Epidermal Growth Factor-2 (HER-2)

It is a transmembrane tyrosine kinase receptor belonging to the epidermal factor receptors (EGFRs) family. It is encoded by Erb B2/HER-2-neu oncogene. HER-2 or neu and EGFR are similar with 80% of homology in the amino acid and of three tyrosine kinase domains [17]. This family consists of four transmembrane receptors kinase namely HER-1 which is also called EGFR, HER-2, HER-3, and HER-4. Alteration in activity and expression of the HER family is observed in breast cancer. About 25% of breast cancer has shown overexpression of HER2 receptors. EGFR is highly expressed in triple-negative breast cancer (TNBC) [18]. These receptors are located on the cell membranes and are found in various tissues. HER-2 receptors and HER family is responsible for normal breast growth and development. Dysregulation in the expression of HER-2 is linked with breast cancer. The mechanism of HER-2 receptor overexpression is due to the amplification of the HER-2 gene. This amplification of the HER-2 gene then results in overexpression of HER-2 receptors and disturb normal control mechanism, which ultimately result into the formation of tumor cell. HER-2 protein levels are 100 times more in the cancerous cells than in normal cells. In breast cancer patients about 92% of cases are because of HER-2 overexpression. HER-2 is now considered to be a good target for novel treatment, which is specific for HER overexpression breast cancer [19].

10.4 Various Types of Breast Cancer Detected Under Hormone Receptor Breast Cancer

10.4.1 Estrogen Receptor (ER) Positive

This type of breast cancer has receptors that allow it to develop by using the hormone estrogen. Antiestrogen hormone treatment (endocrine therapy) can stop cancer cells from growing [20]. Depending on the expression of estrogen-receptor protein we have two distinct types of estrogen-positive breast cancer, which are "strongly ER-positive and weakly ER-positive cancer" [21]. Gene RNA expression further categorized the ER-positive breast cancer as "luminal A and luminal B" type. Estrogen and ER are required for the synthesis of the PR gene is a protein-coding gene that is an estrogen-regulated gene [22].

10.4.2 Progesterone Receptor (PR) Positive

This sort of breast cancer has receptors that empower it to develop by using the hormone called progesterone. Treatment with endocrine therapy blocks the growth of cancer cells [2]. Progesterone-positive breast cancer is found to be more frequent in individuals of age 30–50 years. Patients with PR-positive breast cancer shows more aggressive tumor characteristics including negative lymph node and smaller tumor size. PR-positive group of patients has shown high proportions of ERBB2-positive tumors [21].

10.4.3 Hormone Receptor (HR) Negative

This type of cancer is devoid of HRs, endocrine therapies that impede hormones in the body will have no effect. It represents about 30% of all breast cancers. HR-negative breast cancer has a more complex clinical course. HR-negative tumors are poorly differentiated, associated with a high recurrence rate, and of high histological grade [23].

Breast cancer cells are separated after a biopsy or surgery and are scrutinized to check if they own estrogen or PR proteins. The hormones estrogen and progesterone when bound to the HRs motivate the cancer cells to grow [20].

Owing to whether or not these receptors are present, breast cancers are classified as HR-positive breast cancer or HR-negative breast cancer. When deciding on treatment alternatives, it's critical to know the HR status.

Receptors are proteins found within or on the surface of cells that may bind to certain chemicals in the blood. Normal breast cells and certain breast cancer cells have estrogen and PRs, and these hormones are required for the cells to develop. One, both, or none of these receptors may be present in breast cancer cells.

Both of the HRs ought to be examined in all invasive breast tumors. Either on a biopsy sample or once the tumor is removed surgically. At least one of these receptors is found in almost seven out of ten breast tumors. Older women have a larger proportion than younger women.

10.5 Detection

McCarty and Colleagues 1985 proposed the H-score semiquantitative scoring system. It is calculated by the following the formula [24].

$$H \, score = \sum(\text{Percent of tumor cells staining} \times \\ \text{ordinal value corresponding to the intensity level})$$

The maximum possible score is 300. A score which is less than 1 is considered negative. The score which is less than 100 is considered weakly positive, a score which is between 101 and 200 is considered moderately positive, and a score between 201 and 300 is considered as strongly positive.

Allred and colleagues come up with the Allred score, which is calculated as follows [24].

$$Allred \, score = \text{proportion score} + \text{intensity score}$$

The result is interpreted based on two scores which are the proportion of positive staining cell score (0–5 scale) and staining intensity of tumor cell score (0–3 scale). These two scores are summed up together for the final score. If the final score is 0–2 it is considered negative and a score between 3 and 8 is considered positive [24].

Japanese investigators introduced **a modified J-score**. They evaluated only the number of positive cells and do not consider staining intensity. If the J-score is between 1 and 2, the HR status is negative; if the J-score is 3, the HR status is positive. However, in the Western world, this method was not well-received [25].

American society of clinical oncology and college of American pathologist guidelines were announced in 2010 to enhance the reliability of Immunohistochemical ER and PR testing in breast cancer.

A test called **an immunohistochemistry** (IHC) test is used now most often to find out if cancer cells have estrogen and PRs. Test results are interpreted as the tumor is HR-positive if not less than 1% of the cells tested have either estrogen or PRs. Otherwise, the test will indicate the tumor is HR-negative.

Several New Gene-Classified Tests have been Identified for Determining Hormone Status.

Oncotype Dx is one of the molecular assays with 21-gene signature which consist of 16 cancer genes and 5 reference genes. It is a quantitative test that is based on a reverse transcriptase-polymerase chain reaction (qRT-PCR) test, which is carried out on formalin-fixed paracrine-embedded tissues that generate scores ranging from 0 to 100. Patients with less than 18 scores are classified as low risk of disease recurrence, patients with scores of 18–30 with intermediate, and patients with more than 31 scores have a high risk of disease recurrence [26].

HR-positive (or hormone-positive) breast cancer cells have either estrogen (ER) or progesterone (PR) receptors or both. About 70% of breast cancers are due to the expression of these two receptors [27].

HR-negative (or hormone-negative) breast cancers neither possess estrogen nor PRs.

TNBC cells do not express estrogen or PRs and also do not produce any or too much of the protein called HER-2 (Human Epidermal Growth Factor receptor 2).

10.6 Therapy

Surgery, Radiation, Chemotherapy, and Hormone therapy are the mainstays of breast cancer treatment. Hormone/Endocrine therapy is extremely beneficial in patients with hormone-receptor–positive tumors. Hormone therapy has low morbidity, leading to a notable reduction in mortality for breast cancer. Hormone therapy is usually considered an excellent choice for patients with estrogen-receptor–positive cancers and nonlife-threatening advanced diseases, or for older patients who are not fit for combative chemotherapy treatments. Hormone therapy is broadly classified as ER-positive, PR-positive, triple-negative, and HER-2 positive (Figure 10.1).

Hormonal therapy works by lowering the amount of estrogen in the body or blocking estrogen from attaching to the breast cancer cells. Hormone therapy for breast cancer shall not be confounded with menopausal hormone therapy (MHT) – treatment with estrogen alone or in combination with progesterone to ease and relieve symptoms. Hormone therapy is used for the treatment of hormone-sensitive metastatic breast cancer.

The main types of hormonal therapy used include:

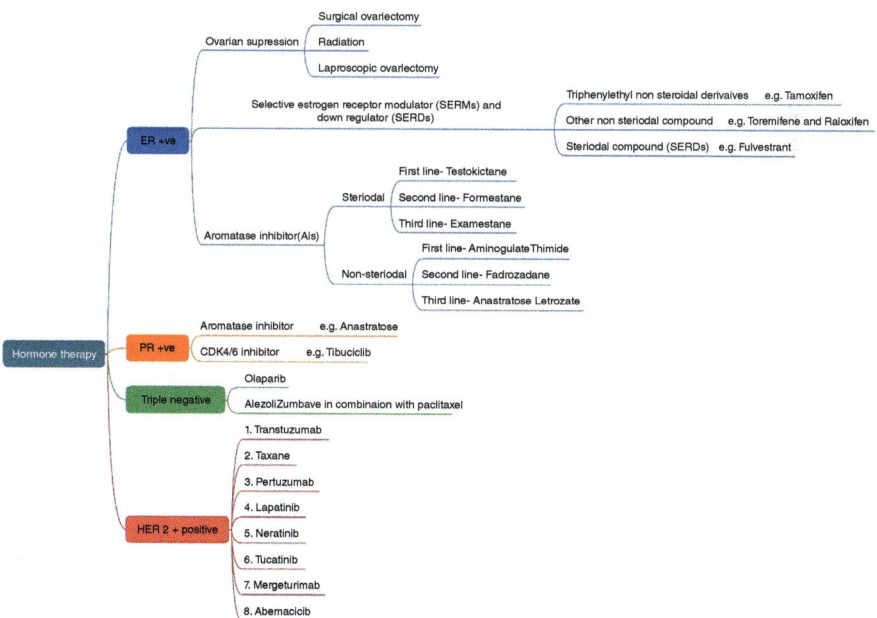

Figure. 10.1 Classification of drugs used in hormone therapy.

10.6.1 Selective Estrogen-Receptor Response Modulators (SERMs)

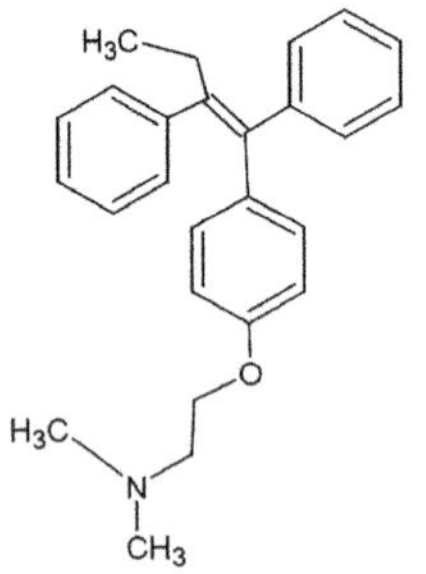

Figure. 10.2 Structure of Tamoxifen.

These types of therapeutic agents chunk the effects of estrogen in the breast tissue by binding to the ERs in breast cells. Tamoxifen (Figure 10.2) is the selective estrogen-receptor response modulators (SERMs) of choice, used to treat breast cancer [28]. These are a category of the therapeutic agents used for the prevention and treatment of diseases such as breast cancer and osteoporosis [29]. SERMs show tissue-selective activity. It acts as an agonist in the liver, bone, and cardiovascular tissue whereas it acts as an antagonist in breast and brain tissue as well it shows mixed agonist–antagonist activity in the uterus, other than Tamoxifen (Figure 10.2), raloxifene, toremifene, and several new chemotherapeutic agents are now in use as SERMs. New SERMs currently being developed include EM-652, LY-353,381, GW-5638, and SP500263, which have potential for chemoprevention [30]. SERMs act by three interactive mechanisms and they are differential ER expression and gene activation through ERE interaction, differential ER conformation upon ligand binding. SERMs induce unique changes in receptor confirmation, which is responsible for required pharmacological action in targeted tissue. SERMs plug the estrogen binding site on the receptors and block estrogen action and thus preventing estrogen to gain access to ERs [30].

10.6.2 Aromatase Inhibitors

Aromatase inhibitors inhibit estrogen production in postmenopausal women. Aromatase inhibitors operate by inhibiting the enzyme aromatase, which converts testosterone into estrogen in minute concentrations in the body. Aromatase inhibitors are further classified as steroidal and nonsteroidal aromatase inhibitors. The aromatase enzyme is a member of the P450 family product of the CYP19 gene. This enzyme is highly expressed in granulosa cells of the ovarian follicles and placenta. Steroidal aromatase inhibitors bind on the same site on aromatase moiety irreversibly and are converted to reactive intermediates by aromatase. Nonsteroidal aromatase inhibitors bind to the heme group of enzymes reversibly by basic nitrogen atom [31]. Aromatase inhibitors include therapeutic agents such as Arimidex (anastrozole) (Figure 10.3), Aromasin (exemestane) (Figure 10.4), and Femara (letrozole) (Figure 10.5) [32].

10.6.3 Estrogen-Receptor Down Regulators (ERDs)

Estrogen receptor down regulators, called ERDs seal the effects of estrogen in breast tissue. These are different in pharmacological activity and molecular activity from selective ER modulators. Though they both act by inhibiting the estrogen activity ERDs show the significant downstream effect [33]. ERDs act as antioestrogen by disrupting ERαcofactor interaction and give rise to receptor turnover in cells. ERDs are generally prescribed for patients who are resistant to SERMs [34]. Faslodex (chemical name: Fulvestrant) (Figure 10.6) is the drug of choice, which can be used as monotherapy for the treatment of postmenopausal women tested with hormone-receptor-positive breast cancer. To improve the pharmacological and pharmaceutical properties of ERDs development of new ERDs is in progress. GW 5638 new compound of ERDs discovered by Willson and colleagues, which contains a shorter ally carboxylic group rather than a tertiary amino side chain [35]. GW 5638 is structurally similar to the ERα ligand-binding domain and hence can be bound to it. Other new ERDs include GW-7604, ZK-703, and ZK-253, which are under development for the treatment of breast cancer [36].

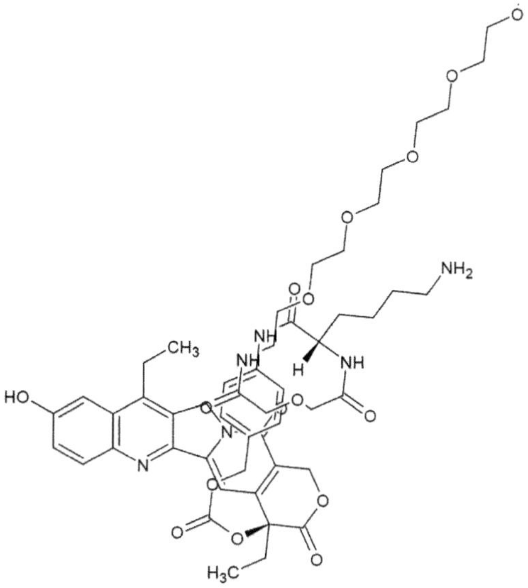

Figure. 10.3 Structure of Anastrozole.

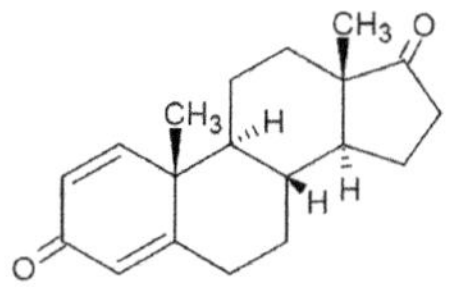

Figure. 10.4 Structure of Exemestane.

10.6.4 Luteinizing Hormone-Releasing Hormone Agents (LHRH)

Luteinizing hormone-releasing hormones (LHRHs) cause the ovaries to stop working and cease generating estrogen, downregulating the amount of estrogen available to support the growth of hormone-receptor-positive breast cancer. LHRH agents include LHRHs agonist and antagonist [37]. Examples include Zoladex (chemical name: goserelin), Lupron

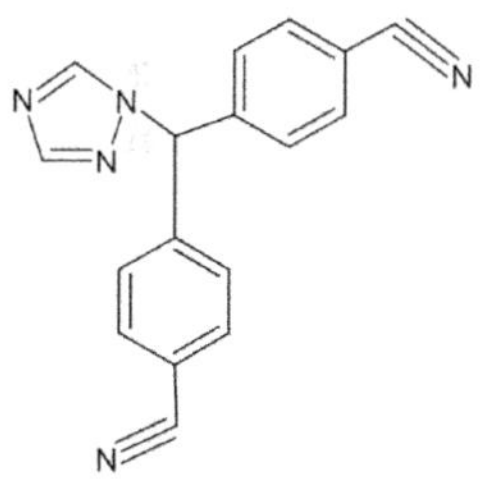

Figure. 10.5 Structure of Letrozole.

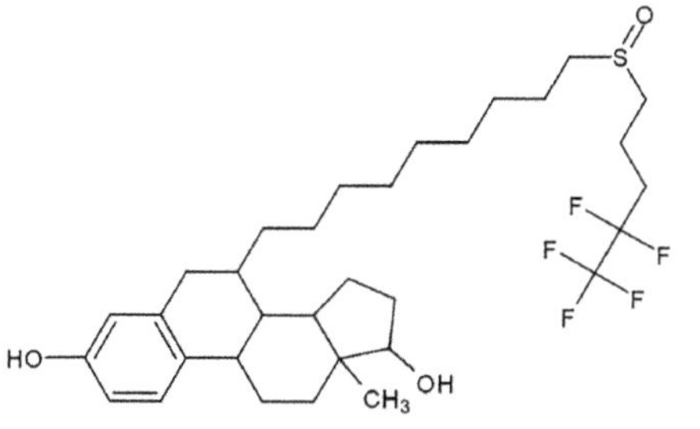

Figure. 10.6 Structure of Fulversant.

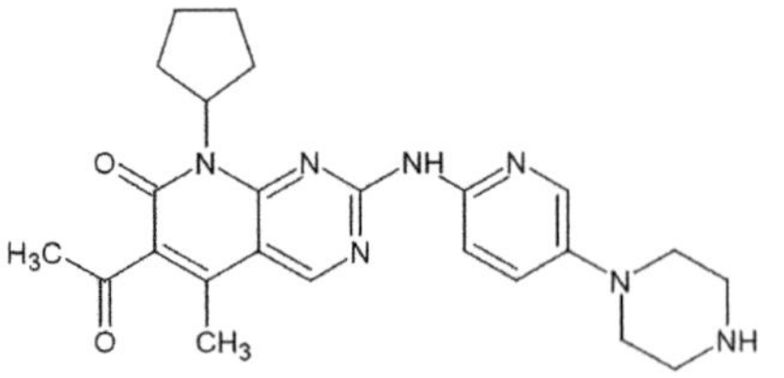

Figure. 10.7 Structure of Palbociclib.

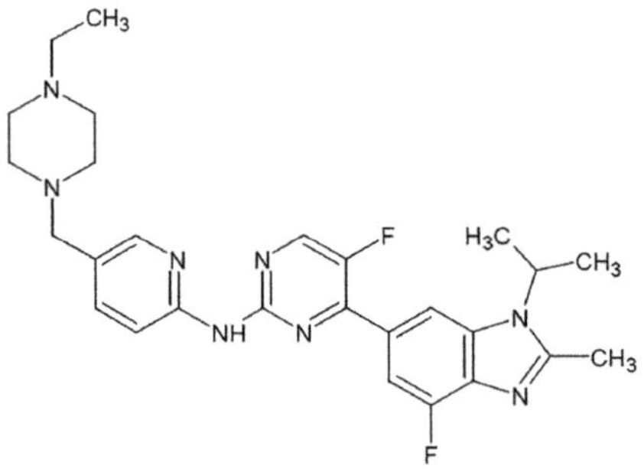

Figure. 10.8 Structure of Abemaciclib.

(chemical name: leuprolide), and Trelstar (chemical name: Triptorelin).

Palbociclib (Ibrance) (Figure 10.7) and Abemaciclib (Verzenio) (Figure 10.8) are two other medications utilized, both of which block two cyclin-dependent kinases (CDK4 and CDK6), which appear to encourage the proliferation of HR-positive breast cancer cells. Palbociclib is also authorized for use in conjunction with fulvestrant in the treatment of HR-positive breast cancer in postmenopausal women [28].

Alternatively, ovarian suppression can be carried out by oophorectomy, LHRH analogs, and chemotherapy drugs. Other types of hormone therapies that are rarely used are megestrol acetate, androgens, and high doses of estrogen.

10.7 Limitations of Hormone Therapy

The complexity of hormone therapy, which might get worse at the start of therapy is a major part of the concern. These side effects have limited the application of hormone therapy. There are several drugs that are approved for the therapy but they show certain side effects. Following are some drugs enlisted with their side effects.

10.7.1 Tamoxifen

It is a SERMs class of drug. Side effects associated with this drug include bone loss in premenopausal women, risk of a blood clot, sleep disorder, vaginal dryness, loss of sexual interest, risk of venous thromboembolic events, depression, hot splashes, uterine sarcoma, etc.

10.7.2 Raloxifene

It is a nonsteroidal SERMs class of drugs, which shows the side effect of suppression of ovaries, loss of bone, depression, and loss of sensuality.

10.7.3 Aromatase Inhibitors

These classes of drugs show side effects like heart failure, hypercholesterolemia, joint pain, bone loss, risk of heart attack and angina, etc.

10.7.4 Fulvestrant

It is a steroidal SERDs class of drug. It shows some side effects musculoskeletal pain, joint pain, gastrointestinal disturbances like nausea, vomiting, and constipation, breathing problem, headache, hot splashes, shortness of breath, anorexia, etc.

10.8 Triple-Negative Breast Cancer

Around 15–20% of all breast cancer patients are diagnosed with TNBC. TNBC is expounded as breast cells short of an expression of the, ER, PR and HER2. TNBC is a heterogeneous disease (having several etiologies) diagnosed by IHC and is characterized by tumors showing an absence of ER, PR expression and does not overexpress human epidermal growth factor receptor 2 (HER-2). Cancer with ER and or PR IHC expression of 0 and HER 2 IHC expression of 0 and 1+ or with short of gene amplification is defined as TNBC [38]. Next-generation sequencing (NGS) has given away a host of molecular features of TNBC, including high rates of *TP*53 (gene to produce tumor protein p53) mutations, PI3K (phosphoinositide 3-kinase) and MEK pathway activation (protein kinase), and similar genetically to serous ovarian cancers, including upregulation of the *BRCA* pathway [39]. Identified genetic susceptibilities of TNBC give rise to promising therapeutic approaches, including DNA-damaging agents (i.e. platinum salts and poly adenosine diphosphate–ribose polymerase inhibitors), as well as immunotherapy [40].

10.8.1 Clinical History of Triple-Negative Breast Cancer

It was observed from studies that the clinical features of TNBCs are different from all other forms of breast cancer [41]. Patients with TNBC have an increased likelihood of reappearance and mortality ratio within five years of diagnosis as compared to other breast cancer patients. The qualitative pattern of recurrence is also different. It has been found that the rate of recurrence was high within three years and then it gets mild. Though adverse effects of TNBC are mild it has more aggressive clinical features as compared to other forms of breast cancer [42].

To compare the clinical features of TNBC with nontriple-negative breast cancer, Janet Yeh et al. Carried out the study in January 2010. About 1964 patients with invasive breast cancer were studied out of which 190 patients had TNBC and 1774 had nontriple-negative breast cancer. The results showed the highest percentage of BRCA mutation carriers as compared to nontriple-negative breast cancer. It was also observed that 10.6–30.9% of TNBC patients showed the carrier rate of BRCA 1 and BRCA2 Mutation [43].

10.8.2 Imaging Characteristics/Features of Triple-Negative Breast Cancer

Mammographic findings of TNBC were reported commonly as a mass without calcification, mass with speculated or ill-defined margin, and mass with irregular shape. Ultrasound characteristics of TNBC were reported commonly as hypoechoic masses with asymmetric shape and nonrestrictive margin. Other characteristics of TNBC include parallel mass orientation with unanticipated surface boundary. Due to the aggressive nature and fast growth of TNBC, it is less likely to be detected by mammography or ultrasound. It is mostly detected clinically either by physically performed breast exam and clinical symptoms, such as breast pain or nipple discharge, or self-breast exam [44].

10.8.3 Subtypes of TNBC

10.8.3.1 Basal-Like Subtype: (i) BL1 (ii) BL 2

The basal-like subtype was discovered by first-generation cDNA microassay [36]. Basal-like tumors are classified under TNBC because of the absence of ER, PR, and HER2 receptors expressions. Basal-like breast cancer is characterized by a gene expression profile similar to the basal myoepithelial layer of a normal breast. Approximately 70–75% of TNBC are basal like and rest 25–30% of TNBC do not exhibit basal-like subtypes [45].

Basal-like subtype is further classified as BL1 and BL2. BL1 is characterized by the enhancement of cell division components, pathways, and cell cycle. Proliferation pathway goes along with elevated DNA damage which is to a great extent supported by the observation of nuclear Ki67 staining and high expression of Ki67 mRNA. BL2 subtype is characterized by enriching growth factor receptors such as EGFR, EPHA2, and MET [46].

10.8.3.2 Claudine Low Subtype

Claudine's low subtype of TNBC is identified by gene expression profile and associated with low survival rate with low recurrence rate. This subtype is different from another subtype of TNBC as it shows lower expression of markers of luminal epithelial cell differentiation, epithelial cell differentiation, epithelial cell–cell adhesion, and elevated levels of genes that are responsible for host defence, immunity, and tumor cell invasiveness [47].

10.8.3.3 Immunomodulatory Subtype (IM)

The immunomodulatory (IM) subtype is an augment for factors responsible for the immune cell process including immune cell signaling. This subtype has a good prognosis, has a high expression of STAT genes, and evince activation of STAT transcription [48].

10.8.3.4 Mesenchymal Subtype (M)

The mesenchymal (M) subtype is responsible for pathways and components involved in the cell differentiation pathway, cell motility, and extracellular receptor interaction [48].

10.8.3.5 Mesenchymal Stem-Like Subtype (MSL)

Mesenchymal stem-like (MSL) subtype shows a similar biological process as that of mesenchymal type. This subtype involves cellular differentiation, growth pathway, and cell motility but this subtype is unique for its gene-representing components and growth factor signaling pathway which includes EGFR, ERK1/2, calcium signaling, and G-protein coupled receptors. This subtype possesses genes that are exclusive to adipocytes and osteocytes [48].

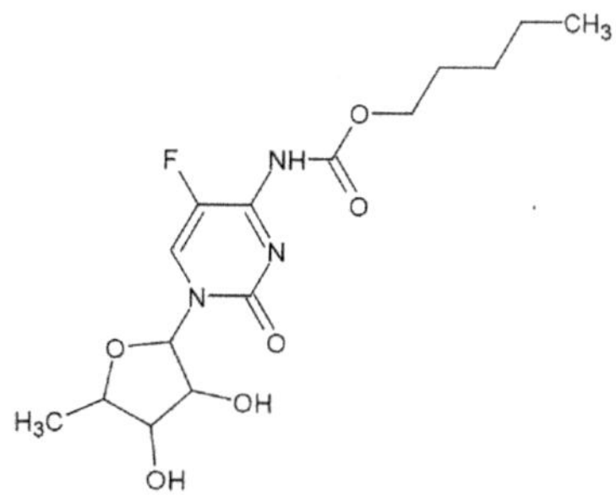

10.8.3.6 Luminal Androgen Receptor Subtype (LAR)

The luminal androgen receptor (LAR) subtype is different than other subtypes of TNBC. This subtype is responsible for hormonally related pathways including porphyrins metabolism, androgen/estrogen metabolism, and steroid synthesis. This subtype shows ERα-negative staining but exhibits ER, AR ErbB4, and proactive signaling. This subtype shows AR-expression 10-fold higher than another subtype of TNBC [46].

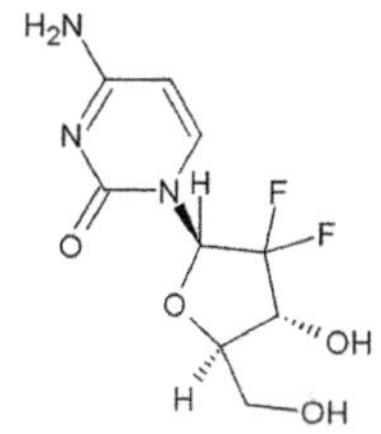

Figure. 10.9 Structure of Capecitabine.

Figure. 10.10 Structure of Gemcitabine.

10.8.4 Treatment of Triple-Negative Breast Cancer

Stage I–III TNBC: It can be treated by surgery, radiation, and mastectomy.

Stage IV TNBC: it can be treated with chemotherapy, e.g. anthracycline class of drug, Taxen, Capecitabine (Figure 10.9), Gemcitabine (Figure 10.10)

10.8.5 Advance TNBC

It can be treated with immunotherapy, immunoconjugates, and platinum salts

Immunotherapy and immunoconjugates: Immunotherapy was found to possess an exciting opportunity for the treatment of TNBC. Immunotherapy in the form of monoclonal antibodies acts as the checkpoint inhibitors. These immunomodulators can increase the immune response against tumor by inhibiting immune-regulating proteins that downregulates the immune system.

Immune therapy includes immune checkpoint inhibitors, antibody–drug conjugate, and recurrent TNBC [49]

Immune checkpoint inhibitors: PD-1, its ligands PD-L1, and cytotoxic associated protein 4 which are involved in T-lymphocyte activity, e.g. Atezolizumab, Pembrolizumab, etc.

Antibody–drug conjugate: e.g. Sacituzumab Govitecan.

Currently, the FDA has approved pembrolizumab (Keytruda®) combined with chemotherapy for the treatment of patients with metastatic TNBC whose tumors express PD-L1. Current research is oriented toward developing novel treatments for early and advanced TNBC. Ongoing studies on variable responses of therapeutic agents in diverse microenvironments of the tumors necessitate the development of biomarkers to identify patients who will respond well to a given therapy. Clinical trials investigating newer targeted therapies and combination strategies reflect further attempts to understand molecular and immunological aspects of TNBC. Studies have shown androgen receptors linked to luminal type TNBC in a few cases which respond to antiandrogen drugs.

10.8.6 Pharmacogenomics

Pharmacogenetics is the probe into the variableness of drug response due to heredity. It is also associated with gene–drug interaction. Pharmacogenomics has significant application in oncology due to life-threatening side effects and low therapeutic index of chemotherapeutic agents [50]. The absence of estrogen, progesterone, and HER-2 receptors in TNBC elicits a good response to chemotherapy as compared to other options. The recurrence and metastatic rate of TNBC are higher than that of non-triple negative breast cancer. Impact of genomic and molecular heterogeneity has shown to affect the patients receiving the treatment for TNBC. Based on the expression of genes and several other features of cancer probable targets have been recognizing for the therapeutic treatment of breast cancer.

Pharmacogenetics have reported variation in metabolism of most of the drugs based on the CYP2D6 genotype. Based on this scientists have separated patients into four metabolic phenotypes, which are ultrarapid, poor, extensive, and intermediate [51].

Patient with TNBC has shown a good response to cisplatin treatment with altered BRCA1 expression due to methylation and p53 mutation [52].

Patients with high expression of CD73 have shown resistance to doxorubicin treatment [53].

Genomic alteration in breast tumors involves loss of PTEN and RB1, overexpression of Myc and H1F1-α, and mutation OF p53. Network TCGA. Comprehensive molecular portraits of human breast tumors.

Additional amplification of genes has been identified such as MCL1, AKT1, CDK4, JAK2, and EGFR, and loss of mutation in BRCA1/2, CDKN2A, and ATM, which has been shown to have an impact on targeted hormone therapy of breast cancer [54].

Various pharmacogenomics markers have been identified to alter and improve response to various drugs used against HR-positive and TNBC. CYP2D6 is a potential biomarker identified for tamoxifen because of its major contribution to the metabolism of the drugs in men. Other non-CYP2D6 pharmacogenomic biomarkers have also been identifying for tamoxifen which includes CYP2B6*6, CYP2C19*2, CYP3A4 *1B, and CYP3A5*3[55] (Table 10.1).

If it was possible to predict which patients can benefit from treatment and which patients may agonize from the chemotherapy and hormone therapy-related

Table 10.1 Genes that can influence the outcome of breast cancer.

Serial No.	Therapeutic agent	Genes	Propose outcome
1	Aromatase inhibitors	CYP19A1, ESR1	Due to toxicity risk of discontinuation
		CYP19	Poor response to letrozole
2	Doxorubicin	MPO	Improved survival
		NOS	Shorter progression rate
		CBR3 11A	Increased neutropenia
3	Cyclophosphamide	GSTP1	Low risk of hematologic toxicity
		CYP2B6	Shorter progression rate
4	Tamoxifen	UGT2B15	Risk of recurrence
		SULT1A1	Risk of recurrence
		ESR2	Decrease hot flashes
		CYP2D6	Increase risk of discontinuation
5	Trastuzumab	Fc gamma RIIIa-158V/V	Progression-free survival
6	Paclitaxel	CYP1B1*3	Progression-free survival
		CYP2C*8	Increased risk of neuropathy
		ABCB1	Increase efficacy

Source: Adapted from Westbrook and Stearns [56].

toxicity then care for cancer patients would considerably improve. Here are some well-studied examples of drug–gene pharmacogenomics pairs. Dihydropyrimidine dehydrogenase genotype and fluoropyrimidine, Thymidylate synthetase, Methylene tetrahydrofolate reductase, Thiopurine S-methyltransferase (TPMT) and thiopurine, Uridine diphosphate glucuronosyltransferase (UGT) genotype and irinotecan, Glutathione S-transferases gene polymorphism and platinum compounds, Excision repair cross-complementing group 1 (ERCC1), Excision repair cross-complementing group 2 (ERCC2), ATP-binding cassettes (ABCB1, ABCC2, ABCG2), X-ray cross-complementing group 1 (XRCC1) [50]. Many of the drugs as monotherapy and a few as combination therapy are under clinical trials at different phases (Table 10.2).

10.9 Conclusion

The importance and significance of steroidal hormones and the receptors associated with them cannot be exaggerated in the treatment of breast cancer. Tamoxifen though the traditional drug of choice for hormone therapy for breast cancer, its use is becoming limited due to adverse side effects as well as hormone resistance in breast cancer patients. TNBC is the one that affects the immunity of the patients

Table 10.2 Ongoing clinical trials in TNBC.

Serial No.	Target	Drug	Phase
1	Cyclin-dependent kinase	Abemaciclib	Phase II
		Trilaciclib	Phase II
		Dinaciclib + Epirubicin	Phase I
2	Growth factors and angiogenesis (EGFR)	Erlotinib + Chemotherapy	Phase II
		Icotinib	Phase II
		Cetuximab + Ixabepilone	Phase II
3	HER 2	Trastuzumab + Paclitaxel + Cyclophosphamide	Phase II
4	VEGFR2	Apatinib + Vinorelbine	Phase II
5	PARP	Olaparib + Carboplatin + Paclitaxel	Phase I
		Olaparib + Cediranib	Phase II
		Olaparib	Phase II
		Rucaparib + Cisplatin	Phase II

Source: Level et al. [57]/John Wiley & Sons.

and leads to a worst prognosis than any other type of cancer. Hence it is very important to have accurate tools to diagnose the TNBC and provide the correct treatment for the patients. As a result of this new therapeutic agents are under clinical trials with fewer side effects, and better therapeutic efficacy and the ones which can bypass the resistance to hormone treatment are being investigated. With increased specialization and a targeted drug delivery system it is possible to improve breast cancer care.

References

1 Sun, S.X., Bostanci, Z., Kass, R.B. et al. (2018). *Breast Physiology: Normal and Abnormal Development and Function*, 5e. Elsevier Inc.
2 Chun, C. (2021). A comprehensive guide to breast cancer. *Healthline* 66 (1): 1–20. https://www.healthline.com/health/breast-cancer.
3 Bruce, D.F. (2019). Invasive breast cancer: symptoms, treatments, prognosis. *webmd* 1–5.
4 Breastcancer.org. (2018). Non-invasive or invasive breast cancer. Https://Www .Breastcancer.Org/Symptoms/Diagnosis/Invasive, pp. 2–4. https://www.breastcancer .org/symptoms/diagnosis/invasive %0Afile:/// C:/Users/pc/Downloads/Documents/ RESEARCH BANK/types of BCA/non inva.pdf.
5 Feng, Y., Spezia, M., Huang, S. et al. (2018). Breast cancer development and progression: risk factors, cancer stem cells, signaling pathways, genomics, and molecular pathogenesis. *Genes Dis.* 5 (2): 77–106. https://doi.org/10.1016/j.gendis.2018.05.001.

6 Snowdon, C.T. and Ziegler, T.E. (2009). Reproductive hormones. *Handb. Psychophysiol.* 319–346. https://doi.org/10.1017/cbo9780511546396.014.

7 Javed, S. and Lteif, A. (2013). Development of the human breast. *Semin. Plast. Surg.* 27: 5–12.

8 Beatty, T. (2013). Everything you need to know about UDIs. *Med. Device Diagnostic Ind.* Nov: 1–9. https://doi.org/10.1037/a0038309.

9 Khan, S.A. (2020). Progesterone exposure and breast cancer risk-addressing barriers. *JAMA Netw. Open* 3 (4): e203608. https://doi.org/10.1001/jamanetworkopen .2020.3608.

10 Nagy, B., Szekeres-Barthó, J., Kovács, G.L. et al. (2021). Key to life: physiological role and clinical implications of progesterone. *Int. J. Mol. Sci.* 22 (20): https://doi .org/10.3390/ijms222011039.

11 Chen, C., Gong, X., Yang, X. et al. (2019). The roles of estrogen and estrogen receptors in gastrointestinal disease (Review). *Oncol. Lett.* 18 (6): 5673–5680. https://doi.org/10.3892/ol.2019.10983.

12 Yaşar, P., Ayaz, G., User, S.D. et al. (2017). Molecular mechanism of estrogen–estrogen receptor signaling. *Reprod. Med. Biol.* 16 (1): 4–20. https://doi.org/10.1002/ rmb2.12006.

13 Björnström, L. and Sjöberg, M. (2005). Mechanisms of estrogen receptor signaling: convergence of genomic and nongenomic actions on target genes. *Mol. Endocrinol.* 19 (4): 833–842. https://doi.org/10.1210/me.2004-0486.

14 Sommer, S. and Fuqua, S.A.W. (2001). Estrogen receptor and breast cancer. *Semin. Cancer Biol.* 11 (5): 339–352. https://doi.org/10.1006/scbi.2001.0389.

15 Horwitz, K.B. and Mcguire, W.L. (1975). *Specific Progesterone Receptors in Human Breast Cancer.* Elsevier.

16 Graham, J.D. and Clarke, C.L. (1997). Physiological action of progesterone in target tissues. *Endocr. Rev.* 18: 502–519.

17 Saule, S., Lagrou, C., Rommens, C. et al. (1979). Three new types of viral oncogene of cellular origin specific for haematopoietic cell transformation. *Macmillan J. Ltd.* 281 (11): 452–455.

18 Nuciforo, P., Radosevic-Robin, N., Ng, T., and Scaltriti, M. (2015). Quantification of HER family receptors in breast cancer. *Breast Cancer Res.* 17 (1): 1–12. https://doi .org/10.1186/s13058-015-0561-8.

19 Yarden, Y. (2001). Biology of HER2 and its importance in breast cancer. *Oncology* 61 (Suppl 2): 1–13.

20 American Cancer Society (2021). "Understanding a Breast Cancer Diagnosis, Atlanta.," Https://Www.Cancer.Org/Cancer/Breast-Cancer/About/Types-of-Breast-Cancer.Html#References, pp. 1–36.

21 Badve, S. and Nakshatri, H. (2009). Oestrogen-receptor-positive breast cancer: towards bridging histopathological and molecular classifications. *J. Clin. Pathol.* 62: 6–12. https://doi.org/10.1136/jcp.2008.059899.

22 Cleator, S.J., Ahamed, E., Coombes, R.C., and Palmieri, C. (2009). A 2009 update on the treatment of patients with hormone receptor-positive breast cancer. *Clin. Breast Cancer* 9 (Suppl 1): S6–S17. https://doi.org/10.3816/CBC.2009.s.001.

23 Nahleh, Z. (2008). Androgen receptor as a target for the treatment of hormone receptor-negative breast cancer: an unchartered territory. *Future Med.* 4: 15–21.

24 Phillips, T., Murray, G., Wakamiya, K. et al. (2007). Development of standard estrogen and progesterone receptor immunohistochemical assays for selection of patients for antihormonal therapy. *Appl. Immunohistochem Mol. Morphol.* 15 (3): 325–331.

25 Kurosumi, M. (2007). Immunohistochemical Assessment of Hormone Receptor Status Using a New Scoring System (J-Score) in Breast Cancer. Springer

26 Paik, S., Shak, S., Tang, G. et al. (2004). "A multigene assay to predict recurrence of tamoxifen-treated, node-negative breast cancer. *N. Engl. J. Med.* 351 (27): 2817–2826.

27 Lim, E., Metzger-filho, O., Winer, E.P. et al. (2012). The natural history of hormone receptor–positive breast cancer. *Oncology* 26 (8): 688–694.

28 Krauss, K. and Stickeler, E. (2020). Endocrine therapy in early breast cancer. *Breast Care* 15 (4): 337–346. https://doi.org/10.1159/000509362.

29 National Cancer Institute (2021). Hormone therapy for breast cancer what are hormones and hormone receptors? pp. 1–10. https://www.cancer.gov/types/breast/breast-hormone-therapy-fact-sheet.

30 Lewis, J.S. and Jordan, V.C. (2005). Selective estrogen receptor modulators (SERMs): mechanisms of anticarcinogenesis and drug resistance. *Mutat. Res.* 591: 247–263. https://doi.org/10.1016/j.mrfmmm.2005.02.028.

31 Smith, I.E. and Dowsett, M. (2003). Aromatase inhibitors in breast cancer. *NEJM* 2431–2442.

32 Miller, W.R. (2003). Aromatase inhibitors: mechanism of action and role in the treatment of breast cancer breast cancer. *Semin. Oncol.* 7754 (04): 3–11. https://doi.org/10.1016/S0093-7754(04)00302-6.

33 Hussain, S.A., Williams, S., Stevens, A., and Rea, D.W. (2004). Endocrine therapy for early breast cancer. *Expert Rev. Anticancer Ther.* 4 (5): 877–888. https://doi.org/10.1586/14737140.4.5.877.

34 Wittmann, B.M., Sherk, A., and McDonnell, D.P. (2007). Definition of functionally important mechanistic differences among selective estrogen receptor down-regulators. *Cancer Res.* 67 (19): 9549–9560. https://doi.org/10.1158/0008-5472.CAN-07-1590.

35 Willson, T.M., Henke, B.R., Momtahen, T.M. et al. (1994). 3-[4-(1,2-diphenylbut-1-enyl)phenyl]acrylic acid: a non-steroidal estrogen with functional selectivity for bone over uterus in rats. *J. Med. Chem.* 37: 1550–1552.

36 Wang, D.Y., Jiang, Z., Ben-David, Y. et al. (2019). Molecular stratification within triple-negative breast cancer subtypes. *Sci. Rep.* 9 (1): 1–10. https://doi.org/10.1038/s41598-019-55710-w.

37 Liu, S.V., Liu, S., and Pinski, J. (2011). Luteinizing hormone-releasing hormone receptor targeted agents for prostate cancer. *Expert Opin. Investig. Drugs* 20 (6): 769–778. https://doi.org/10.1517/13543784.2011.574611.

38 Lehmann, B.D., Bauer, J.A., Chen, X. et al. (2011). Identification of human triple-negative breast cancer subtypes and preclinical models for selection of targeted therapies. *J. Clin. Invest.* 121 (7): 2750–2767. https://doi.org/10.1172/JCI45014.

39 Griffiths, C.L. and Olin, J.L. (2012). Triple negative breast cancer: a brief review of its characteristics and treatment options. *J. Pharm. Pract.* 25 (3): 319–323. https://doi.org/10.1177/0897190012442062.

40 Wahba, H.A. and El-Hadaad, H.A. (2015). Current approaches in treatment of triple-negative breast cancer. *Cancer Biol. Med.* 12 (2): 106–116. https://doi.org/10.7497/j.issn.2095-3941.2015.0030.

41 Anders, C.K., Abramson, V., Tan, T., and Dent, R. (2016). The evolution of triple-negative breast cancer: from biology to novel therapeutics. *Am. Soc. Clin. Oncol. Educ. B.* 36: 34–42. https://doi.org/10.14694/edbk_159135.

42 Dent, R., Trudeau, M., Pritchard, K.I. et al. (2007). Triple-negative breast cancer: clinical features and patterns of recurrence. *Clin. Cancer Res.* 13 (15): 4429–4434. https://doi.org/10.1158/1078-0432.CCR-06-3045.

43 Yeh, J., Chun, J., Schwartz, S. et al. (2017). Clinical characteristics in patients with triple negative breast cancer. *Int. J. Breast Cancer* 2017: 21–23. https://doi.org/10.1155/2017/1796145.

44 Krizmanich-Conniff, K., Paramagul, C., Patterson, S.K. et al. (2012). Triple receptor-negative breast cancer: imaging and clinical characteristics. *AJR Am. J. Roentgenol.* 199 (2): 458–464.

45 Hubalek, M. and Müller, H. (2017). Biological subtypes of triple-negative breast cancer. *Breast Care* 12: 8–14. https://doi.org/10.1159/000455820.

46 Lehmann, B.D. (2011). Identification of human triple-negative breast cancer subtypes and preclinical models for selection of targeted therapies. *J. Clin. Invest.* 121 (7): 1–18. https://doi.org/10.1172/JCI45014DS1.

47 Dias, K., Dvorkin-Gheva, A., Hallett, R.M. et al. (2017). Claudin-low breast cancer; clinical & pathological characteristics. *PLoS One:* 1–17. https://doi.org/10.1371/journal.pone.0168669.

48 Burstein, M.D., Tsimelzon, A., Poage, G.M. et al. (2015). Comprehensive genomic analysis identifies novel subtypes and targets of triple-negative breast cancer. *Clin. Cancer Res.* 21 (7): 1688–1698. https://doi.org/10.1158/1078-0432.CCR-14-0432. Epub 2014 Sep 10. PMID: 25208879; PMCID: PMC4362882.

49 Herrera Juarez, M., Tolosa Ortega, P., Sanchez De Torre, A., and Ciruelos Gil, E. (2020). Biology of the triple-negative breast cancer: immunohistochemical, RNA, and DNA features. *Breast Care* 15 (3): 208–216. https://doi.org/10.1159/000508758.

50 Miteva-marcheva, N.N., Ivanov, H.Y., Dimitrov, D.K., and Stoyanova, V.K. (2020). Application of pharmacogenetics in oncology. *Biomark. Res.* 4: 1–10.

51 Hicks, J., Swen, J., and Gaedigk, A. (2014). Challenges in CYP2D6 phenotype assignment from genotype data: a critical assessment and call for standardization. *Curr. Drug Metab.* 15 (2): 218–232. https://doi.org/10.2174/1389200215666140202215316.

52 Silver, D.P., Richardson, A.L., Eklund, A.C. et al. (2010). Efficacy of neoadjuvant cisplatin in triple-negative breast cancer. *J. Clin. Oncol.* 28 (7): 1145–1153. https://doi.org/10.1200/JCO.2009.22.4725.

53 Loi, S., Pommey, S., Haibe-Kains, B. et al. (2013). CD73 promotes anthracycline resistance and poor prognosis in triple-negative breast cancer. *Proc. Natl. Acad. Sci. U. S. A.* 110 (27): 11091–11096. https://doi.org/10.1073/pnas.1222251110.

54 Balko, J.M., Giltnane, M., Schwarz, K.J. et al. (2014). Actionable therapeutic targets. *Cancer Discov.* 4 (2): 232–245. https://doi.org/10.1158/2159-8290.CD-13-0286 .Molecular.

55 Slanař, O., Hronová, K., Bartošová, O., and Šíma, M. (2021). Recent advances in the personalized treatment of estrogen receptor-positive breast cancer with tamoxifen: a focus on pharmacogenomics. *Expert Opin. Drug Metab. Toxicol.* 17 (3): 307–321. https://doi.org/10.1080/17425255.2021.1865310.

56 Westbrook, K. and Stearns, V. (2013). Pharmacology & therapeutics pharmacogenomics of breast cancer therapy: an update. *Pharmacol. Ther.* 139 (1): 1–11. https://doi.org/10.1016/j.pharmthera.2013.03.001.

57 ClinicalTrials.gov Search Results (accessed 07 May 2022).

11

Donor–Acceptor-Based Heterocyclic Compounds as Chemotherapy and Photothermal Agents in Treatment of Breast Cancer Cell

Vinayak M. Adimule[1], Sheetal R. Batakurki[2], Maya M. Pai[2] and Santosh Nandi[3]

[1] *Angadi Institute of Technology and Management (AITM), Department of Chemistry, Savagaon Road, Belagavi, Karnataka, 590009, India*
[2] *M. S. Ramaiah University of Applied Sciences, Department of Chemistry, Bangalore, Karnataka, 560054, India*
[3] *KLE Technological University Dr. M. S. Sheshgiri College of Engineering and Technology, Chemistry Section, Department of Engineering Science and Humanities, Angol Main Road, Udyambag, Belagavi 590008, Karnataka, India*

11.1 Introduction

Breast cancer has been one of the major causes in the women with 50 plus years of age. As per the WHO report around 7.8 million women are diagnosed with breast cancer and reports indicate that 2.7 million mortalities are recorded for the year 2020 [1]. The major causes of the breast cancer are obesity, post and premenopausal hormone effects, exposure to longtime electromagnetic radiations, alcohol and tobacco dependence, irregular menstrual cycles, poor nutritious food, physical activity, etc. [2–7]. Breast cancer cell lines are of different types depending upon the location of cancerous cell lines [8]. Figure 11.1 indicates the types of breast cancer. Treatment of the breast cancer included both noninvasive techniques and invasive techniques [9–11] that depends upon the type of cancer cell line, infectious stage, site of cancer cell lines, premedical conditions of the patients, etc. Treatment includes radiation, hormonal, targeted, radio, immune, and chemotherapy. All these therapies involve either inorganic compounds or organic molecules containing oxygen, nitrogen, and sulfur. Many heterocyclic compounds [12] possessing hydrogen bond donors and acceptors substituents for the privileged molecules such as indole, quinoline, and arylpiperazines are used for photothermal agents, chemotherapeutical agents, progesterone inhibitors, estrogen enhancers, etc. [13–16]. Doxorubicin, paclitaxel, mifamurtide, vincristine, and cytarabine are approved

Drug and Therapy Development for Triple Negative Breast Cancer, First Edition.
Edited by Pravin Kendrekar, Vinayak Adimule, and Tara Hurst.
© 2023 WILEY-VCH GmbH. Published 2023 by WILEY-VCH GmbH.

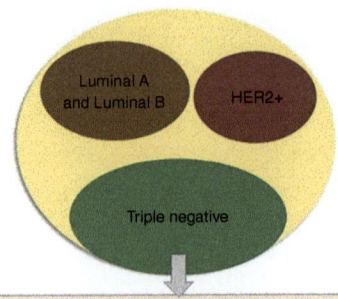

Figure 11.1 Types of breast cancer and different types of triple-negative cancer cell lines.

active ingredients used as anticancer agents to treat the breast cancer cells that are approved by FDA, EMA, NMPA, and KFDA [17–20]. Derivatives of these active ingredients or addition of certain linkers having hydrophobic and hydrophilic to increase the solubility of drug, bioavailability, biocompatibility, reduce the side effects, etc., are reported in the literature [21].

11.2 Causes for Breast Cancer

From many years identification of breast cancer with its risk factors in women has become the talk of the town. However, until recently, the reason for smoking cigarettes was not thought to be a factor. Researchers are looking for a comparable factor for the rise in the number of female smokers after seeing an increase in the prevalence of lung and breast cancer in women in recent decades. Researchers have been studying the link between smoking and breast cancer for nearly 20 years, but only 22 articles were published by the late 1980s [22]. A weak relationship, a lack of relationship, a supportive influence, depression, insomnia, etc. have all been proposed by various research [23–26]. Women who were exposed to cigarette smoke as children or who were married to smokers are more likely to get breast cancer. Breast cancer risk increases with earlier smoking initiation. Breast cancer risk increased in women who began smoking between the ages of 10 and 14 years. Women with a family history of breast cancer, ovarian cancer, or both are more likely to develop the disease [27]. Nicotine present as the major compound in the cigarette is reported to trigger the breast cancer [28] and also suppress the treatment responses to many standard anticancer agents [27, 29, 30].

11.3 Imaging and Screening of Breast Cancer

Breast cancer therapy success depends on an early diagnosis. T3 tumors, which are largely the result of delayed diagnosis, have a 10-year survival of less than 60% while T1 tumors with a size of less than 2 cm have an average 85% 10-year survival rate [31].

Breast cancer is ultimately found using four diagnostic techniques: clinical examination, mammography, breast ultrasound, and breast magnetic resonance imaging (MRI) [32–34]. Today artificial intelligence, machine learning, and deep learning algorithms have provided a much faster, safer, and high accuracy data in identifying the stages of breast cancer [35–37]. The easiest method of diagnosis is clinical examination [38]. It is a straightforward method of early detection that can identify cancers measuring 1–2 cm or more, depending on their location and breast size [39]. It continues to

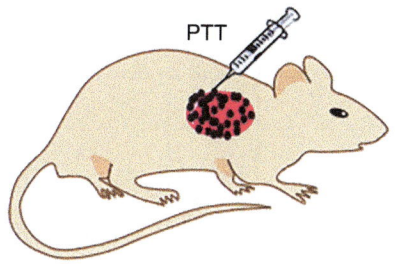

Figure 11.2 Illustration of PTT induced inhibition of cancer growth.

be the most typical method for discovering breast malignancies for the first time by the affected woman. But a breast exam presupposes an in-depth knowledge of progressive extent of breast cancer [40] and moreover in many instances the clinical examination may take long way leading to depression in patients [41]. Recently, high-power lasers are necessary for effective tumor ablation but employing them also raises questions regarding safety and the cost of the necessary equipment. These limitations have hindered the use of Photothermal therapy (PTT) devices in clinical settings. Figure 11.2 illustrates the PTT-induced inhibition of cancer growth on mice model. There has been a lot of research done to enhance the photothermal effects and tumor targetability of PTT agents and get rid of the biological obstacles to treating cancer [42]. The simultaneous use of PDT or PTT with nanoparticles in the treatment of breast cancer has been investigated by many research teams. Our research focus is on how to create simple compounds with excellent PTT performance.

11.4 Photothermal Therapy (PTT)

In recent years use of PTT is widely preferred noninvasive technique to treat the cancerous cell lines, developed by Li and coworkers [43]. PTT involves the use of optical radiation of 700–2000 nm NIR wavelength range to generating the heat upto 41–47 °C, to kill cancerous cells and clinically this is termed as hyperthermia. The greater advantage of phototherapy is to irradiate primary and secondary lymph node of initial phase of metastatic stage. Several photosensitizers such as smaragdyrin, indocyanine green (ICG), polypyrrole, heptamethine cyanine, methyl-β-cyclodextrin, phinazine cyanine, pyrazino 2,3-g quinoxaline, triphenylamine-perylene diimide, diketopyrrole poly phenyl amine, and 2-dicyanomethylenethiazole [44–53] are well reported in literature and are indicated in Table 11.1. Inspite of several advantages, there are certain limitations such as nonspecificity of the photothermal agents and all physicochemical parameters including pH and temperature. The first-generation PTT agents were inorganic materials, and metals with high conductivity were used. Due to low biocompatibility, toxicity, and target specificity and to increase the efficiency of heat conversion, surface modifications of inorganic materials were carried out [54–56]. Surface modifications of the inorganic materials are carried out by using biocompatible polymers such as poly(lactic acid) [57], polyethylene glycol [58], poly(lactic-*co*-glycolic

Table 11.1 Few of the photosensitizers that are screened against various breast cancer cell lines.

Chemical structure	Heterocycle	Breast cancer cell line	Activity	Type	References
	Smaragdyrin	MDAMB-231 cell	IC$_{50}$ cell viability ~90%	PT	[44]
	Indocyanine green	M4M2	ICG (100 μM)	PTT	[45]
	Polypyrrole	MCF-7	Cell viability down to 56.6% at 100 μg ml^{-1} The photocytotoxicity of MPDAPPy-0.97	PTT	[46]
	Heptamethine cyanine	ES2	ICG (10 μM)	PTT	[47]

Table 11.1 (Continued)

Chemical structure	Heterocycle	Breast cancer cell line	Activity	Type	References
	Methyl-β-cyclodextrin	HTB-38	ICG-CD (730 nm)	PTT	[48]
	Polypyrrole	MCF7	PPy-80(500 μg ml^{-1})	PTT	[49]

R = CH$_3$ or H

Table 11.1 (Continued)

Chemical structure	Heterocycle	Breast cancer cell line	Activity	Type	References
	Phinazine cyanine	4T1	PH-1@HSA (200 ml, 500 mM)	PTT	[50]
	Pyrazino 2,3-g quinoxaline	4T1	NPs: PQs-TPA NPs (20 μg ml^{-1})	PQs-TPAPTT	[51]
	Triphenylamine-perylene diimide	A549	PA intensities at 635 nm of different PIT NPs concentrations (ii) THF, NaOH, TsCI, 93%	PIT	[52]

Table 11.1 (Continued)

Chemical structure	Heterocycle	Breast cancer cell line	Activity	Type	References
	Diketopyrrole poly phenyl amine	HCT-116	100 μl aliquot of DPP-TPA NPs (40 μg ml^{-1})	PTT	[53]
	2-Dicyanomethylenethiazole	4T1	TPTHM NPs (100 ml, 2.5 mg kg^{-1})/water	PTT	[50]

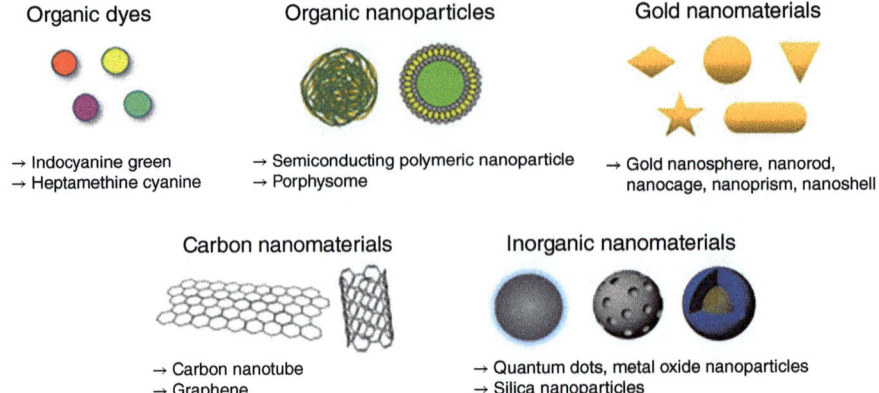

Figure 11.3 Different materials used for photothermal therapy. Source: Adapted from Adimule et al. [63].

acid) [59], (PLA) [60], poly(ε-caprolactone) (PCL) [61] poly(amino acid) (PAA) chitosan [62], and enhanced anticancer activity of drug along with easy biodegradability. Further to increase the photostability conversion efficiency, thermal stability and biodegradability of nano metal-based materials were used as PTT agents. An ideal photothermal agent would specifically target tumors and have a high photothermal conversion efficiency (PCE) without affecting biological tissue's chromophores' absorption. In recent years many nanomaterials such as organic nanoparticles, gold nanomaterials carbon nanomaterials, and inorganic nanomaterials are screened against inhibition of breast cancer cell lines [63] and are indicated in Figure 11.3.

11.5 Acceptor–Donor-Based Heterocyclic Compounds

A donor–acceptor (D–A) molecule upon exposure to light absorbs certain energy and moves from lower energy state (S_0) to a higher energy state, which is generally a singlet state (S_n), this is an unstable state and hence immediately gets converted to lowest excited singlet state (S_1) [64]. The Jablonski diagram is presented in the Figure 11.4 that depicts the PTT, these S_1 state will release energy that aid in generation of reactive oxygen species (ROS) whereas in photodynamic therapy the singlet state will jump to higher triplet state through intersystem crossing (ISC) [65]. This triplet state is also unstable and hence lands up to lower triplet state (T_1) thereby generates ROS ($^3O_2$ to ROS), which will kill the molecular oxygen present abundantly found in cancer cell. The triplet state oxygen can generate ROS through two mechanisms either through charge transfer or energy transfer. Oxygen radicals are generated through charge transfer mechanism and singlet oxygen through energy transfer. Both will aid in killing the molecular oxygen. However, the generated singlet oxygen can lead to form oxides, particularly nitric oxide, which are again harmful to the cancer cells.

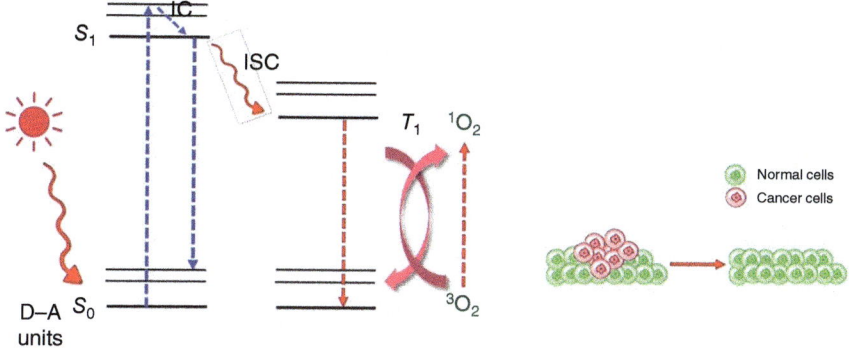

Figure 11.4 Illustration of mechanism of PDT using Jablonski diagram.

A donor–acceptor architecture of molecules is designed by selecting the appropriate molecules with high HOMO levels as donors and acceptor with low LUMO energy levels. Such donor–acceptor molecules are separated by the conjugated system that results in lowering the bandgap energy and increasing the absorption to longer wavelength [66, 67]. By varying the donors and acceptors or increasing the chain length between D–A molecules, various physical and chemical properties can be achieved that can be designed by modifying either donors or acceptors or increasing the chain length to achieve the required electronic properties that can be used as fluorescence materials and photosensitizing agents [68, 69].

Triphenylamine (TPA) as electron–donor and benzo[1,2-c:4,5-c′]bis([1,2,5]thiadiazole) (BBT) as electron acceptor are popular D–A units in constructing NIR-II fluorophores [70, 71].

Kai Sun et al. have designed a new class of electron donor–acceptor complex using aryl radicals by photoactivation of arenes via thianthrenation [72]. Further, the arylation of quinoxalin-2(1H)-one radicals were generated upon exposure to UV light through electron-transfer process through formation of EDA complex mechanism. Total of around 102 N-methylquinoxalin-2(1H)-one molecules were designed and synthesized containing electron–donating, electron-withdrawing and neutral functional groups such as –Me, –Et, –Bn, –OAc, –Ph, –OMe, –OPh, –F, –Cl, –Br, –CN, –CHO, –COOMe, –COCH$_3$, further o-vanillin, p-vanillin, zingerone, and ibuprofen were also used in arylation resulting in expected product with high yield and were screened for PTT applications. However, heterocyclic compounds nitrogen such as indole pyridine, quinoline, and oxygen-containing benzoxazole failed to be used as PTT agents.

Yu Kaiwu et al. [73] designed an usual acceptor–donor pattern of molecules TPA-2TCC and TPA-2TCP starting from TPA and (E)-2-cyano-3-(thiophen-2-yl)acrylic acid and (Z)-2-(4-(pyridin-4-yl)phenyl)-3-(thiophen-2-yl) acrylonitriles as efficient photostable photosensitizers generating ROS resulting in high competent apoptosis of 4T1 breast cancer cells. The unique pattern of acceptor–π–donor–π–acceptor (A–π–D–π–A) structure present in these two molecules may lead to practice in PDT applications.

Leilei Shi et al. [74] designed and synthesized new quinoline-based QpyNHOH molecules as histone deacetylases (HDAC) inhibitors. QpyNHOH molecules were screened against MCF-7, MDA-MB-231, and HUVEC cells and were found that QpyNHOH can be a potent candidature for the HDAC inhibitors.

Shao et al. [75] have designed donor–acceptor–donor concept using TPA with benzobis(thiadiazole) as acceptor with thiophene bridge as donor (CSM0), (CSM1) (CSM2). The synthesized molecules (CSMs) were further converted to nanoparticles by coprecipitation method using Pluronic F127 micelles. CSMN2 showed a broad adsorption between 700 and 1200 nm in NIR-I and NIR-II regions.

11.6 Examples of Organic-Based Donor–Acceptor

11.6.1 Indocyanine

ICG is an FDA-approved [76] photosensitizer for near infrared clinical imaging [59] green is widely used as chemotherapeutic agent since six decades [60]. In spite of long-term usage, ICG has certain limitations such as poor aqueous stability in vitro, concentration-dependent aggregation, rapid elimination from the body, and lack of target specificity[61] hinders the use of ICG alone. Thus, to improve the above limitations, ICT have been employed along with the polymers, such as poly-lactic acid [77], reduced graphene [78], polydopamine [79].

The delivery of the drug to the target site is the foremost requirement of any drug to perform its action. In recent years, various nanoparticles are used for efficient drug delivery, which not only aids in delivery but also increases the bioavailability of the drug, in many instances, they provide synergistic effects to inhibit the cancerous cell lines, increases the immune power of the cells, helps in healing, and energize the healthy cells and tissue surrounding the target site. Polydopamine nanoparticles were used for loading ICG [80].

Further, it was observed that reduced graphene along with polydopamine enhanced the optical absorption of ICG in NIR wavelength thereby improving complete remission of 4T1 breast subcutaneous model [76]. ICG was adsorbed on the rGO layer containing polydopamine and was synthesized via Michael addition reaction. Due to the water solubility, bioavailability properties of rGO and biocompatibility, strong adhesive properties of dopamine, well-established chemotherapeutic drug etc., all leading to the formation of polymeric layers adhere to the ICG resulting in increased photothermal effect of ICG–PDA–rGO making it as efficient material of choice for chemo-photothermal therapy.

A nano formulation of ICG with 7-Ethyl-10-hydroxycamptothecin (SN38) enhanced the π–π stacking and hydrophobic interactions and was used to inhibit the BCap37 cells. The synergistic chemo-photo therapy of ICG was achieved by self-assembling with the SN38 drug and resulting in high percentage of heat conversion efficiency compared to free ICG [81].

ICG was encapsulated in the halloysite nanotubes containing metalorganic framework of Zeolitic imidazolate framework-8 along with standard anticancer doxorubicin hydrochloride was used as photosensitizer to irradiate and kill MDA-MB-231 cells. Further, hyalouric acid was also coated with the biomaterial HNTs@ZIF-8@DOX@ICG, which increased the apoptosis of CD44-positive MDA-MB-231 cells [82].

Novel Dye Conjugate (nDC) was synthesized using 17β-estradiol with ICG dye. The dye showed photothermal absorption at 757 and 787 nm in DMSO solvent and the dye showed a specific binding affinity with ER-positive cell lines (MCF-7) whereas no specific binding site was observed with ER-negative cell lines (MDA MB 231) [83].

11.7 Polymers-Based Agents

Hybrid D–A polymer particles (H-DAPPs) are generally used for photoacoustic and fluorescence imaging due to the readily available excitons, which improves the potency of drug molecules against cancer cells. Poly[(9,9-dihexylfluorene)-co-2,1,3-benzothiadiazole-co-4,7-di(thiophen-2-yl)-2,1,3-benzothiadiazole] (PFBTDBT10) was used as fluorescent imaging along with poly[4,4-bis(2-ethylhexyl)-cyclopenta[2,1-b;3,4-b′]dithiophene-2,6- diyl-alt-2,1,3-benzoselenadiazole-4,7-diyl] (PCPDTBSe) as photosensitizers generated the heat required to kill the BALB/c CL.7 and, EO771, or 4T1 (murine breast cancers) cells [68]. Figure 11.5 depicts the hybrid D–A polymeric materials for breast cancer treatment.

11.7.1 Phthalocyanine

The most researched photosensitive was porphyrin, a planar ring molecule. It was utilized to destroy paramecium after being conjugated with a metallic iron to create haematophophylin. Tetrapyrrolic macrocycles called phthalocyanines have nitrogen atoms connecting the different pyrrole units. In 1907, a molecule of the third generation was created from porphyrin. When 1,2-cyanobenzamide's characteristics were being studied, porphyrin was conjugated to the benzene ring and connected via aza nitrogen's rather than methine carbons [84]. Due to this chemical alteration, the absorption spectrum of phthalocyanine shifts from 630 nm (for its precursor porphyrin) to roughly 680 nm [85]. With enhanced spectroscopic and photochemical capabilities, phthalocyanine replicates the biological significance of the precursor molecule porphyrin and so boosts its specificity for cancer targets. The phthalocyanine molecule contains a core cavity that can house 63 different elemental ions, improved tissue penetration, and an intense absorption band in the red or near-infrared portion of the electromagnetic spectrum, among other special properties [86]. The phthalocyanine ring's spectral, photophysical, and photochemical behavior is improved by the

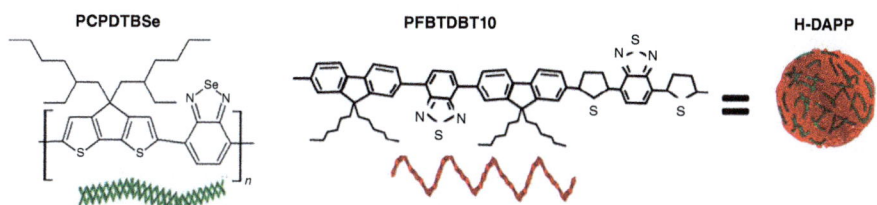

Figure 11.5 Hybrid donor–acceptor polymeric materials for breast cancer treatment. Source: Graham-Gurysh et al. [68].

presence of substituents. As a result, fuming sulfuric acid is occasionally replaced for phthalocyanine photosensitizer in photodynamic therapy to improve its ability to cross the water-based circulatory system and other organic solvents [87].

Phthalocyanines that contain sulfur have demonstrated improved tissue permeability, optimal potency, and reduced toxicity [88]. Phthalocyanine is a versatile chemical with applications outside of biology. Due to its capacity to combine with metal cations to generate colorful complexes, it can also be utilized industrially. Because of its great photochemical stability, phthalocyanine was utilized as a light-resistant pigment dye in the textile and paper industries as early as the 1930s. Its expanded applications include serving as an electronic catalyst, a photosensitizer in photodynamic therapy, and a photoconducting agent in photocopiers [89]. Also used commercially are recording layer dyes for CD-R and DVD-R optical storage discs, photovoltaic materials for solar cells, photoconductors for xenography, functional polymers, liquid crystals, non-optics, and nanotechnology. These applications were all made possible by their distinctive optical and redox characteristics, which may be adjusted by applying the necessary peripheral substitutions [90]. When diamagnetic metals like zinc, silicon, or aluminum are included into the macrocycle of phthalocyanine, it improves the compound's triplet state and singlet oxygen quantum yield when exposed to visible light. Due to their innate fluorescence capabilities, photosensitizers are beneficial for more than just singlet oxygen generation in the treatment of cancer [91]. They can also be used for neoplastic tissue diagnostics and in situ dosimetry [89, 92].

11.8 Conclusion

Over the past 20 years, there have been significant advancements in the diagnosis and treatment of breast cancer. The development of photodynamic therapy, which uses the administration of a photosensitizer and light of a particular wavelength, has been facilitated by extensive study into the treatment of breast cancer. This therapy is more targeted and less harmful. The photosensitizer is activated by light, which sets off a sequence of photobiological and photochemical processes that result in cancer cell destruction. Various donor–acceptor heterocyclic compounds are being synthesized and used as photosensitizers, photodynamic therapeutic agents, and chemotherapeutical agents. Termination of molecular oxygen present inside the cancer cells can be terminated by the release of ROS by the small organic D–A molecules. Introduction of different types of electron donor and acceptors bridge by small heterocyclic moieties can increase the wavelength absorption in NIR-I and NIR-II regions leading to more efficient photosensitizers in PTT and PDT applications. However, with several examples of breast cancer relapse and medication resistance, single agents for therapy have successfully demonstrated that they are insufficient due to the heterogeneity of metastatic breast cancer. Combination therapy and its synergistic effect may enhance the apoptosis of breast cancer cells with decreased toxicity, which may be a novel technique to minimize these negative consequences. This review article gives an insight in designing new small organic donor–acceptor heterocyclic moieties that can serve as photosensitizers as well as chemotherapeutical agents leading to combined effect in stopping the breast cancer cells.

References

1 Hamed, G., Marey, M.A., Amin, S.E., and Tolba, M.F. (2020). Deep learning in breast cancer detection and classification. In: *The International Conference on Artificial Intelligence and Computer Vision* (8 April 2020), pp. 322–333. Cham: Springer.

2 Hjerkind, K.V., Johansson, A.L., Trewin, C.B. et al. (2022). Incidence of breast cancer subtypes in immigrant and non-immigrant women in Norway. *Breast Cancer Res.* 24 (1): 1–4.

3 Elovainio, M., Komulainen, K., Lipsanen, J. et al. (2022). Long-term cumulative light exposure from the natural environment and sleep: a cohort study. *J. Sleep Res.* 31 (3): e13511.

4 Ren, Q., Luo, F., Ge, S., and Chen, P. (2023). Major depression disorder may causally associate with the increased breast cancer risk: evidence from two-sample Mendelian randomization analyses. *Cancer Med.* 12 (2): 1984–1996.

5 Soldato, D., Havas, J., Crane, T.E. et al. (2022). Coffee and tea consumption, patient-reported, and clinical outcomes in a longitudinal study of patients with breast cancer. *Cancer* 128 (19): 3552–3563.

6 Mohan Gift, M.D., Pattnaik, B., Nandi, S.S. et al. (2022). Determination of prohibition mechanism of cationic polymer/SiO_2 composite as inhibitor in water using drilling fluid. *Mater. Today Proc.* https://doi.org/10.1016/j.matpr.2022.08.171.

7 Keri, P.B., R.S., Adimule, V., Kendrekar, P. et al. (2022). The nano-based catalyst for the synthesis of benzimidazoles. *Top. Catal.* https://doi.org/10.1007/s11244-022-01562-0.

8 Tan, P.H., Ellis, I., Allison, K. et al. (2020). WHO classification of tumours editorial board the 2019 World Health Organization classification of tumours of the breast. *Histopathology* 77 (2): 181–185.

9 Vinayak, A., Santosh, N., Basappa, Y. et al. (2021). Enhanced photoluminescence properties of $Gd_{(x-1)}$ Srx O: CdO nanocores and their study of optical, structural, and morphological characteristics. *Mater. Today Chem.* 20 (3): 100438.

10 Ma, X., Gong, J., Hu, F. et al. (2023). Pretreatment multiparametric MRI-based radiomics analysis for the diagnosis of breast phyllodes tumors. *J. Magn. Reson. Imaging* 57 (2): 633–645.

11 Dahan, M., Cortet, M., Lafon, C., and Padilla, F. (2022). Combination of focused ultrasound, immunotherapy, and chemotherapy: new perspectives in breast cancer therapy. *J. Ultrasound Med.* 42(3): 559–573.

12 Adimule, V., Nandi, S.S., and Yallur, B.C. (2022). Devices and sensors based on additively manufactured shape-memory of hybrid nanocomposites. In: *Shape Memory Composites Based on Polymers and Metals for 4D Printing* (ed. M.R. Maurya, K.K. Sadasivuni, J.J. Cabibihan, et al.). Cham: Springer https://doi.org/10.1007/978-3-030-94114-7_15.

13 Atmaca, H., Ilhan, S., Korkmaz, E., and Zora, M. (2022). Endoplasmic reticulum stress-induced apoptotic effects of novel 1-pyrroline (3,4-dihydro-2H-pyrrole) derivatives on breast cancer cells. *Chem. Biodivers.* 19 (7): e202200123.

14 El Rabeeb, S.I., El Deeb, M.A., Sarg, M.T., and Hassan, A.Y. (2022). Imidazo [1,2,4] triazolone and fused imidazo [1,2,4] triazolone derivatives: synthesis, in vitro anticancer screening, CDK2 inhibitory activity, and molecular modeling studies. *J. Heterocycl. Chem.* 59 (12): 2207–2224.

15 Dutta, K., Majumdar, A.G., Kushwah, N. et al. (2022). Synthesis of novel indole-oxadiazole molecular hybrids by a regioselective C-3 sulfenylation of indole with 1,3,4-oxadiazole-2-thiols using iodine-dimethyl sulfoxide and their anticancer properties. *J. Heterocycl. Chem.* 59(12): 2165–2176.

16 Singh, R., Kumar, R., Pandrala, M. et al. (2021). Facile synthesis of C6-substituted benz [4,5] imidazo [1,2-a] quinoxaline derivatives and their anticancer evaluation. *Arch. Pharm.* 354 (7): 2000393.

17 Adimule, V., Yallur, B.C., Batakurki, S.R., and Nandi, S.S. (2022). Synthesis, morphology and enhanced optical properties of novel $Gd_xCo_3O_4$ nanostructures. *Adv. Mater. Res.* 1173: 71–82. https://doi.org/10.4028/p-3pkhf6.

18 Hwang, T.J., Kesselheim, A.S., Tibau, A. et al. (2022). Clinical benefit and expedited approval of cancer drugs in the United States, European Union, Switzerland, Japan, Canada, and Australia. *JCO Oncol. Pract.* 18 (9): e1522–e1532.

19 Nandi, S.S., Adimule, V., and Yallur, B.C. (2022). Synthesis, structural and optical properties of co doped Sm_2O_3 nanostructures. *Adv. Mater. Res.* 1173: 59–69. Trans Tech Publications, Ltd.

20 Wang, W. and Sun, Q. (2020). Novel targeted drugs approved by the NMPA and FDA in 2019. *Signal Transduct. Target Ther.* 5 (1): 65. https://doi.org/10.1038/s41392-020-0164-4. PMID: 32385226; PMCID: PMC7205905.

21 Shaikh, N.M., Adimule, V., Bagihalli, G.B. et al. (2022). A novel mixed Ag–Pd nanoparticles supported on SBA silica through [DMAP-TMSP-DABCO]OH basic ionic liquid for Suzuki coupling reaction. *Top. Catal.* https://doi.org/10.1007/s11244-022-01586-6.

22 Malvezzi, M., Carioli, G., Bertuccio, P. et al. (2019). European cancer mortality predictions for the year 2019 with focus on breast cancer. *Ann. Oncol.* 30 (5): 781–787.

23 Adimule, V., Yallur, B.C., Pai, M.M. et al. (2022). Biogenic synthesis of magnetic palladium nanoparticles decorated over reduced graphene oxide using *Piper betle* petiole extract (Pd-rGO@Fe_3O_4 NPs) as heterogeneous hybrid nanocatalyst for applications in Suzuki-Miyaura coupling reactions of biphenyl compounds. *Top. Catal.* 1–14.

24 Di Meglio, A., Gbenou, A.S., Martin, E. et al. (2021). Unhealthy behaviors after breast cancer: capitalizing on a teachable moment to promote lifestyle improvements. *Cancer* 127 (15): 2774–2787.

25 Bean, H.R., Diggens, J., Ftanou, M. et al. (2021). Insomnia and fatigue symptom trajectories in breast cancer: a longitudinal cohort study. *Behav. Sleep Med.* 19 (6): 814–827.

26 Woldeamanuel, Y.W., Blayney, D.W., Jo, B. et al. (2021). Headache outcomes of a sleep behavioral intervention in breast cancer survivors: secondary analysis of a randomized clinical trial. *Cancer* 127 (23): 4492–4503.

27 Adimule, V., Medapa, S., Rao, P.K., and Kumar, L.S. (2014). Synthesis of Schiff bases of 5-[5-(4-fluorophenyl) thiophen-2-yl]-1,3,4-thiadiazol-2-amine and its anticancer activity. *Int. J. Adv. Pharm. Sci. 5* (1): 1761–1768.

28 Khodabandeh, Z., Valilo, M., Velaei, K., and Pirpour Tazehkand, A. (2022). The potential role of nicotine in breast cancer initiation, development, angiogenesis, invasion, metastasis, and resistance to therapy. *Breast Cancer* 29 (5): 778–789.

29 Adimule, V., Batakurki, S., Yallur, B.C. et al. (2022). Enhanced photoluminescence, optical, structural properties of ZrO_2-incorporated Sm_2O_3:Co_3O_4 nanocomposite and their applications in photocatalytic degradation of methylene blue. *J. Mater. Res.* 37: 2396–2405.

30 Aebi, S., Davidson, T., Gruber, G., and Cardoso, F. (2011). Primary breast cancer: ESMO clinical practice guidelines for diagnosis, treatment and follow-up. *Ann. Oncol.* 22: vi12–vi24.

31 Becker, S. (2015). A historic and scientific review of breast cancer: the next global healthcare challenge. *Int. J. Gynecol. Obstet.* 131: S36–S39.

32 He, Z., Chen, Z., Tan, M. et al. (2020). A review on methods for diagnosis of breast cancer cells and tissues. *Cell Prolif.* 53 (7): e12822.

33 Ginsburg, O., Yip, C.H., Brooks, A. et al. (2020). Breast cancer early detection: a phased approach to implementation. *Cancer* 15 (126): 2379–2393.

34 Menko, F.H., Monkhorst, K., Hogervorst, F.B. et al. (2022). Challenges in breast cancer genetic testing. A call for novel forms of multidisciplinary care and long-term evaluation. *Crit. Rev. Oncol. Hematol.* 5: 103642.

35 Batakurki, S.R., Adimule, V., Pai, M.M. et al. (2022). Synthesis of Cs-Ag/Fe_2O_3 nanoparticles using *Vitis labrusca* rachis extract as green hybrid nanocatalyst for the reduction of arylnitro compounds. *Top. Catal.* 1–14. https://doi.org/10.1007/s11244-022-01593-7.

36 Boeri, C., Chiappa, C., Galli, F. et al. (2020). Machine learning techniques in breast cancer prognosis prediction: a primary evaluation. *Cancer Med.* 9 (9): 3234–3243.

37 Allugunti, V.R. (2022). Breast cancer detection based on thermographic images using machine learning and deep learning algorithms. *Int. J. Eng. Comput. Sci.* 4 (1): 49–56.

38 McCart Reed, A.E., Kalita-De Croft, P., Kutasovic, J.R. et al. (2019). Recent advances in breast cancer research impacting clinical diagnostic practice. *J. Pathol.* 247 (5): 552–562.

39 Plewes, D.B., Bishop, J., Samani, A., and Sciarretta, J. (2000). Visualization and quantification of breast cancer biomechanical properties with magnetic resonance elastography. *Phys. Med. Biol.* 45 (6): 1591.

40 Shaikh, N.M., Sawant, A.D., Bagihalli, G.B. et al. (2022). Highly active mixed Au–Pd nanoparticles supported on RHA silica through immobilised ionic liquid for Suzuki coupling reaction. *Top. Catal.* 1–10.

41 Vinayak, A., Sudha, M., Lalita, K.S., and Kumar, R.P. (2015). Synthesis, characterization and cytotoxic evaluation of novel derivatives of 1-[2-(aryl substituted)-5-(4′-Fluoro-3-methyl biphenyl-4-yl)-[1,3,4] oxadiazole-3-yl]-ethanone. *Arch. Appl. Sci. Res. 7* (5): 4–8.

42 Liu, Y., Bhattarai, P., Dai, Z., and Chen, X. (2019). Photothermal therapy and photoacoustic imaging via nanotheranostics in fighting cancer. *Chem. Soc. Rev.* 48 (7): 2053–2108.

43 Li, B., Hao, G., Sun, B. et al. (2020). Engineering a therapy-induced "immunogenic cancer cell death" amplifier to boost systemic tumor elimination. *Adv. Funct. Mater.* 30 (22): 1909745.

44 Laxman, K., Reddy, B.P., Robinson, A. et al. (2020). Cell-penetrating peptide-conjugated BF2-oxasmaragdyrins as NIRF imaging and photothermal agents. *ChemMedChem* 15 (19): 1783–1787.

45 Fan, H., Chen, S., Du, Z. et al. (2022). New indocyanine green therapeutic fluorescence nanoprobes assisted high-efficient photothermal therapy for cervical cancer. *Dyes Pigm.* 1 (200): 110174.

46 Huang, J., Wang, S., Xing, Y. et al. (2019). Interface-hybridization-enhanced photothermal performance of polypyrrole/polydopamine heterojunctions on porous nanoparticles. *Macromol. Rapid Commun.* 40 (19): 1900263.

47 St. Lorenz, A., Buabeng, E.R., Taratula, O. et al. (2021). Near-infrared heptamethine cyanine dyes for nanoparticle-based photoacoustic imaging and photothermal therapy. *J. Med. Chem.* 64 (12): 8798–8805.

48 Kadapure, S.A., Kadapure, P., Nandi, S., and Shet, A. Overview on catalyst and co-solvents for sustainable biodiesel production. *Proc. Inst. Civ. Eng. Energy* 1–9.

49 Bhattarai, D.P., Tiwari, A.P., Maharjan, B. et al. (2019). Sacrificial template-based synthetic approach of polypyrrole hollow fibers for photothermal therapy. *J. Colloid Interface Sci.* 15 (534): 447–458.

50 Yan, Y., Chen, J., Yang, Z. et al. (2018). NIR organic dyes based on phenazine-cyanine for photoacoustic imaging-guided photothermal therapy. *J. Mater. Chem. B* 6 (45): 7420–7426.

51 Zhang, L.P., Kang, L., Li, X. et al. (2021). Pyrazino [2,3-g] quinoxaline-based nanoparticles as near-infrared phototheranostic agents for efficient photoacoustic-imaging-guided photothermal therapy. *ACS Appl. Nano Mater.* 4 (2): 2019-29.

52 Li, H., Yue, L., Li, L. et al. (2021). Triphenylamine-perylene diimide conjugate-based organic nanoparticles for photoacoustic imaging and cancer phototherapy. *Colloids Surf., B* 205: 111841.

53 Cai, Y., Liang, P., Tang, Q. et al. (2017). Diketopyrrolopyrrole–triphenylamine organic nanoparticles as multifunctional reagents for photoacoustic imaging-guided photodynamic/photothermal synergistic tumor therapy. *ACS Nano* 11 (1): 1054–1063.

54 Adimule, V.M., Nandi, S.S., Kerur, S.S. et al. (2022). Recent advances in the one-pot synthesis of coumarin derivatives from different starting materials using nanoparticles: a review. *Top. Catal.* 1–31.

55 Vlachopoulos, A., Karlioti, G., Balla, E. et al. (2022). Poly(lactic acid)-based microparticles for drug delivery applications: an overview of recent advances. *Pharmaceutics* 14 (2): 359.

56 Vinayak, A., Sudha, M., Jaadeesha, A.H. et al. (2014). Synthesis, characterization of some novel 1,3,4-oxadiazole compounds containing 8-hydroxy quinolone moiety as potential antibacterial and anticancer agents. *Int. J. Pharm. Res.* 4 (4): 180–185.

57 Espinoza, S.M., Patil, H.I., San Martin Martinez, E. et al. (2020). Poly-ε-caprolactone (PCL), a promising polymer for pharmaceutical and biomedical applications: focus on nanomedicine in cancer. *Int. J. Polym. Mater. Polym. Biomater.* 69 (2): 85–126.

58 Landsman, M.L., Kwant, G., Mook, G.A., and Zijlstra, W.G. (1976). Light-absorbing properties, stability, and spectral stabilization of indocyanine green. *J. Appl. Physiol.* 40 (4): 575–583.

59 Shirata, C., Kaneko, J., Inagaki, Y. et al. (2017). Near-infrared photothermal/photodynamic therapy with indocyanine green induces apoptosis of hepatocellular carcinoma cells through oxidative stress. *Sci. Rep.* 7 (1): 1–8.

60 Schaafsma, B.E., Mieog, J.S., Hutteman, M. et al. (2011). The clinical use of indocyanine green as a near-infrared fluorescent contrast agent for image-guided oncologic surgery. *J. Surg. Oncol.* 104 (3): 323–332.

61 Zheng, C., Zheng, M., Gong, P. et al. (2012). Indocyanine green-loaded biodegradable tumor targeting nanoprobes for in vitro and in vivo imaging. *Biomaterials* 33 (22): 5603–5609.

62 Yin, L., Li, Q., Mrdenovic, S. et al. (2022). KRT13 promotes stemness and drives metastasis in breast cancer through a plakoglobin/c-Myc signaling pathway. *Breast Cancer Res.* 24 (1): 1–3.

63 Adimule, V., Revaiah, R.G., Nandi, S.S., and Jagadeesha, A.H. (2021). Synthesis, characterization of Cr doped TeO_2 nanostructures and its application as EGFET pH sensor. *Electroanalysis* 33: 579.

64 Kong, X., Nir, E., Hamadani, K., and Weiss, S. (2007). Photobleaching pathways in single-molecule FRET experiments. *J. Am. Chem. Soc.* 129 (15): 4643–4654.

65 Guo, M., Mao, H., Li, Y. et al. (2014). Dual imaging-guided photothermal/photodynamic therapy using micelles. *Biomaterials* 35: 4656–4666.

66 Guo, B., Huang, Z., Shi, Q. et al. (2020). Organic small molecule based photothermal agents with molecular rotors for malignant breast cancer therapy. *Adv. Funct. Mater.* 30 (5): 1907093.

67 Yap, J.E., Zhang, L., Lovegrove, J.T. et al. (2020). Visible light—responsive drug delivery nanoparticle via donor–acceptor stenhouse adducts (DASA). *Macromol. Rapid Commun.* 41 (21): 2000236.

68 Graham-Gurysh, E., Kelkar, S., McCabe-Lankford, E. et al. (2018). Hybrid donor-acceptor polymer particles with amplified energy transfer for detection and on-demand treatment of breast cancer. *ACS Appl. Mater. Interf.* 10 (9): 7697–7703.

69 Li, J., Ou, H., Li, J. et al. (2021). Large π-extended donor-acceptor polymers for highly efficient in vivo near-infrared photoacoustic imaging and photothermal tumor therapy. *Sci. China Chem.* 64 (12): 2180–2192.

70 Antaris, A.L., Chen, H., Cheng, K. et al. (2016). *Nat. Mater.* 15 (2): 235–242.

71 Suryavanshi, A., Adimule, V., and Nandi, S.S. (2020). Synthesis, impedance, and current–voltage characteristics of strontium-manganese titanate hybrid nanoparticles. *Macromol. Symp.* 392: 2000002.

72 Sun, K., Shi, A., Liu, Y. et al. (2022). A general electron donor–acceptor complex for photoactivation of arenes via thianthrenation. *Chem. Sci.* 13 (19): 5659–5666.

73 Yu, K., Pan, J., Tian, M. et al. (2022). Unusual electron donor-acceptor sequenced NIR AIEgen for highly efficient mitochondria-targeted cancer cell photodynamic therapy. *Chem. Asian J.* 17(17), e202200571.

74 Shi, L., Zhang, P., Liu, X. et al. (2022). Activity-based photosensitizer to reverse hypoxia and oxidative resistance for tumor photodynamic eradication. *Adv. Mater.* 2206659. https://doi.org/10.1002/adma.202206659.

75 Shao, W., Wei, Q., Wang, S. et al. (2020). Molecular engineering of D–A–D conjugated small molecule nanoparticles for high performance NIR-II photothermal therapy. *Mater. Horiz.* 7: 1379–1386.

76 Hu, D., Zhang, J., Gao, G. et al. (2016). Indocyanine green-loaded polydopamine-reduced graphene oxide nanocomposites with amplifying photoacoustic and photothermal effects for cancer theranostics. *Theranostics* 6 (7): 1043.

77 Xiao, L., Wang, B., Yang, G., and Gauthier, M. (2012). Poly(lactic acid)-based biomaterials: synthesis, modification and applications. *Biomed. Sci. Eng. Technol.* 11: 247–282.

78 Shahrokhian, S. and Salimian, R. (2018). Ultrasensitive detection of cancer biomarkers using conducting polymer/electrochemically reduced graphene oxide-based biosensor: application toward BRCA1 sensing. *Sens. Actuat. B Chem.* (266): 160–169.

79 Jin, A., Wang, Y., Lin, K., and Jiang, L. (2020). Nanoparticles modified by polydopamine: Working as "drug" carriers. *Bioact. Mater.* 5 (3): 522–541.

80 Xu, F., Liu, M., Li, X. et al. (2018). Loading of indocyanine green within polydopamine-coated laponite nanodisks for targeted cancer photothermal and photodynamic therapy. *Nanomaterials* 8 (5): 347.

81 Hu, S., Dong, C., Wang, J. et al. (2020). Assemblies of indocyanine green and chemotherapeutic drug to cure established tumors by synergistic chemo-photo therapy. *J. Controlled Release* 324: 250–259.

82 Long, Y., Feng, Y., He, Y. et al. (2022). Hyaluronic acid modified halloysite nanotubes decorated with ZIF-8 nanoparticles as dual chemo-and photothermal anticancer agents. *ACS Appl. Nano Mater.* 5 (4): 5813–5825.

83 Pillai, V.J. and Jose, I. (2022). Near infrared molecular imaging of breast cancer cell lines using a novel ER targeted fluorescent dye.

84 Keri, R., Patil, M., Brahmkhatri, V.P. et al. (2022). Copper(II)-β-cyclodextrin promoted Kabachnik-fields reaction: an efficient one-pot synthesis of α-aminophosphonates. *Top. Catal.* https://doi.org/10.1007/s11244-021-01556-4.

85 Mensing, J.P., Wisitsoraat, A., Tuantranont, A., and Kerdcharoen, T. (2013). Inkjet-printed sol–gel films containing metal phthalocyanines/porphyrins for opto-electronic nose applications. *Sens. Actuators, B* 176: 428–436.

86 Li, C., Chen, G., Zhang, Y. et al. (2020). Advanced fluorescence imaging technology in the near-infrared-II window for biomedical applications. *J. Am. Chem. Soc.* 142 (35): 14789–14804.

87 Ethirajan, M., Chen, Y., Joshi, P., and Pandey, R.K. (2011). The role of porphyrin chemistry in tumor imaging and photodynamic therapy. *Chem. Soc. Rev.* 40 (1): 340–362.

88 Lo, P.C., Rodríguez-Morgade, M.S., Pandey, R.K. et al. (2020). The unique features and promises of phthalocyanines as advanced photosensitisers for photodynamic therapy of cancer. *Chem. Soc. Rev.* 49 (4): 1041–1056.

89 Aniogo, E.C., George, B.P., and Abrahamse, H. (2017). Phthalocyanine induced phototherapy coupled with Doxorubicin; a promising novel treatment for breast cancer. *Exp. Rev. Anticancer Ther.* 17 (8): 693–702.

90 Baygu, Y., Soganci, T., Kabay, N. et al. (2020). Phthalocyanine-cored conductive polymer design: effect of substitution pattern and chalcogen nature on optical and electrical properties of Zn(II)-phthalocyanine–cored polycarbazoles. *Mater. Today Chem.* 18: 100360.

91 Vinayak, A., Sudha, M., Rao, K., and Lalita, K.S. (2014). Synthesis of n-{[5-(2,4-dichlorophenyl)-1,3,4-oxadiazol-2-yl] methyl} amine derivatives as anticancer precursors. *Int. J. Med. Chem. Anal.* 4: 231–235.

92 Andersson-Engels, S., Af Klinteberg, C., Svanberg, K., and Svanberg, S. (1997). In vivo fluorescence imaging for tissue diagnostics. *Phys. Med. Biol.* 42 (5): 815.

Part IV

Regulatory, Clinical Aspects and Case Studies

12

An Insight into Drug Regulatory Affairs and the Procedures

Shaik A. Begum[1, 2] and Joshna Rani S[2]

[1] *Nirmala College of Pharmacy, Department of Pharmacology & Pharmacy Pratice, Mangalagiri, Andhra Pradesh, 522503, India*
[2] *Institute of Pharmaceutical technology, Sri Padmavati Mahila Viswa Vidyalayam, Tirupati, Andhra Pradesh, 517502, India*

12.1 Endpoints of Clinical Trials for the Approval of Cancer Drugs and Biologics

Clinical trial endpoints serve a variety of purposes. In conventional oncology drug development, early phase clinical trials investigate safety and identify evidence of biological treatment efficacy, such as tumor shrinking. Later-phase efficacy trials frequently use endpoints to determine whether a medicine has a therapeutic benefit, such as increased survival or symptom improvement [1].

12.2 Statutory and Regulatory Requirements for Effectiveness

A 1962 amendment to the Federal Food, Drug, and Cosmetic Act mandated that new medications demonstrate efficacy (FD&C Act). This law stipulates that considerable evidence of efficacy must be derived from adequate and well-controlled clinical trials. Biological goods must also be safe, pure, and effective, according to the Public Health Service Act [2].

Not only have major clinical outcomes (e.g. enhanced survival, symptomatic improvement) been used to justify drug approval, but also effects on surrogate endpoints that are known to predict clinical benefit (e.g. blood pressure).

The 1992 accelerated approval regulations (21 CFR part 314, subpart H, and 21 CFR part 601, subpart E) allow for the use of additional endpoints for the approval of drugs or biological products intended to treat serious or life-threatening diseases and that generally demonstrate an improvement over available therapy or

Drug and Therapy Development for Triple Negative Breast Cancer, First Edition.
Edited by Pravin Kendrekar, Vinayak Adimule, and Tara Hurst.
© 2023 WILEY-VCH GmbH. Published 2023 by WILEY-VCH GmbH.

provide therapy where none exists [3]. In this instance, the Food and Drug Administration (FDA) may approve clearance based on a surrogate endpoint result that is fairly expected to anticipate therapeutic benefit. Surrogate endpoints, such as blood pressure for cardiovascular disease, are less well-established than traditional endpoints.

The FDA may also grant accelerated approval based on an effect on a clinical endpoint that can be measured before irreversible morbidity or mortality that is reasonably likely to predict an effect on irreversible morbidity or mortality or other clinical endpoints [4].

Benefits (i.e. an intermediate clinical endpoint). The accelerated approval process allows a medicine to be approved quickly. Regulations that require the company to conduct clinical research in order to validate and fully characterize the clinical benefit. The FDA may withdraw approval of a medicine or indication if, among other reasons, post-marketing studies fail to establish clinical benefit or if the applicant fails to execute the requisite studies with appropriate diligence. The term "conventional approval" refers to the long-standing method of drug approval based on clinical benefit or an effect on a surrogate endpoint that predicts clinical benefit. This is distinct from accelerated approval, which is associated with the use of a surrogate or intermediate clinical endpoint that is reasonably likely to predict clinical benefit to support drug approval [5].

12.2.1 Endpoints Supporting Previous Oncology Approvals

To gain conventional approval, applicants must provide direct evidence of clinical benefit or improvement in a surrogate endpoint known to predict clinical benefit. In oncology, survival improvement is considered an appropriate metric of therapeutic benefit. Other endpoints have also been utilized by sponsors to secure approval for cancer treatments. The FDA used to approve cancer medications based on tumor assessments obtained by radiological studies or physical examinations. In the early 1980s, the FDA decided that cancer medicine approval should be based on more direct evidence of clinical benefits, such as increased survival, quality of life, physical functioning, or tumor-related symptoms, after consulting with the Oncologic Drugs Advisory Committee (ODAC) [6]).

Over time, there have been larger increase in tumor reduction and tumor growth delay, and tumor measuring endpoints have been used to support both traditional and accelerated approval. Disease-free survival (DFS) improvements have facilitated therapeutic approval in specific adjuvant contexts when a large proportion of patients were predicted to have cancer symptoms at the time of recurrence. Traditional approval in leukemia has been supported by durable complete response (CR), which is related to decreased infection, hemorrhage, and blood product support. Traditional approval in certain cancers has been based on a considerable improvement in progression-free survival (PFS) or a high, verified durable ORR, but the degree of effect, treatment of tumor-related symptoms, and drug toxicity should all be taken into account. ORR has been used as an endpoint in randomized trials for hormonal drugs for breast cancer (BC) and a single-arm trial for a drug for

ROS1-positive metastatic non-small cell lung cancer. Traditional approval has been supported in several clinical settings by improvements in tumor-related symptoms, an improved ORR, and adequate response duration [7].

12.2.2 Endpoints Based on Tumor Assessments

Several endpoints based on tumor evaluations are discussed in this section. Radiographic tumor assessments directly quantify disease components in many cancer forms, and tumor measures are frequently used to guide treatment decisions in clinical practice. As a result, tumor-based outcomes are thought to be more therapeutically useful than other markers. DFS (including EFS), ORR, CR, TTP, and PFS are examples of tumor-measure-based goals that can be used to support traditional or accelerated approval [8].

Two judgments must be taken when choosing a tumor evaluation endpoint. To begin, determine whether the endpoint can accommodate either speedy or traditional approval. Second, any risk of endpoint bias or uncertainty must be assessed. Data from a second study may be required to support drug applications that depend solely on tumor assessment endpoints as confirmation of efficacy. The precision with which tumors are measured varies depending on the tumor's surroundings. In areas where there are no delineated margins, tumor measures utilized in ORR estimates can be inaccurate (e.g. pleural or peritoneal mesothelioma, pancreatic cancer, and brain tumors) [9].

12.2.3 Clinical Practice Guideline for the Diagnosis, Staging, and Treatment of Patients with Metastatic Breast Cancer

Despite the fact that metastatic breast cancer (MBC) is an incurable disease, appropriate treatment strategies have been found to improve survival rates. Systemic therapy is the gold standard in MBC, however, depending on the patient's illness condition, it may be supplemented with locoregional therapies (LRTs) [10]. As a result, a multidisciplinary team (MDT) is required for optimal management. Even if current targeted medicines may lead to subtype changes in the future, as seen by the first tumor-agnostic approvals, these guidelines are based on biological subtypes of BC [11].

Regardless of the patient's age, treatment decisions must be made, but comorbidities, patient characteristics, and patient preferences must all be considered as part of a collaborative decision-making process. A comprehensive geriatric assessment may provide valuable information in the case of aged patients. Supportive care should always be a component of the treatment strategy, and bringing in experienced palliative care early can help with symptom control [12].

If the disease-free interval (DFI) is at least 12 months after the last drug administration and no residual toxicities are apparent, rechallenge with drugs previously used in early breast cancer (EBC) is a potential option. Approved biosimilars can be used instead of the original drug in all approved indications. This patient population should be encouraged to seek clinical trials early in the course of their illness, with a preference for enrollment in a clinical research if one is available in each line of therapy [13].

12.2.4 Cancer Drug and Diagnostic Regulation by the FDA

The FDA is revamping the way it reviews oncology products. The Office Of Oncology Drug Productss (OODPs) had three divisions that reviewed oncology products at its inception: biologics, drugs, and imaging [14].

Office of Hematology and Oncology Products (OHOPs), with a new structure encompassing four divisions:

1) The Division of Hematology Products (DHPs)
2) The Division of Hematology Oncology Toxicology (DHOT)
3) The Division of Oncology Products 1 (DOP 1)
4) The Division of Oncology Products 2 (DOP 2)

Review personnel will specialize in various oncologic diseases (e.g. BC, gastrointestinal cancer, and melanoma) within the two divisions of oncology goods, similar to how academic comprehensive cancer centers do it presently. The purpose of this reorganization was to improve FDA staff career development and foster more consistency in review by boosting interaction between FDA workers and outside academic experts. In addition, the FDA has formed the DHOT, a toxicology–pharmacology division that will focus on mechanisms of action as well as routine toxicology, in response to the rising need to mix unapproved treatments. Topic leaders, such as individuals with experience in clinical pharmacology or biomarkers, are also needed [15].

The FDA's Oncology Program helps to coordinate activities within the agency, such as meetings with the CBER (Center for Biologics Evaluation and Research) on tumor vaccines and cellular therapies, or meetings with the CDRH (Center for Devices and Radiologic Health) on in vitro diagnostics and other cancer-related products, as well as cross-center meetings to discuss oncology drug development. One of the current important issues in oncology drug development is determining how to properly handle the multiple possible therapeutic combinations and, from a regulatory perspective, obtaining clearance of the combination of no licensed candidate medicines. He also talked about drug-diagnostic combinations and how to examine biomarkers that predict treatment response so that tailored medicines may be provided [4]. FDA also advised that biomarkers be utilized more frequently to anticipate and prevent cancer medication toxicity.

Traditional cancer clinical trial endpoints, such as overall survival, may not be sufficient, especially when evaluating the value of newer targeted therapies, and should be replaced by other biologically significant and predictive of outcome endpoints that have not been used as a basis for regulatory review or approval [16].

One of these potential endpoints could be pathologic CR.

People may be able to comprehend the therapeutic value of some of these biological endpoints, which could help speed up the development of novel drugs.

In the HPV (human papillomavirus) vaccine trials, researchers had enough knowledge of the biology of cervical cancer to use moderate and high-grade dysplasia as an endpoint.

12.2.5 Considerations for Clinical Trial Design and Analysis

The FDA approves drugs based on "sufficient and well-controlled investigations," as stated by the regulations (21 CFR 314.126). A quantifiable assessment of the drug's effect as well as a meaningful comparison to a control group is required in studies. The most reliable technique to establish efficacy in randomized controlled trials is to show a statistically significant improvement in a clinically meaningful endpoint (i.e. superiority).

12.2.6 Single-Arm Studies

In instances when there is no approved medication and large tumor regressions can be presumed to be attributable to the tested medicine, the FDA has occasionally viewed ORR and response duration found in single-arm studies as significant evidence for quick approval. Response rates have been used for traditional approval in cases like acute leukemia, where CRs have been connected to fewer transfusions, fewer infections, and a higher life expectancy. Overall survival, DFS (including EFS), TTP, and PFS are all time-to-event outcomes that single-arm trials are unable to assess well. Due to the variability in the natural history of various forms of cancer, a randomized study is essential to investigate time-to-event endpoints [17].

12.2.7 Randomized Studies Designed to Demonstrate Noninferiority

A noninferiority (NI) trial establishes the efficacy of a novel treatment by demonstrating that it is not less effective by a predetermined amount than a typical regimen (the active control) (NI margin). This NI margin should be a clinically acceptable loss that is not more than the effect of the active control medication. The normal regimen should include a well-defined therapeutic benefit (survival benefit) [18]. If the new treatment is unsuccessful by more than the NI margin, it will be assumed that it is useless.

NI studies use externally controlled (historical) data to determine the active control's treatment impact magnitude. In most cancer trials, this effect is not clearly documented. In NI experiments, consistency is also assumed. According to this hypothesis, the active-control effect has remained consistent between the externally controlled trial and the current analysis. This assumes that patient population characteristics, supportive care measures, and evaluation procedures are consistent between the current study and the externally controlled data from which the active-control effect was obtained [19].

To determine the amount of the active-control treatment impact, a full meta-analysis of externally controlled studies should be used. These trials should be able to reliably replicate the active-control effect when compared to the placebo arm. In NI trials, determining the active-control effect and selecting the quantity of effect (NI margin) to be preserved are both tough challenges. NI experiments have a higher sample size than superiority trials and often entail replication of clinical trial results. Furthermore, subsequent drugs and crossover to the active-control arm can throw off any NI analysis. In NI research, endpoints other than overall survival and ORR are problematic [20].

12.3 Clinical Trial Design Considerations

The protocol and SAP should include the methods for assessing, measuring, and analyzing the endpoint(s).To avoid systematic bias, visits, and radiological examinations should be symmetric between the two study arms. To the extent practicable, the FDA and the application should agree on the following elements in advance, and the applicant should complete the items before the research begins [21]:

- The study designs
- The definition of progression
- The SAP
- The methodology for handling missing data and censoring methods
- The data to be recorded on the Case Report Form (CRF)
- The operating procedures of an independent endpoint review committee (IRC)

12.4 Clinical Trial Analysis Issues

Missing data can make endpoint analysis difficult. For objectives dependent on tumor assessments, the procedure should design a suitable assessment visit for each patient (i.e. a visit when all scheduled tumor assessments have been done). The analytic plan should include a comparison of the adequacy of follow-up in each treatment arm. The approach for analyzing incomplete and/or missing follow-up visits, as well as filtering mechanisms, should be included in the protocol. To assess the results' robustness, the analysis strategy should include the primary analysis and one or more sensitivity tests. Although any study with missing data can be difficult, similar results in the primary and sensitivity analyses can help to improve the results.

When appropriate, the number of deaths in patients who have been lost to follow-up for an extended period of time should be included in the evaluation. An imbalance in such deaths, for example, could skew a PFS calculation by overestimating PFS in the therapy arm with less follow-up [20].

12.5 Use of Pathological Complete Response as an Endpoint to Support Accelerated Approval in Neoadjuvant Treatment of High-Risk Early-Stage Breast Cancer

The FDA's accelerated approval program aims to expedite the development of drugs for the treatment of serious or life-threatening disorders that provide a significant therapeutic advantage over existing treatments. Despite recent advancements in adjuvant systemic therapy for BC, we recognize that some high-risk or poor prognosis subsets of early-stage BC patients still have a significant unmet medical need. Consideration of PCR as an acceptable study endpoint for accelerated approval in the neoadjuvant setting could drive industry innovation and accelerate the development of novel treatments for high-risk early-stage BC [21].

The FDA may grant accelerated approval on a determination that the product has an effect on a surrogate endpoint that is reasonably likely to predict clinical benefit, or on a clinical endpoint that can be measured earlier than an effect on irreversible morbidity or mortality, as amended by the "Food and Drug Administration Safety and Innovation Act of 2012," according to Section 506(c) of the Federal Food, Drug, and Cosmetic Act (FD&C Act), 21 U.S.C. 356(b), as

The accelerated approval regulations also state that "where there is uncertainty as to the relation of the surrogate endpoint to clinical benefit, or of the observed clinical benefit to ultimate outcome, approval under this section will be subject to the requirement that the sponsor study the drug further, to verify and describe its clinical benefit." Post marketing studies are usually studies that are already in progress. When such investigations are required, they must be adequate and well-controlled. Any such studies must be carried out with care by the sponsor.

12.6 Developing Treatments for Premenopausal Women with Breast Cancer

This section of the chapter provides guidance to companies that make CDER and CBER-regulated medicines or biological products for the treatment of BC. Premenopausal women, as defined by serum hormonal levels, should be included in BC clinical trials, according to this advice (including but not limited to follicle-stimulating hormone and estradiol). The issues of premenopausal women with BC in terms of conception and fertility preservation are not addressed in this chapter of the industry guidance.

Various clinical studies assessing the efficacy of drugs that alter the hormonal axis for the treatment of hormone receptor (HR)-positive BC have excluded premenopausal women. Separate trials have been carried out in certain cases to establish the advantage in this patient population, causing delays in the availability of these medicines for premenopausal women with HR-positive BC.

In premenopausal and postmenopausal women with BC, chemotherapy, immunotherapy, and targeted therapy (drugs that function outside of the hormonal axis) were all equally effective.

FDA believes hormonal medications given to premenopausal women with appropriate estrogen suppression will have the same efficacy and safety profile as hormonal therapies given to postmenopausal women, based on a review of nonclinical, clinical pharmacology, and clinical literature. BC oncology product development programs will include premenopausal women, resulting in more complete clinical information to aid clinical decision-making and timely delivery of safe and efficacious medicines to this patient population.

12.7 Recommendations by FDA

BC medicine development programs should explore including premenopausal women. FDA encourages sponsors to meet with CDER and CBER early in the development process to discuss their BC medicine development goals.

- According to the FDA, both premenopausal women 4 with appropriate estrogen suppression and postmenopausal women 4 with acceptable estrogen suppression should be equally eligible and included in clinical investigations for hormonal axis modifying drugs or combinations.
- If there are concerns about efficacy and/or safety, stratifying randomization based on menopausal status at trial entry may be necessary. An assessment of the weight of evidence for reproductive toxicity, including published literature and current nonclinical data, should be supplied to allow FDA to determine if reproductive toxicity studies are required for an indication that would include premenopausal women [1].
- As part of the clinical development plan, data on the clinical impacts (e.g. bone health, cardiac health) of BC medications in all patients should be gathered and analyzed generally, as well as by stratification (if available) based on menopausal status at study enrollment. FDA may require or request agreement from the sponsor to perform post-marketing studies to investigate additional long-term clinical consequences if justified by applicable law or regulation.
- To assist the assessment of short- and long-term effects of these medicines across clinically relevant subgroups of patients, clinical studies testing medications in premenopausal women with BC should reflect the racial and ethnic diversity of this patient population.
- Refer to FDA recommendations for methodologies and approaches for collecting patient experience data throughout the development program (e.g. to better inform tolerability of a drug).
- While fertility and fertility preservation are outside the scope of this recommendation, a gynecologist should be consulted as needed throughout trial planning and monitoring.

12.7.1 Access to Experimental Cancer Drugs

The US Food and Medicine Administration has cleared an experimental drug for human testing after it has been studied in the lab and on animals (FDA) [21]. However, such a medicine cannot be advertised, sold, or prescribed at this time. "Investigational medications" is another term for experimental drugs.

12.7.2 How to Get a Hold of an Experimental Drug

Clinical trials, improved access, or the right to try may make experimental medications available.

12.7.3 Access to More Information (Compassionate Use)

The use of an experimental drug outside of clinical trials to treat persons with serious or life-threatening disorders is known as expanded access, or compassionate usage. The FDA is in charge of it. You might be eligible to get an experimental medicine through extended access if you match specific conditions. To be eligible, you must:

- you have a life-threatening illness
- having no access to standard therapies
- be ineligible to participate in a clinical trial

The only way to get a medicine through extended access is with your doctor's assistance. He or she will have to do the following:

- Inquire with the company that is creating and testing the experimental medicine if they are willing to make it available to you. If the corporation considers that providing the drug is in its best interests or if there isn't enough drug available, it may refuse to do so.
- If the drug manufacturer agrees to furnish the medicine, submit an application to the FDA. For more information, your doctor should visit the FDA's Expanded Access: Information for Physicians page.

If you are treated with a medicine through extended access, there are several safeguards in place to ensure your safety:

- The use must be approved by an Institutional Review Board or its representative.
- The use must be approved by the FDA.
- You'll go through an informed consent process to make sure you're aware of the dangers and advantages of taking the experimental medicine.

12.8 What is Right to Try?

Right to try, like expanded access, allows persons with serious or life-threatening diseases to utilize an experimental treatment outside of clinical trials [21]. It is not, however, regulated by the FDA, unlike extended access. You may be eligible to acquire an experimental medicine through right to try if you meet specific conditions. These are some of the criteria:

- You are suffering from a life-threatening sickness.
- You are not eligible for any usual therapies.
- You are not eligible for the drug's clinical trial.

The medicine you are looking for must also meet specific requirements. These are some of the criteria:

- A Phase 1 clinical trial has previously been conducted on it.
- The FDA has not cleared it for any use.
- Its owner has sought FDA approval or is conducting a clinical trial to further investigate the medicine before requesting approval.

The only way to get a medicine through the right-to-try program is with your doctor's aid. To learn more, he or she will need to speak with the firm that is creating and testing the experimental medicine.

- If the drug is eligible for usage under the right to try program, and
- if the firm is willing to offer it (which it is not required to do)

12.9 Examples of Drugs Approved for Breast Cancer

In March 2022, the FDA approved therapies for patients with resettable NSCLC, BRCA-mutated, HER2-negative BC, unresectable or metastatic melanoma, advanced endometrial carcinoma with high microsatellite instability and mismatch repair deficiency, and metastatic castration-resistant prostate cancer [22]. The FDA also granted an indication for a companion diagnostic assay to detect BRCA mutations in high-risk EBC.

The agency granted various fast-track designations, orphan drug classifications, and priority reviews for a number of cancer types, including gastrointestinal cancer, lymphoma, melanoma, multiple myeloma, leukemia, and more.

In March, the CYAD-101 trial in patients with metastatic colorectal cancer mCRC and the NEON-2 trial in patients with advanced solid tumors were both put on hold. The FDA also released updated cancer clinical trial standards, which are in accordance with President Biden's 2016 Cancer Moonshot initiative.

The FDA approved fam-trastuzumab deruxtecan-nxki (Enhertu, Daiichi Sankyo, Inc.) on 4 May 2022 for adult patients with unresectable or metastatic HER2-positive BC who have received a prior anti-HER2-based regimen either in the metastatic setting or in the neoadjuvant or adjuvant setting and have developed disease recurrence during or within six months.

Adult patients with unresectable or metastatic HER2-positive BC who have received two or more prior anti-HER2-based regimens in the metastatic setting gained accelerated approval for fam-trastuzumab deruxtecan-nxki in December 2019. The confirmatory trial for the speedy approval came after that.

DESTINY-Breast03 (NCT03529110) was a multicenter, open-label, randomized trial that enrolled 524 patients with HER2-positive, unresectable, and/or MBC who had previously received trastuzumab and taxane therapy for metastatic disease or had disease recurrence during or within six months of finishing neoadjuvant or adjuvant therapy. Patients were randomly assigned to receive Enhertu or ado-trastuzumab emtansine intravenously every three weeks until intolerable toxicity or disease progression. The study was stratified by HR status, past pertuzumab treatment, and visceral illness history.

FDA approvals					
Generic name	**Brand name**	**Tumor type**	**MOA**	**Manufacturer**	**Approval**
Filgrastim-ayow (biosimilar)	Releuko	Supportive care: neutropenia	Leukocyte growth factor	Kashiv Biosciences	2/25/22
Lutetium Lu 177 vipivotide tetraxetan	Pluvicto	Prostate cancer	PSMA-targeted radioligand therapy	Novartis	3/23/22
Nivolumab	Opdivo + platinum-doublet chemotherapy	NSCLC	PD-1 inhibitor	Bristol Myers Squibb	3/4/22
Nivolumab and relatlimab-rmbw (biologic)	Opdualag™	Melanoma	PD-1 inhibitor and LAG-3 blocker	Bristol Myers Squibb	3/18/22
Olaparib	Lynparza	Early breast cancer	PARP inhibitor	AstraZeneca	3/11/22
Pembrolizumab	Keytruda	Endometrial carcinoma	PD-1 inhibitor	Merck	3/21/22

Source: Adapted from FDA (2023).

References

1 Food and Drug Administration (2019). *2018 New Drug Therapy Approvals.* Washington, DC, USA: Food and Drug Administration.

2 O'Shaughnessy, J.A., Wittes, R.E., Burke, G. et al. (1991). Commentary concerning demonstration of safety and efficacy of investigational anticancer agents in clinical trials. *J. Clin. Oncol.* 9: 2225–2232.

3 Adimule, V., Yallur, B.C., and Gowda, A.H.J. (2022). Chapter 14 – Advanced sensors based on carbon nanomaterials. In: *Carbon Nanomaterials-Based Sensors* (ed. J.G. Manjunatha and C.M. Hussain), 259–268. Elsevier https://doi.org/10.1016/B978-0-323-91174-0.00004-4.

4 Martinalbo, J., Bowen, D., Camarero, J. et al. (2016). Early market access of cancer drugs in the EU. *Ann. Oncol.* 27: 96–105. https://doi.org/10.1093/annonc/mdv506.

5 Eisenhauer, E.A., Therasse, P., Bogaerts, J. et al. (2009). New response evaluationcriteria in solid tumours: revised RECIST guideline (version 1.1). *Eur. J. Cancer* 45 (2): 228–247.

6 Deisseroth, A., Kamiskas, E., Grillo, J. et al. (2012). U.S. food and drug administration approval: ruxolitinib for the treatment of patients with intermediate and high-risk myelofibrosis. *Clin. Cancer Res.* 18 (12): 3212–3217.

7 Adimule, V., Nandi, S.S., and Yallur, B.C. (2022). Devices and sensors based on additively manufactured shape-memory of hybrid nanocomposites. In: *Shape Memory Composites Based on Polymers and Metals for 4D Printing* (ed. M.R. Maurya, K.K. Sadasivuni, J.J. Cabibihan, et al.). Cham: Springer https://doi .org/10.1007/978-3-030-94114-7_15.

8 Huitfeldt, B. and Hummel, J., on behalf of European Federation of Statisticians in the Pharmaceutical Industry (EFSPI) (2011). The draft FDA guideline on non-inferiority clinical trials: a critical review from European pharmaceutical industry statisticians. *Pharm. Stat.* 10: 414–419.

9 Batakurki, S.R., Adimule, V., Pai, M.M. et al. (2022). Synthesis of Cs-Ag/Fe$_2$O$_3$ nanoparticles using *Vitis labrusca* rachis extract as green hybrid nanocatalyst for the reduction of arylnitro compounds. *Top. Catal.* https://doi.org/10.1007/ s11244-022-01593-7.

10 Adimule, V., Medapa, S., Adarsha, H.J. et al. (2014). Synthesis, characterization and in vitro anticancer properties of 1-{5-aryl-2-[5-(4-fluoro-phenyl)-thiophen-2-yl]-[1, 3, 4] oxadiazol-3-yl}-ethanone. *Int. J. Pharm. Res. Rev.* 3 (12): 20–25.

11 Adimule, V., Yallur, B., and Gowda, A. (2022). Crystal structure, morphology, optical and super-capacitor properties of Srx: α-Sb$_2$O$_4$ nanostructures. *Anal. Bioanal. Electrochem.* 14 (1): 1–17.

12 Leo, C.P., Leo, C., and Szucs, T.D. (2020). Breast cancer drug approvals by the US FDA from 1949 to 2018. *Nat. Rev. Drug Discov.* 19: 11. https://doi.org/10.1038/ d41573-019-00201-w.

13 Richey, E.A., Lyons, E.A., Nebeker, J.R. et al. (2009). Accelerated approval of cancer drugs: improved access to therapeutic breakthroughs or early release of unsafe and ineffective drugs? *J. Clin. Oncol.* 27: 4398–4405. https://doi.org/10.1200/ JCO.2008.21.1961.

14 Adimule, V., Nandi, S.S., and Jagadeesha Gowda, A.H. (2021, 2020). Enhanced power conversion efficiency of the P3BT (poly-3-butyl thiophene) doped nanocomposites of Gd-TiO$_3$ as working electrode. In: *Techno-Societal* (ed. P.M. Pawar, R. Balasubramaniam, B.P. Ronge, et al.). Cham: Springer https://doi .org/10.1007/978-3-030-69925-3_6.

15 Dugger, S.A., Platt, A., and Goldstein, D.B. (2018). Drug development in the era of precision medicine. *Nat. Rev. Drug Discov.* 17: 183–196. https://doi.org/10.1038/ nrd.2017.226.

16 Gianni, L., Pienkowski, T., Im, Y.H. et al. (2012). Efficacy and safety of neoadjuvant pertuzumab and trastuzumab in women with locally advanced, inflammatory, or early HER2-positive breast cancer (NeoSphere): a randomisedmulticentre, open-label, phase 2 trial. *Lancet Oncol.* 13: 25–32. https://doi.org/10.1016/S1470-2045(11)70336-9.

17 Temple, R. and Ellenberg, S.S. (2000). Placebo-controlled trials and active-control trials in the evaluation of new treatments - part 1: ethical and scientific issues. *Ann. Intern. Med.* 133: 455–463.

18 De Mattos-Arruda, L., Cortes, J., Santarpia, L. et al. (2013). Circulating tumour cells and cell-free DNA as tools for managing breast cancer. *Nat. Rev. Clin. Oncol.* 10: 377–389.

19 Adimule, V., Kerur, S.S., Chinnam, S. et al. (2022). Guar Gum and its nanocomposites as prospective materials for miscellaneous applications: a short review. *Top. Catal.* https://doi.org/10.1007/s11244-022-01587-5.

20 Dale, D.C. (2003). Poor prognosis in elderly patients with cancer: the role of bias and undertreatment. *J. Support. Oncol.* 1: 11–17.

21 Ring, A., Harder, H., Langridge, C. et al. (2013). Adjuvant chemotherapy in elderly women with breast cancer (AChEW): an observational study identifying MDT perceptions and barriers to decision making. *Ann. Oncol.* 24: 1211–1219.

22 Adimule, V., Vageesha, P., Bagihalli, G. et al. (2019). Synthesis, characterization of hybrid nanomaterials of strontium, yttrium, copper doped with indole Schiff base derivatives possessing dielectric and semiconductor properties. In: *Emerging Research in Electronics, Computer Science and Technology*, Lecture Notes in Electrical Engineering, vol. 545 (ed. V. Sridhar, M. Padma, and K. Rao). Singapore: Springer https://doi.org/10.1007/978-981-13-5802-9_97.

13

A Comprehensive Review of Some Heat-Shock Proteins in the Development and Progression of Human Breast Cancer

Xolani H. Makhoba[1] and Ofentse J. Pooe[2]

[1] *University of Limpopo, Department of Biochemistry, Microbiology and Biotechnology, Turfloop Campus, Sovenga, Eastern Cape 0727, South Africa*
[2] *University of KwaZulu-Natal, School of Life Sciences, Discipline of Biochemistry, Westville Campus, Westville, Durban, KwaZulu-Natal, 3629, South Africa*

13.1 Introduction

Heat-shock proteins (HSPs) were first discovered by the Italian scientist (known as Ritossa) who mistakenly increased the incubation temperature of *Drosophila melanogaster larvae*, resulting in overexpression of special heat protective that were "unknown proteins" at that time [1]. After that incident, many scientists conducted research on these genes whose expression had shown an increase due to temperature. They came to the conclusion that these are molecular chaperones, and their primary function was to stop protein aggregation and cellular harm. Numerous investigations on these molecules were conducted, and the results revealed that HSPs are therefore not only generated by temperature but also by a variety of other factors, including exposure to UV radiation, viruses, toxins, and physiological or environmental assaults [2, 3]. HSPs are classified mainly according to their molecular sizes: namely small HSPs (HSP10, HSP21, HSP27), medium HSPs (HSP60, HSP70), and large HSPs (HSP90, HSP100). HSPs with more than 60 kDa in their sizes are mostly adenosine triphosphate (ATP) dependent and employ their enzymatic activity to induce conformational changes in the newly synthesized proteins by converting ATP to adenosine diphosphate (ADP) through hydrolysis. The compartmentalization of HSPs varies in the cell structure such as the nucleus, mitochondria, chloroplasts, endoplasmic reticulum, and cytosol in all types of prokaryotes and eukaryotes [4, 5].

The major role of HSPs is to protect newly synthesized polypeptides, correct misfolded proteins, and are involved in protein transportation where various activities are taking place or where correctly folded proteins are required for action. HSPs usually

Drug and Therapy Development for Triple Negative Breast Cancer, First Edition.
Edited by Pravin Kendrekar, Vinayak Adimule, and Tara Hurst.
© 2023 WILEY-VCH GmbH. Published 2023 by WILEY-VCH GmbH.

make up about 5–10% of the total protein in most cells but their intracellular concentration may increase within a few minutes under the influence of stressors up to 20 times to suppress or attenuate undesired effects [6]. At physiological conditions, HSPs are entangled with HSP factors such as HSF-1, especially under normal temperatures. However, as a result of increased temperature-induced protein oxidation, HSF-1 dissociates from HSPs, leading to an increase in molecular-like machinery in the cellular system that plays a proteostasis role in cellular function and cell viability [7]. Therefore, controlling the dynamic balance between protein biogenesis, folding, translocation, and protein localization. HSPs are major role players in cellular protection and proteostasis, but their very role can be turned against the same organism in which they are found. Breast cancer is the second leading cause of cancer-related death among women worldwide. Breast cancers are mostly adenocarcinomas that include invasive types such as infiltrating ductal carcinoma (IDC) and noninvasive type, ductal carcinoma in situ (DCIS) [2]. IDC is the most common subtype of breast cancer accounting for 75–80% of all the cases diagnosed [8]. Breast cancer incidence is highest in developed countries and increasing rapidly in developing countries due to inadequate medical support and infrastructure [9]. Therefore, there is a need to characterize a tumor-associated molecule for early detection of breast cancer and for identifying a novel therapeutic target for better cancer treatment. Therefore, this document the roles of some HSPs is discussed and how they cooperate during protein folding stages. How these molecular like machineries can be used as therapeutic treatment of the second most killer of cancer in women can be used as the cheapest treatment or diagnosis of breast cancer, particularly in the developing countries with less resources [10, 11]. Breast cancer affects mostly women, there are few cases of breast cancer in men, about 0.1% of men are affected by breast cancer worldwide. Therefore, there is an urgent need to develop innovative treatment for breast cancer, thus understanding how cancer is developed has been thoroughly studied. In this review, the role of HSPs in breast cancer development will be discussed and how these molecules can be used to come up with cancer treatment will also be covered [12].

13.1.1 Cancer and Its Economic Burden on Human

Nearly half of cancer survivors experience financial distress [13]. Cancer-related financial toxicity, the harmful personal economic burden caused by cancer treatment, affects nearly half of cancer survivors [1] and is present even among those with health insurance [14]. Costs of cancer care are even higher for those with adverse treatment effects [15, 16], such as breast cancer-related lymphedema [13, 14], and comorbidities [15, 17]. Breast cancer-related lymphedema affects up to 35% [17, 18] of the 3.5 million breast cancer survivors in the USA (2016) [18]. Breast cancer-related lymphedema is a chronic inflammatory condition that arises when there is disruption of lymphatic flow due to surgery, adjuvant radiation, and some forms of chemotherapy, infection, obesity, or other trauma to the lymphatic system [17, 19], leading to a buildup of lymphatic fluid in the upper body, especially the arms, breast, and torso [20, 21]. The arm swelling and altered lymphatic function caused by lymphedema may affect a breast cancer survivor's ability to complete activities of daily living and maintain employment, leading to psychosocial distress, secondary comorbidities [19, 22–24], and limited work and career opportunities [25–27]. Previous work has estimated that incremental costs due to lymphedema for US cancer survivors at \$14 877 (excluding cancer-related costs) in the first two years after cancer treatment initiation [25]; however, these estimates are nearly 15 years old, focused on only the short-term costs, and predate the 2010 Affordable Care Act that expanded coverage for cancer-related care and banned refusal of coverage for those who might have a preexisting condition [26]. A patient's decision about whether or not to expend resources on medical care versus other competing needs is driven by out-of-pocket costs. Yet, previous estimates rely on claims and administrative data, which neglect the impact of out-of-pocket costs. Altogether, data on out-of-pocket costs of lymphedema management is lacking for US-based samples and overlooks the long-term impact of cost and indirect costs, such as lost productivity.

13.2 Structure-Functional Features of HSPs

13.2.1 Heat-Shock Protein 40

Newly synthesized proteins are the most vulnerable proteins inside the cellular system, if not protected they can form inclusion bodies and also be exposed to proteolytic enzymes. Therefore, HSP40 recognizes the newly produced polypeptide upon their exit from ribosomes. HSP40 is known for its role as a co-chaperone not necessarily being the main player in assisting the newly produced proteins to fold properly [27]. HSP40 has at least three types, Type I with J-domain, zinc finger motifs, and peptide fragments. While Type II possesses the J-domain and peptide binding fragments. With Type III, the J-domain is surrounded by J-domains on both sides as shown in Figure 13.1 below. The HSP40 co-chaperone role is very important during protein folding in breast cancer development and progression [28].

13.2.2 Heat-Shock Protein 60

HSP60 is amongst the most important molecular chaperones. In cancer cells, it has been suggested that HSP60 plays a vital role in activating human immune system as

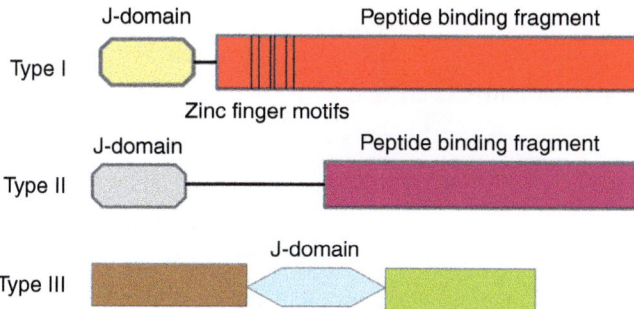

Figure 13.1 Different types of HSP40/DnaJ and its subdomains.

it was found in the extracellular system. For example, when proteins need further folding, HSP60 plays an important role in making sure that the incomplete protein folding is taken care with the help of its co-chaperone called GroES in *Escherichia coli*. Figure 13.2 below show the structure of HSP60 with its lid where protein is enclosed during their folding processes [29, 30].

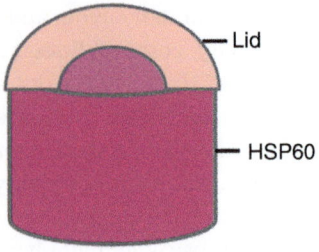

Figure 13.2 HSP60 structure and its lid.

13.2.3 Heat-Shock Protein 70

HSP70 is the major chaperone with various groups almost eight different types of HSP70s have been studied so far suggesting them as main players in protein folding. With the help of HSP40 as its co-chaperone, HSP70 promotes proper folding of the newly synthesized polypeptide without being part of the final product. HSP70 is an ATP-dependent chaperone, it also has three domains: namely, the ATP domain, N-terminal domain, and binding domain [31, 32]. In its ATP form, HSP70 cannot bind to the newly produced proteins but once the ATP gets converted into ADP form, the binding ability of the HSP70 gets activated thus recognizing the newly synthesized protein for proper folding. Upon its folding, the ADP form gets converted back to ATP therefore, HSP70 releases the folded protein (Figure 13.3).

A study conducted shows that HSP70 is overexpressed in breast cancer patients and was involved in the malignant properties of breast cancer. This suggests that HSP70 may be a potential candidate molecule for the development of better breast cancer treatment. Briefly, HSP70-2 expression was detected in the majority of breast cancer patients (83%) irrespective of various histotypes, stages, and grades [33, 34]. HSP70-2 expression was also observed in all breast cancer cells (BT-474, MCF7, MDA-MB-231, and SK-BR-3) used in this study. Depletion of HSP70-2 in MDA-MB-231 and MCF7 cells resulted in a significant reduction in cellular growth, motility, the onset of apoptosis, senescence, and cell cycle arrest as well as reduction of tumor growth in the xenograft model. At the molecular level, downregulation of HSP70-2 resulted in reduced expression of cyclins, cyclin-dependent kinases, antiapoptotic molecules, and mesenchymal markers and enhanced expression of CDK inhibitors, caspases, proapoptotic molecules, and epithelial markers [30]. The figure below summarizes

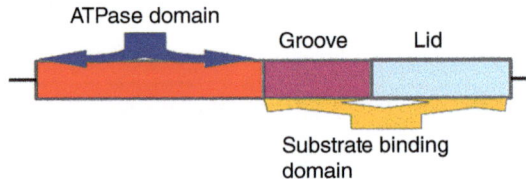

Figure 13.3 The organizational domains of HSP70, with ATPase domain, substrate binding domain with the groove and the lid.

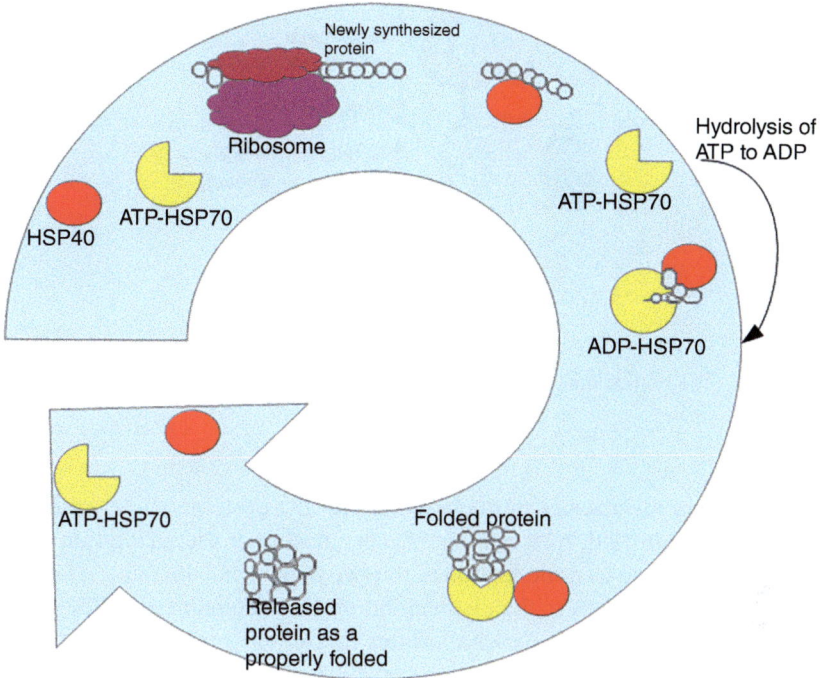

Figure 13.4 Folding processes of the newly synthesized protein in partnership between HSP40 and HSP70.

how HSP70 helps the newly synthesized proteins to fold properly. Upon the protein exit from ribosomes, HSP40 recognizes the protein and therefore activates the hydrolysis of ATP to ADP thus allowing HSP70 to bind to the protein handed over by the very same HSP40. Once the protein has folded properly, the ADP is then converted to ATP, which then allows HSP70 to release the folded protein (Figure 13.4).

13.2.4 Heat-Shock Protein 90

The 90-kDa HSP (HSP90) is a highly ubiquitous and evolutionarily conserved molecular chaperone [35]. It is essential in eukaryotes, where it is involved in the folding, stability, and activation of more than 200 client proteins including many transcription factors, steroid hormone receptors, and protein kinases [36]. In addition, HSP90 stabilizes and activates oncoproteins and therefore is a potential target for drug discovery for the treatment of cancer. HSP90 is a homodimer with each monomer consisting of three domains: an N-terminal domain that possesses an ATP binding site [10]; a middle domain containing residues that participate in binding client proteins; and a C-terminal domain that is essential for dimerization and is also involved in client binding. ATP binding and hydrolysis by HSP90 triggers large conformational changes that are necessary for the cycle of client binding, remodeling, and release [37]. The HSP90 dimer exists in a predominantly open V-shaped

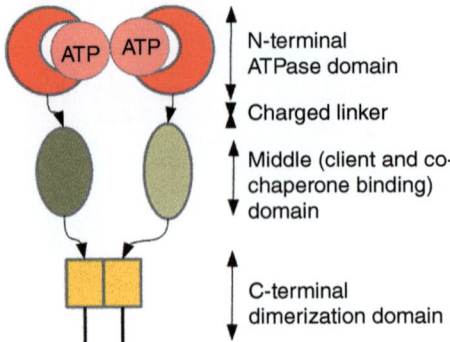

Figure 13.5 The homodimer of HSP90 structure and its domain.

conformation in the absence of nucleotides with the protomers interacting via the C-terminal domain [38]. ATP binding triggers closing of the ATP lid on the ATP-binding site, followed by dimerization of the two N-terminal domains. ATP hydrolysis and ADP release leads to the dissociation of the N-domains and HSP90 returns to the open conformation. Co-chaperones and client protein binding bias the HSP90 chaperone cycle and stabilize or destabilize various conformations of HSP90 (Figure 13.5).

Both HSP70 and HSP90 are some of the most studied molecular chaperones, proteins that themselves are responsible for the folding of other proteins in the cell. HSP70 binds non-native proteins whilst substrates of HSP90 are usually in native-like forms [39]. Proteins that require both HSP70 and HSP90 to fold are thus transferred from HSP70 to HSP90 during the folding process. Eukaryotic HSP90 participates in the conformational regulation of signal transduction molecules, such as tyrosine kinases and steroid hormone receptors. For example, steroid hormone receptors associate with HSP90 in order for them to adopt conformational competence for hormone binding [40]. In eukaryotes, the essential interaction between HSP70 and HSP90 is mediated by the HSP70–HSP90 organizing protein. Both HSP70 and HSP90 possess C-terminally located EEVD motifs that interact with HOP via its tetratricopeptide repeat (TPR) domains, TPR1 and TPR2A motifs, respectively. It is most likely that the HSP70–HSP90 functional partnership in breast cancer cells facilitates the folding of key proteins during the development of cancer cells, possibly those involved in signal transduction [41] (Figure 13.6).

13.3 Conclusion and Future Perspectives

HSPs are the housekeepers of the cells; they take care of newly produced proteins during their folding stages (Table 13.1). In doing so, the cells remain viable and in a good functional activity. With their increase in breast cancer, HSPs seem to be the main role players in the development of cancer cells, proliferation, and growth.

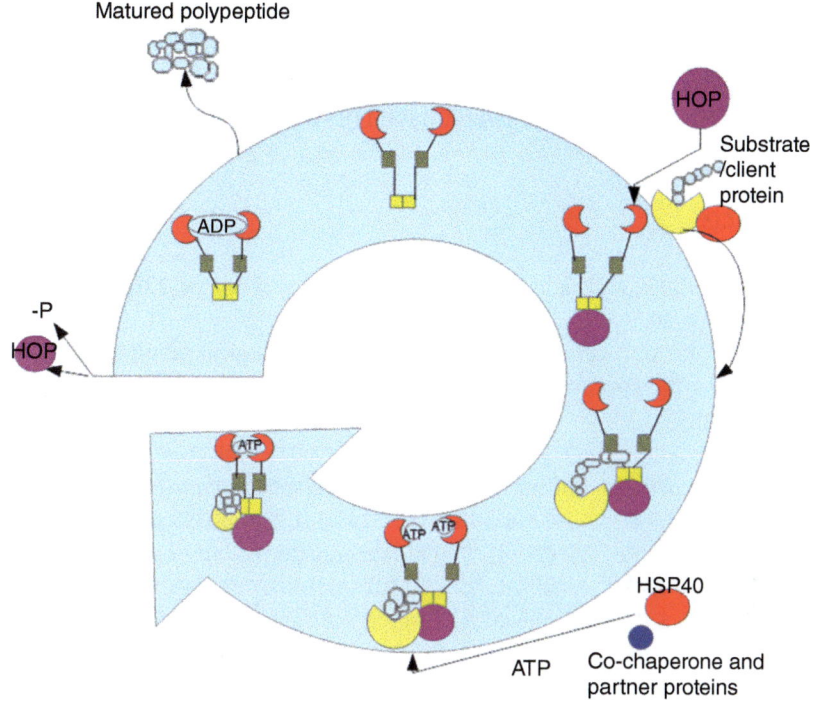

Figure 13.6 Folding of the substrate or client protein presented by HSP70 into HSP90 with the help of co-chaperones.

Table 13.1 Summarizes the sizes and the role of heat-shock proteins during protein folding.

Name	Functional activities	References
HSP40	Recognize the newly produced protein for folding processes and hand them over to HSP70	[1, 2]
HSP60	Holdase and foldase	[14, 17]
HSP70	Folding of the newly synthesized proteins in partnership with HSP40	[5, 7]
HSP90	Assist in the folding process of the newly synthesized polypeptide in partnership with HSP40-HSP70 with heat shock organizing protein (HOP)	[9, 14]
HSP100	Helps newly produced proteins to prevent aggregation	[1, 7]

HSPs in this manner can be used as biomarkers in breast cancer treatment. Additionally, in order to develop a novel treatment for the most deadly form of breast cancer in women worldwide, it is crucial to completely comprehend the HSPs cooperative nature [41]. Especially in developing countries where hospitals and the affordability of medication seem to be the major bottleneck.

Acknowledgments

XHM would like to thank the University of Fort Hare for its SEED grant (C415) and the South African Medical Research Council (SA-MRC-PA19) for financial support.

References

1 Siegel, R., Miller, K.D., and Jemal, A. (2016). Cancer statistics, 2016. *CA Cancer J. Clin.* 66: 7–30.

2 Makki, J. (2015). Diversity of breast carcinoma: histological subtypes and clinical relevance. *Clin. Med. Insights Pathol.* 8: 23–31. https://doi.org/10.4137/CPath.S31563.

3 Yoder, B.J., Wilkinson, E.J., and Massoll, N.A. (2007). Molecular and morphologic distinctions between infiltrating ductal and lobular carcinoma of the breast. *Breast J.* 13: 172–179. https://doi.org/10.1111/j.1524-4741.2007.00393.x.

4 Vineis, P. and Wild, C.P. (2014). Global cancer patterns: causes and prevention. *Lancet* 383: 549–557. https://doi.org/10.1016/S0140-6736(13)62224-2.

5 Calderwood, S.K., Khaleque, M.A., Sawyer, D.B., and Ciocca, D.R. (2006). Heat shock proteins in cancer: chaperones of tumorigenesis. *Trends Biochem. Sci.* 31: 164–172. https://doi.org/10.1016/j.tibs.2006.01.006.

6 Straume, O., Shimamura, T., Lampa, M.J. et al. (2012). Suppression of heat shock protein 27 induces long-term dormancy in human breast cancer. *Proc. Natl. Acad. Sci. U.S.A.* 109: 8699–8704. https://doi.org/10.1073/pnas.1017909109.

7 Cheng, Q., Chang, J.T., Geradts, J. et al. (2012). Amplification and high-level expression of heat shock protein 90 marks aggressive phenotypes of human epidermal growth factor receptor 2 negative breast cancer. *Breast Cancer Res.* 14: R62. https://doi.org/10.1186/bcr3168.

8 Johnson, J.L. (2012). Evolution and function of diverse HSP90 homologs and cochaperone proteins. *Biochim. Biophys. Acta,– Mol. Cell Res.* 1823: 607–613. https://doi.org/10.1016/j.bbamcr.2011.09.020.

9 Nandi, S.S., Suryavanshi, A., Adimule, V., and Yallur, B.C. (2020). Super capacitor characteristics of novel rare earth perovskite nanomaterials of $Sr_{0.5}$, $Cu_{0.4}$, $Y_{0.1}$. *AIP Conf. Proc.* 2274: 020007. https://doi.org/10.1063/5.0022454.

10 Taipale, M., Jarosz, D.F., and Lindquist, S. (2010). HSP90 at the hub of protein homeostasis: emerging mechanistic insights. *Nat. Rev. Mol. Cell Biol.* 11: 515–528. https://doi.org/10.1038/nrm2918.

11 Li, J., Soroka, J., and Buchner, J. (2012). The HSP90 chaperone machinery: conformational dynamics and regulation by co-chaperones. *Biochim. Biophys. Acta* 1823: 624–635. https://doi.org/10.1016/j.bbamcr.2011.09.003.

12 Zuehlke, A. and Johnson, J.L. (2010). HSP90 and co-chaperones twist the functions of diverse client proteins. *Biopolymers* 93: 211–217. https://doi.org/10.1002/bip.21292.

13 Nandi, S.S., Suryavanshi, A., Adimule, V., and Maradur, S.R. (2020). Semiconductor current–voltage characteristics of some novel perovskite ionic nanocomposites of $Sr_{0.5}$, $Cu_{0.4}$, $Y_{0.1}$ and $Sr_{0.5}$, $Mn_{0.5}$ and their electronic sensor applications. *AIP Conf. Proc.* 2274: 020006. https://doi.org/10.1063/5.0022453.

14 Trepel, J., Mollapour, M., Giaccone, G., and Neckers, L. (2010). Targeting the dynamic HSP90 complex in cancer. *Nat. Rev. Cancer* 10: 537–549. https://doi.org/10.1038/nrc2887.

15 Bardwell, J.C. and Craig, E.A. (1988). Ancient heat shock gene is dispensable. *J. Bacteriol.* 170: 2977–2983. https://doi.org/10.1128/jb.170.7.2977-2983.1988.

16 Thomas, J.G. and Baneyx, F. (2000). ClpB and HtpG facilitate de novo protein folding in stressed *Escherichia coli* cells. *Mol. Microbiol.* 36: 1360–1370. https://doi.org/10.1046/j.1365-2958.2000.01951.x.

17 Yosef, I., Goren, M.G., Kiro, R. et al. (2011). High-temperature protein G is essential for activity of the *Escherichia coli* clustered regularly interspaced short palindromic repeats (CRISPR)/Cas system. *Proc. Natl. Acad. Sci. U.S.A.* 108: 20136–20141. https://doi.org/10.1073/pnas.1113519108.

18 Grudniak, A.M., Markowska, K., and Wolska, K.I. (2015). Interactions of *Escherichia coli* molecular chaperone HtpG with DnaA replication initiator DNA. *Cell Stress Chaperones* 20: 951–957. https://doi.org/10.1007/s12192-015-0623-y.

19 Adimule, V., Kerur, S.S., Chinnam, S. et al. (2022). Guar Gum and its nanocomposites as prospective materials for miscellaneous applications: a short review. *Top. Catal.* https://doi.org/10.1007/s11244-022-01587-5.

20 Genest, O., Reidy, M., Street TO et al. (2013). Uncovering a region of heat shock protein 90 important for client binding in *E. coli* and chaperone function in yeast. *Mol. Cell.* 49: 464–473. https://doi.org/10.1016/j.molcel.2012.11.017.

21 Adimule, V., Suryavanshi, A., and Nandi, S.S. (2020). A facile synthesis of poly(3-octyl thiophene):$Ni_{0.4}Sr_{0.6}TiO_3$ hybrid nanocomposites for solar cell applications. *Macromol. Symp.* 392: 2000001. https://doi.org/10.1002/masy.202000001.

22 Southworth, D.R. and Agard, D.A. (2008). Species-dependent ensembles of conserved conformational states define the HSP90 chaperone ATPase cycle. *Mol. Cell.* 32: 631–640. https://doi.org/10.1016/j.molcel.2008.10.024.

23 Graf, C., Stankiewicz, M., Kramer, G., and Mayer, M.P. (2009). Spatially and kinetically resolved changes in the conformational dynamics of the HSP90 chaperone machine. *Embo J.* 28: 602–613. https://doi.org/10.1038/emboj.2008.306.

24 Adimule, V.M., Nandi, S.S., Kerur, S.S. et al. (2022). Recent advances in the one-pot synthesis of coumarin derivatives from different starting materials using nanoparticles: a review. *Top. Catal.* https://doi.org/10.1007/s11244-022-01571-z.

25 Nandi, S.S., Adimule, V., and Yallur, B.C. (2022). Synthesis, structural and optical properties of co doped Sm_2O_3 nanostructures. *Adv. Mater. Res.* 1173: 59–69. Trans Tech Publications, Ltd.

26 Adimule, V., Yallur, B.C., Batakurki, S., and Nandi, S.S. (2022). Synthesis, morphology and enhanced optical properties of novel GdxCo$_3$O$_4$ nanostructures. *Adv. Mater. Res.* 1173, Trans Tech Publications, Ltd.: 71–82.

27 Richter, K., Haslbeck, M., and Buchner, J. (2010). The heat shock response: life on the verge of death. *Mol. Cell.* 40 (2): 253–266. https://doi.org/ 10.1016/j .molcel.2010.10.006.

28 Morimoto, R., Tissieres, A., and Georgopoulos, C. (1990). *Stress Proteins in Biology and Medicine*. New York, NY: Cold Spring Harbor Laboratory Press.

29 Tissières, A., Mitchell, H.K., and Tracy, U.M. (1974). Protein synthesis in salivary glands of *Drosophila melanogaster*: relation to chromosome puffs. *J. Mol. Biol.* 84 (3): 389–398. https://doi.org/ 10.1016/0022-2836(74)90447-1.

30 Adimule, V., Yallur, B.C., Pai, M.M. et al. (2022). Biogenic synthesis of magnetic palladium nanoparticles decorated over reduced graphene oxide using piper betle petiole extract (Pd-rGO@Fe$_3$O$_4$ NPs) as heterogeneous hybrid nanocatalyst for applications in Suzuki-Miyaura coupling reactions of biphenyl compounds. *Top. Catal.* https://doi.org/10.1007/s11244-022-01672-9.

31 Lemaux, P.G., Herendeen, S.L., Bloch, P.L., and Neidhardt, F.C. (1978). Transient rates of synthesis of individual polypeptides in *E. coli* following temperature shifts. *Cell* 13 (3): 427–434. https://doi.org/10.1016/0092-8674(78)90317-3.

32 Adimule, V., Batakurki, S., Yallur, B.C. et al. (2022). Enhanced photoluminescence, optical, structural properties of ZrO$_2$-incorporated Sm$_2$O$_3$:Co$_3$O$_4$ nanocomposite and their applications in photocatalytic degradation of methylene blue. *J. Mater. Res.* https://doi.org/10.1557/s43578-022-00641-y.

33 Peterson, N.S., Moller, G., and Mitchell, H.K. (1979). Genetic mapping of the coding regions for three heat-shock proteins in *Drosophila melanogaster*. *Genetics* 92 (3): 891–902. https://doi.org/10.1093/genetics/92.3.891.

34 McAlister, L. and Finkelstein, D.B. (1980). Heat shock proteins and thermal resistance in yeast. *Biochem. Biophys. Res. Commun.* 93 (3): 819–824. https://doi .org/ 10.1016/0006-291x(80)91150-x.

35 Adimule, V., Jagadeesha Gowda, A.H., Nandi, S.S., and Bowmik, D. (2022). Antimalarial activity of novel class of 1,3-benzoxaborole derivatives containing 1,3,4-oxadiazole moiety. In: *Drug Development for Malaria* (ed. P. Kendrekar). https://doi.org/10.1002/9783527830589.ch12.

36 Young, J.C., Agashe, V.R., Siegers, K., and Hartl, F.U. (2004). Pathways of chaperone-mediated protein folding in the cytosol. *Nat. Rev. Mol. Cell. Biol.* 5 (10): 781–791. https://doi.org/10.1038/nrm1492.

37 Jolly, C. and Morimoto, R.I. (2000). Role of the heat shock response and molecular chaperones in oncogenesis and cell death. *J. Natl. Cancer Inst.* 92 (19): 1564–1572. https://doi.org/10.1093/jnci/92.19.1564.

38 Adimule, V., Batakurki, S., Yallur, B.C. et al. (2022). Samarium-decorated ZrO$_2$@ SnO$_2$ nanostructures, their electrical, optical and enhanced photoluminescence properties. *J. Mater. Sci: Mater. Electron.* 33: 18699–18715. https://doi.org/10.1007/ s10854-022-08718-4.

39 Gomez-Pastor, R., Burchfiel, E.T., and Thiele, D.J. (2017). Regulation of heat shock transcription factors and their roles in physiology and disease. *Nat. Rev. Mol. Cell Biol.* 19 (1): 4–19. https://doi.org/ 10.1038/nrm.2017.73.

40 Kadapure, S.A., Kadapure, P., Nandi, S.S., and Shet, A. (2022). Overview on catalyst and co-solvents for sustainable biodiesel production. *Proc. Inst. Civ. Eng. Energy* 1–9. https://doi.org/10.1680/jener.21.00092.

41 Adimule, V., Nandi, S.S., and Yallur, B.C. (2022). Devices and sensors based on additively manufactured shape-memory of hybrid nanocomposites. In: *Shape Memory Composites Based on Polymers and Metals for 4D Printing* (ed. M.R. Maurya, K.K. Sadasivuni, J.J. Cabibihan, et al.). Cham: Springer https://doi .org/10.1007/978-3-030-94114-7_15.

14

Nanoparticle-Based Therapeutics for Triple Negative Breast Cancer

Isidore A. Egebe and Kamalinder K. Singh

University of Central Lancashire, School of Pharmacy and Biomedical Sciences, Fylde Road, Preston PR1 2HE, United Kingdom

14.1 Breast Cancer: State of Research and Practice

Breast cancer (BC) is a malignant transformation that arise from cells of the mammary glands. It is the most diagnosed cancer worldwide with approximately 2.3 million cases recorded in 2020 and the leading cause of cancer-related mortality in women [1]. BC affect both females and males disproportionately with an estimated 1 in 100 000 incidence for men and 126 in 100 000 for women [2]. As with other cancers, BC follow similar patterns of global variation in incidence. Recent statistics revealed that there are more African American, and Hispanics diagnosed with BC compared to their Caucasian counterparts [3]. Though previously known to be more concentrated in the developed world, recently, there are more cases of BC diagnosed in lower and middle-income countries (LMICs) than in developed countries [3], indicating the influence of social and economic confounding factors in the overall disease burden.

BC survival analysis follows similar disproportionate pattern to incidence around the world. The five-year survival in developed countries goes beyond 80% compared to <70% in India and <50% in South Africa [1]. Although the incidence and mortality were hugely attributed to age and BRCA1/2 mutational status, it is evidenced now that the trajectory has changed to more unpredictable factors. It has been reported that socio-economic transitions, increasing life expectancy, increase in obesity and other health-related factors will contribute to higher BC incidence in developing countries [4]. This is further confounded by delays in presentation, low cancer awareness and fragmented healthcare systems in LMICs as compared to better healthcare access and awareness in developed countries [5, 6]. To this effect, WHO introduced the Global Breast Cancer Initiative in 2021 to address these discrepancies. This commitment to increase BC survival and promote health will focus on early presentation, diagnosis, appropriate treatment and supportive care [7].

Drug and Therapy Development for Triple Negative Breast Cancer, First Edition.
Edited by Pravin Kendrekar, Vinayak Adimule, and Tara Hurst.
© 2023 WILEY-VCH GmbH. Published 2023 by WILEY-VCH GmbH.

Given the global BC statistics anticipating a rise in incidence and mortality, and the commitment to tackling this disease, the focus now is toward proper screening and prevention, diagnosis (subtyping) and treatment. It is established that BC is a heterogenous disease with various molecular subtypes based on estrogen (ER), progesterone (PR), Her-2 and claudin receptors expression status [8]. This has informed scientists and oncologists further to group BC into major different subtypes with the aim of personalizing treatment and improving clinical outcomes. The major BC subtypes includes luminal A, luminal B, basal-like, Her-2 enriched, triple negative BC, and minor subtypes: claudin-low, molecular apocrine and luminal Her-2 [8]. Treatment failures and disease recurrence has warranted reliance on the mentioned subtypes for patient stratification and employment of personalized and targeted treatments.

Selecting a treatment regime for BC is based on multiple factors, however, main treatment types are surgery, chemotherapy, radiotherapy, immunotherapy, molecularly targeted therapies and recent employment of nanoparticle (NP)-based therapeutics [8]. Surgical intervention in BC as for all other non-metastatic resectable tumors is thought to be the most effective treatment type as it can take off whole tumor mass in a single attempt. However, late diagnosis with presentation of metastases and relapse has rendered surgery inappropriate for tackling some forms of BC [9]. This has caused the combination of either surgery with chemotherapy or radiation therapy for BC treatment. Chemotherapy, which is readily available, can target all forms or stages of BC and is cost effective. This is usually used to treat BC but in most cases it encounters recurrence as a result of resistance programmes switched on by tumor cells [10]. And the undesired effects that come with chemotherapy have forced scientists and clinicians to consider other potent but cost-effective interventions.

Radiation therapy, immunotherapy and other targeted therapies have been widely applied in BC treatment. Because of the potency of radiation therapies in combination with immunotherapy in targeting BC metastases and advanced-stage disease, it has been employed at all stages of treatment including as adjuvant therapies [11]. Although treatment outcomes are favorable, the overall survival of BC globally still need to be improved and the major challenge faced with targeted treatment in BC is cost-effectiveness.

The reviewed treatment strategies for BC have been successful but also faced challenges in terms of improving overall survival and cure. NP-based therapeutics have been long applied in BC treatment, to improve on resistance to chemotherapeutics and other forms of treatments, with focus on targeting the tumor microenvironment and heterogeneity. The advent of NP-based therapeutics has brought hopes on improving curative approaches to diseases including cancers. In BC, it has great potential as it can both be used as a treatment and diagnostic tool in what is known as theranostics [12]. Numerous nanocarriers have been used to deliver chemotherapy, small molecules, and gene therapies for BC therapeutics and diagnostics (Figure 14.1). Amongst them includes but are not limited to lipid-based nano formulations [13], solid lipid nanoparticles (SLNs) [14], liposomal formulations [15], lipid polymer hybrid NPs (LPH-NP) [16], polymeric nanoparticles [17], polymeric micelles [18], inorganic NPs such as carbon nanotubes and mesoporous silica NPs [19, 20]. Some FDA-approved and drugs undergoing clinical trials for BC include Doxil [21], Abraxane [22], Myocet [23], NK-105 [24] and have been summarised in Table 14.1.

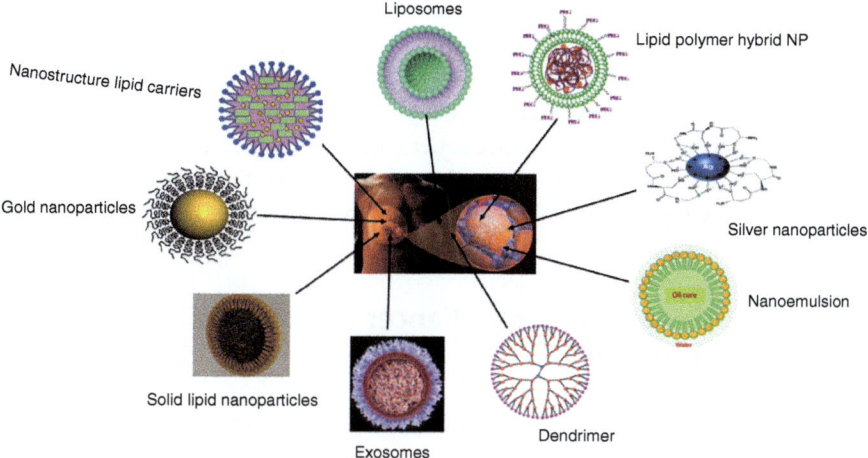

Figure 14.1 Structural representation of various nanoparticle systems utilized in breast cancer treatment. NP = nanoparticles.

Table 14.1 Nanomedicines FDA approved and undergoing clinical trials for breast cancer [25].

Drug name	Nanoparticle-drug combination	Product status
Abraxane	Nanoparticle albumin-bound paclitaxel	Approved in 2005 [22]
Doxil	PEGylated liposomal/doxorubicin hydrochloride	Approved in 1995 [21]
Myocet	Non-PEGylated liposomal/doxorubicin	Phase III [26]
NK-105	PEG-polyaspartate/paclitaxel	Phase III [24]
Genexol-PM	PEG-poly(D, L-lactide)/paclitaxel	Phase III [27]
LEP-ETU	Liposomal paclitaxel	Phase II [28]
NK-102	PEG-polyglutamic acid/SN-38	Phase II [23]
Xyotax	Paclitaxel poliglumex	Phase II [29]
ThermoDox	Heat-activated liposomal-doxorubicin	Phase I/II [30]
Liposomal annamycin	Liposome/semi-synthetic doxorubicin analogue annamycin	Phase I/II [31]
Rexin-G	Protein targeted conjugate phospholipid/ miRNA-122	Phase I/II [32]
SPI-077	Stealth liposomal cisplatin	Phase I [33]
S-CKD602	PEGylated liposomal/CKD602	Phase I [34]
Nanoxel	PEG-poly(D, L-lactide)/docetaxel	Phase I [35]
BIND-014	PEG-polylactic-co-glycolic acid/docetaxel	Phase I [36]

In spite of the successes recorded so far with advancement in research, advocacy for screening and prevention of BC, still remain a powerful approach to mitigate the impact of this disease [37]. As outlined above, deciding a proper treatment for BC is based on the underlying subtype. Additionally, it has been shown that hormone receptors and Her-2 positive BC subtypes are better targeted therapeutically as compared to receptor-negative subtypes [38]. This makes triple negative BC even harder to target and so, will be the next focus of this chapter with exploration of nanotherapeutics.

14.2 Triple Negative Breast Cancer (TNBC) and Treatment Approaches

Unlike other BC subtypes, triple negative breast cancer (TNBC) does not express either of the most common receptors (ER, PR, Her-2) associated with major subtypes of BC. This has made it very challenging to target TNBC with hormone or molecular targeted therapies and till date, there are no standardized treatment for TNBC [39]. Therefore, chemotherapy remains the main systemic treatment and often records poor outcomes as metastatic lesions often lead to recurrence [40]. Combination therapy is considered the main approach to tackle TNBC, however, bevacizumab has been recently combined with chemotherapy for treating TNBC in some countries but recorded no significant increase in patient survival [41], warranting more research into possible treatment regimens.

Epidemiologically, it has been shown that TNBC affects premenopausal woman in a larger proportion and they account for 15–20% of all BC cases [42]. Outcomes in TNBC are poorer as compared to other subtypes, with a 40% five-year survival rate, and approximately 46% of patients develops metastases [43]. Additionally, it has only about 13.5 median survival time after metastasis, 25% recurrence after surgery, a relapse time of 19–40 months compared to 35–67 months for non-TNBC patients and the mortality rate is 75% within three months post recurrence [44, 45]. As a result, TNBC is classified as the most challenging BC subtype and one of the hardest cancers to treat generally.

To understand the molecular heterogeneity of TNBC, to design more rationale drugs to improve patient outcomes, Lehmann et al. performed a deep gene expression profiling on 587 TNBC samples from patients and identified 6 subtypes. They included mesenchymal (M), mesenchymal stem-like (MSL), immunomodulatory (IM), luminal androgen receptor (LAR), basal-like 1 (BL1) and basal-like 2 (BL2) subtypes [46]. Genetic profiling was also performed on existing TNBC cell lines and classified them into six subtypes, providing a matched-up cellular model for effective treatment of TNBC clinically [47]. Although nonspecific, analysis of this type has provided some clue on how patients presenting with TNBC should be stratified further and treated with various combination drug therapies.

The absence of hormone receptors and other molecular target on TNBC, which are present on other BC subtypes and makes them easily "druggable," contributes to the poor treatment outcomes recorded for TNBC. Patients with TNBC are often diagnosed with metastasis, prone to relapse leading to poorer prognosis. This has come down to chemotherapeutic agents being the only treatment option. An interesting recent finding from literature reveals that the application of neoadjuvant chemotherapy treatment regimens for TNBC showed a significantly higher disease remission percentage than in hormone receptor-positive BC thus improving the prognosis in patients with TNBC [39]. According to the national comprehensive cancer network guidelines, recommendations have been made to use drug combination for TNBC based on cisplatin, fluorouracil, cyclophosphamide, taxane and anthracycline. Therefore, currently, cyclophosphamide + epirubicin + fluorouracil + paclitaxel/docetaxel (CEF-T), taxel/docetaxel + Adriamycin + cyclophosphamide (AC), methotrexate + cyclophosphamide + fluorouracil (MCF), and cyclophosphamide + Adriamycin + fluorouracil (CAF) are the adjuvant therapies advised to use for TNBC treatment [39].

Chemotherapeutic agents are still the most commonly used drugs for treating many types of cancers to date. As discussed above, they are recommended to use in combination with TNBC. However, the advent of nanotherapeutics has changed the trajectory with hopes of targeting refractory tumors like TNBC to improve treatment outcomes. NP-based therapeutics have now been applied in treating TNBC with successes recorded and even better outcomes expected. Targeting TNBC with NP-based therapeutics can either be done passively or actively. Passive targeting involves exploiting leaky vasculature around tumors and poor lymphatic drainage that increase uptake of NPs into tumors by the enhanced permeability and retention (EPR) effect. Actively, based on the nanoparticle construct, ligands are appended on their surfaces which are specific to the expressed surface proteins on tumor cells only, thus directing the chemotherapeutic drug only to tumor cells. Based on this understanding, the advances of NP-based therapeutics for TNBC will be discussed below.

14.3 Nanoparticle Therapeutics for TNBC

NPs are substances or materials with sizes within the nanoscale and are generally defined to be <100 nm. NP-based therapeutics is a combination of a nanoparticles and a drug which can be a peptide, small molecule, siRNA, miRNA, chemical substances, etc. The role of the NPs in most cases is to carry these substances with potential therapeutic effect to systemic or defined locations in the body. It is worth mentioning that with NPs acting as the delivery vehicle, they have the potential of controlling the release of these drugs in various amount and concentrations in the body based on the targeted location. Additionally, their properties can be modified to increase the half-life and bioavailability of a therapeutic substance by preventing degradation, and immune and renal clearance.

Broadly, NPs can perform two major activities in a biological system. It is applicable in diagnostics where it is employed to carry and deliver substances with fluorescence properties thus whole, or certain body parts can be imaged. In disease treatment it can deliver a whole range of substances with therapeutic effect and fluorescent substances for guided surgeries [48].

Due to TNBC being aggressive, heterogeneous, and lacking targeting receptors, the main treatment approach remains chemotherapy. A large amount of systemic chemotherapy used in treating patients with TNBC, brings about diverse toxicities and they are resisted by tumor cells in many cases [13]. The need for treatment regimens that selectively target TNBC to reduce toxicity and resistance is on the rise. NP-based therapeutics have the potential to meet these requirements and so they have been applied in treating TNBC [49–51]. A range of NP-based systems have been utilized or undergoing preclinical and clinical trials for TNBC treatment. Table 14.2 summarises the clinical trials being carried out on NP-based therapeutics for TNBC.

Table 14.2   Summary of clinical trials on nanoparticle-based therapeutics for TNBC.

Study type	Treatment type/name	Phase	Trial reference/number
Neoadjuvant therapy for locally advanced or inflammatory TNBC using Carboplatin and paclitaxel albumin stabilized NPs before surgery	Laboratory-based analysis. Drugs: carboplatin, paclitaxel albumin formulation	II	(NCT01525966) [52]
Evaluation of nab-paclitaxel and CORT125134 in patients with solid tumors	Nab-paclitaxel and CORT125132	I/II	NCT02762981 [53]
Paclitaxel liposome weekly and tri-weekly injection for metastatic breast cancer	Paclitaxel liposome ampoule	IV	NCT02142790 [54]
Veliparib treatment for previously treatment unresponsive/refractory malignant solid tumors	Veliparib	I	NCT00892736 [55]
Neoadjuvant therapy of nab-paclitaxel/pembrolizumab followed by epirubicin/pembrolizumab/cyclophosphamide for TNBC	Pembrolizumab, cyclophosphamide, epirubicin, nab-paclitaxel	II	NCT03289819 [56]
Carboplatin + AZD2281 for breast and ovarian cancers treatment	Carboplatin and AZD2281	I	NCT01445418 [57]
Safety and efficacy evaluation of Genexol-PM compared to Genexol for recurrent or metastatic breast cancer treatment	Genexol and Genexol-PM	III	NCT00876486 [27]

14.3.1 Metallic Nanoparticles

Metallic NPs are inorganic in nature and can be grouped into four main types: metal ions, metal oxide, metal sulfide and bimetallic NPs. The metal ions include gold, copper, silver, titanium, platinum, magnesium, and zinc NPs [58]; and the metal oxides are titanium oxide, silver oxide and zinc oxide NPs widely employed in drug trials [59]. Metal sulfide NPs include silver sulfide, copper sulfide and iron sulfide that have excellent antimicrobial activities, so they are widely applicable in medical sciences. The last group is the bimetallic NPs that are a combination of two metals, e.g., copper sulfide, silver sulfide, iron Sulfide [58]. Amongst the metallic NPs, gold and silver NPs have been extensively utilized in biology including cancer for diagnosis, treatment and drug delivery [60].

14.3.1.1 Gold Nanoparticles (AuNPs)

AuNPs are used in cancer therapy due to their low toxicities and ability to bond with anti-tumor agents [61]. A recent combination of AuNPs and paclitaxel (PTX) in an *in vivo* model increased drug delivery efficiency in targeting cells thus decreasing BC cell proliferation. Vemuri et al. supported these findings by combining PTX, curcumin with AuNPs and assessing their anti-metastatic effects in different *in vitro* and *in vivo* models of TNBC. They found that AuNPs decreased expression of VEGF, cyclin-D1, and stat-3 genes and upregulated caspase-9 for apoptosis. The TNBC tumor shrank significantly in the treatment group using AuNPs-drug combination as compared to the control group that received curcumin-PTX only [62]. Another study by Kong et al. found little or no toxic effect of AuNPs on healthy cells but reduced cancer cells migration and invasion by decreasing energy rate thus killing the cells [63].

14.3.1.2 Silver Nanoparticles (AgNPs)

This type of inorganic NPs have been used widely due to their antimicrobial properties and cytotoxic effects on tumor cells. They have been used to target BC brain cancers, hematological malignancies, colon and liver cancers, etc [64, 65].

In a study by Swanner et al., it was shown that both normal BC cells and TNBC cells uptake silver nanoparticles (AgNPs) effectively, however, it was cytotoxic only to the transformed cells because of their rapid degradation of these particles. The stress levels of malignant cells were shown to increase after particle internalization due to increased synthesis of antioxidants with the effect felt at the endoplasmic reticulum, but no changes were observed in non-malignant cells [66]. The same team proved using *in vivo* model that AgNPs can be administered systemically and that was evidenced by a reduction in the size of TNBC in murine model, but there have not been much clinical trials for TNBC patients using AgNPs yet.

14.3.2 Dendrimers

Dendrimers are symmetrical and overly branched NPs with size between 2 and 10 nm in diameter (Figure 14.1). They can be monodisperse and homogeneous which is a unique property which makes them good nano-delivery system for cancer cells targeting [58].

Liu and colleagues performed a study on cancer therapy with this delivery vehicle, focusing on ligand-receptor mediated endocytosis. In their study, MDR1 gene was targeted for silencing in MCF-7 cells by loading siRNA onto PAMAM dendrimers. They generated nanocomplex called dendriplexes with a modified phospholipid PAMAM-siMDR1 complex. The complex showed an increased siMDR1 uptake leading to a gene expression silencing evidenced by decreased expression of p-glycoprotein (p-gp) and there was also accumulation of doxorubicin (DOX) intracellularly [67].

Chittasupho and colleagues, in a study, targeted and inhibited the CXCR4 on the surface of BT-549 TNBC cells using poly(amidoamine) (PAMAM) dendrimers that encapsulated DOX modified with LFC-131 peptide [68]. Other researchers have used poly(propylene imine) (PPI) dendrimers for drug delivery in MCF-7 cells for BC therapy including a study by Kaur and colleagues in which they created a folate conjugated PPI dendrimers that delivers methotrexate (MTX) to MCF-7 with high efficiency [69].

14.3.3 Lipid-Based Nanoparticles (LNPs)

Structurally, LNPs are made up of lipids that are biocompatible and biodegradable which include phospholipids, triglycerides, cholesterol amongst others. Due to their non-toxic properties and the limited inclusion of organic solvents, contributes to LNPs being better option for drug delivery in TNBC as compared to inorganic and polymeric NPs [70, 71]. Various LNPs have been developed for TNBC treatment. They include liposomes, nanoemulsions (NEs), SLNs, nanostructured lipid carriers (NLCs), LPH-NP. A comparison of various LNPs with respect to their advantages and limitations have been compiled in Table 14.3.

14.3.3.1 Liposomes
Liposomes are observed to be a vesicular-type drug carrier system (Figure 14.1) first discovered in 1963 by Alec Bangham who spontaneously dispersed phospholipids in aqueous medium [79]. Liposomes are classified as either uncharged, positively, or negatively charged, and zwitterionic (amphiphilic). The lipids used for positively charged liposomes formation includes *N*-[1-(2,3-dioleyloxy) propyl]-*N*, *N*, *N*-triethylammonium (DOTA) and 1,2-dioleoyl-3-trimethylammoniopropane (DOTAP). Whereas for negatively charged liposomes, phosphatic acid, phosphatidylglycerol, phosphatidylinositol and dicetylphosphate are used. Amphiphilic liposomes composed of phosphatidylethanolamine, phosphatidylcholine, etc. [80].

Charged lipids were shown to possess better drug entrapment properties and some of their components does induce cell death in cancer cells including BC [79]. For TNBC treatment, Guo and colleagues developed a dual-targeting liposome made up of 1,2-dioleoyl-*sn*-glycero-3-phosphocholine (DOPC) and 1,2-distearoyl-*sn*-glycero-3-phosphoethanolamine-*N*-[carboxy(polyethylene glycol)-2000] (DSPE-PEG-COOH) that encapsulated DOX and functionalized on the surface with

Table 14.3 Advantages and limitations of lipid-based nanoparticles in TNBC treatment.

Lipid-based nanoparticles	Positive aspects	Shortcomings	References
Liposomes	Used in drug resistant tumors, lipids compatible with parenteral administration, systemic circulation time prolonged by stealth coating, easily formulated as pegylated liposomes, easily altered pharmacokinetics and biodistribution, has a reduced cardiac toxicity and effective in multidrug resistant tumors.	Attached ligands in active targeted forms may disrupt their stability. Has general stability problems.	[72, 73]
Solid lipid NPs	Possess long-term stability properties, reduced systemic toxicities, possible to load hydrophilic and lipophilic drugs onto, could be scaled up and protects its payload from degradation.	Low drug loading capacity, high incidence of polymorphic transitions, drug expulsion issues, possibility of agglomeration.	[74]
Nanostructured lipid carriers	High water content of NLC dispersion, possibility of drug expulsion during storage, increased drug payload, complement limitations with SLNs.	Contaminated by metal during probe-sonication, high volume of aqueous phase used brings in possibility of dilution of dispersion, can't be applicable in thermos-labile drugs delivery.	[75, 76]
Lipid polymer hybrid NPs	Better serum stability, deliver both hydrophilic and lipophilic drugs, good particle size, allows for surface modification, high drug payload, drug release pattern amenable.	Limited industrial scale production, not cost effective due to aseptic processing, cumbersome lab-scale preparation.	[77, 78]

antibodies against ICAM1 and EGFR for selective TNBC targeting. Their set-up showed an enhanced internalization on MDA-MB-231 (42.7%) and MDA-MB-436 (60.9%) with *in vitro* proliferation inhibition of up to 40% [81] rationalizing dual complementary targeting liposomes as an effective system for targeting TNBC.

The slug gene activates the TGF-β1/Smad pathway, promoting proliferation and invasion of TNBC. Yang et al. synthesized a tLyp-1-peptide modified liposomes made up of DSPE-PEG 2000 that encapsulated miRNA to target and silence this gene. The liposome-miRNA complex showed better TNBC uptake (48.79 ± 0.42) compared to miRNA only (3.69 ± 0.08). Also, the complex showed inhibition of cancer cell proliferation far better than miRNA singly, indicating an overall anti-cancer activity with the usage of DSPE-PEG2000 lipid as a drug nanocarrier [82]. Chen and coauthors generated a detachable immune liposome (ILips) on an immunochemotherapeutic principle that delivered anti-CD47 and PTX into TNBC cells. DOPE was the main lipid in this set-up and their results showed that ILips enhanced the expression of CD-47 in response to MMP2, polarizing macrophages from M2 subtype to M1 thus driving TNBC engulfment and activation of cytotoxic T lymphocytes [83].

Another study by Alawak and colleagues built a thermos-responsive liposome that encapsulated DOX to target TNBC. The liposomes were functionalized with MAB 1031 antibody by covalent coupling (LipTS–GD–MAB), targeting ADAM8 overexpressed in TNBC tumor bearers. Their results showed a significantly increased internalization compared to DOX liposome and toxicity analysis showed up to 80% cellular viability [84]. Additionally, an azadiradione-loaded liposome (AZD-lipo) for TNBC treatment was developed by El-Senduny and coauthors. Their results revealed that AZD-lipo possesses enhanced anti-tumor properties and good bioavailability. This was evidenced by a reduction in proliferative and neo-vascularisation proteins expression in TNBC including D1, COX-2, surviving, VEGF-A compared to free AZD [85].

14.3.3.2 Nanoemulsions (NEs)

NE are made up of two immiscible liquids in a monophasic equilibrium by adding surfactants and co-surfactants (Figure 14.1). Kim and colleagues targeted the LPC and LPAR1 receptors on TNBC by synthesizing decitabine (DAC) and Panobinostat (PAN) NE which was further functionalized with lysophosphatidylcholine and lysophosphatidic acid. They showed that DAC-PAN-LNEs restored CDH 1/E-cadherin and simultaneously inhibited FOXM1 expression, suppressing TNBC cells growth. In addition, MDA-MB-231 cells viability was decreased by 55% supporting NEs to possess therapeutic properties against TNBC [86]. In another study by Xu et al., a NE was developed from soya lecithin and Kolliphor HS15 known as puerarin nanoemulsion (NanoPue) which can be administered orally and is potent against TNBC. Their results revealed that NanoPue inhibited cancer-associated fibroblasts by sixfold and promoted infiltration of cytotoxic T cells into the tumor by twofold hence enhancing the chemotherapeutic effect of nanopaclitaxel in desmoplastic TNBC [87].

Additionally, Han and colleagues developed NE of elemene, a sesquiterpene compound extracted from the rhizome of Curcuma herbs (E-NE) with its lipid phase composed of soybean phospholipid and cholesterol which could treat TNBC and inhibit lung metastasis. The E-NE was found to slow angiogenesis, NLRP3 inflammasomes, IL-1β, and destabilized HIF-1α by successfully scavenging ROS [88]. Also, Saraiva and coauthors, fabricated edelfolsine NE (ET-NEs) from phosphatidylcholine and Miglyol 812 and observed that they possessed *in vitro* and *in vivo* anti-cancer activities. ET-NEs were found to traverse the physiological barriers of xenografted MDA-MB-231 cells into zebrafish embryos hence hindered proliferation of tumor cells confirmed by confocal microscopy [89]. Self-nanoemulsifying systems (SNELS) generating an isotropic thermodynamically stable system have also shown enhanced anticancer activity when tested in DMBA induced preclinical tumor mice model and clathrin mediated endocytosis by MDA-MB-231 cells. Apoptosis assay (65% cell death) and cell cycle distribution (47% inhibition at G2/M phase) further corroborated the cytotoxicity of docetaxol loaded SNELS towards cancerous cells [90].

14.3.3.3 Solid Lipid Nanoparticles (SLNs)

SLNs are colloidal NPs with an outer aqueous phase and inner lipid phase stabilized by a mixture of surfactants [91] (Figure 14.1). It was found by research that its mechanism of drug carriage is by solubilizing lipophilic chemotherapeutic agents within its lipidic matrix or developing an enriched shell for drugs surrounding its lipid core [92, 93]. The most commonly used SLNs for BC treatment include fatty components including soybean lecithin, monoglycerides (glyceryl monostearate [GMS]), triglycerides (glyceryl behenate, trimyristin), and fatty acids (palmitic acid, stearic acid [SA]), etc. [94].

For TNBC treatment, Eskiler and colleagues fabricated a BMN-673 (Talozoparib) loaded SLNs with GMS being the solid lipid phase and Tween 80 the surfactant. SLNs increased the therapeutic index of the drug and was potent against BRCA1 mutated and sensitive TNBC. BMN-673-SLNs significantly decreased HCC1937 and HCC1937-R cells with minimal damage to MCF-10A cells as compared to naked BMN-673. BMN-673-SLNs also significantly caused toxicities in TNBC cells via DNA double strand break, G2/M cell cycle arrest and cleaved moieties of PARP [95]. In an another study by Siddhartha et al., di-allyl-disulfide (DADS) loaded SLNs were synthesized using pluronic F-68 and soy lecithin as surfactant mix with palmitic acid as solid lipid phase, and further tagged with RAGE antibody to increase uptake by TNBC cells with effective targeting. Their results showed that DADS-RAGE-SLNs increase apoptosis and cytotoxicity by 61.8% compared to 15% by DADS only. Also, DADS-RAGE-SLNs were internalized by cells via receptor-mediated endocytosis, effectively bypassing P-gp efflux pump compared to naked DADS [20].

Further, Kothari et al. designed a docetaxel (DTX)-α lipoic acid (ALA) SLNs using stearic acid (SA), GMS, Compritol ATO 888 (glyceryl behenate) and Tween 80 surfactant to target TNBC. From their results, DTX-ALA-SLNs caused significant cytotoxicity in 4T1 cells compared to ALA-SLNs, DTX-SLNs and naked drugs. Moreover, there was 32% apoptosis induced by DTX-ALA-SLNs as opposed to 11% by DTX singly [96]. Furthermore, Pindiprolu and coauthors fabricated niclosamide SLNs (Niclo-SLNs) using Tween 80, pluronic F-68 surfactants and stearyl amine as lipid phase for TNBC targeting. There was significant cytotoxicity induced by Niclo-SLNs and enhanced cellular uptake (77.06%) at cell cycle's G0/G1 phase as opposed to 69.50% for free Niclo. The enhanced cellular internalization of Niclo-SLNs was proposed to have resulted from their ability to escape the cells efflux pump thus increasing drug absorption around the cells [97].

14.3.3.4 Nanostructured Lipid Carriers (NLCs)

NLCs are a mixture of solid and liquid lipid which is stabilized in the aqueous phase by one or more surfactants [98]. They are classified as second-generation lipidic NPs and their ideal drug carrying capabilities are structured by incorporating solid and liquid lipids in the lipid matrix to form a amorphous structure or crystal imperfections (Figure 14.1). Inclusion of liquid lipid facilitates better drug loading into the lipid matrix and reduced drug expulsion [99].

Commonly used liquid lipids for NLCs manufacture are polyoxyl castor oil, labrafil WL 2609 BS, oleic acid, labrafil PG, labrafil M2125 Cs, olive oil, glyceryl tridecanoate, etc. And the solid lipids used include Compritol ATO 888, stearic acid, Precirol, glyceryl tripalmitate, GMS, etc. NLCs are highly tolerable and bio-compatible, have better EPR effect, less drug leakage, and decreased gelation, making it a good choice for anti-cancer drugs delivery.

NLCs have been researched upon extensively, pre-clinically to target TNBC. Pedro et al. developed PTX-loaded NLCs (PTX-NLCs) by combining Compritol ATO 888, MCT, soya lecithin and Tween 80. They observed PTX-NLCs to cause significant cytotoxicity and anti-clonogenic effect on MDA-MB-231 cells compared to PTX singly. Their cell viability analysis showed that there was 56% viability with PTX as compared to 38% with PTX-NLCs. Also, there was 1.5- and 1.7-fold accumulation of PTX-NLCs at 0.5 and 1.5 hours respectively around tumor site in mice harboring tumors as compared to only PTX [100].

Zhang and coauthors also showed that PTX coupled with NLCs targeted TNBC with enhanced uptake in MDA-MB-231 cells [101]. Lages et al. proved the concept further by also showing that DOX and α-tocopherol succinate NLCs were effective against TNBC by improving survival in murine model and preventing metastasis to other organs [102]. Additionally, Gadag et al., showed that resveratrol-loaded NLCs were potent against TNBC and caused reduced cell viability in MDA-MB-231 cells with the possibility to increase bioavailability via intradermal administration [103]. And recently, Gilani et al. developed luteolin-loaded NLCs and used them to treat TNBC cells. They found that MDA-MB-231 cells viability was reduced and the NLCs released its payload slowly over 24 hours with enhanced mucoadhesion, gastro-intestinal stablity and increased intestinal uptake [104].

14.3.3.5 Lipid Polymer Hybrid Nanoparticles (LPH-NPs)

LPH-NPs are a combination of polymeric NPs and LNPs in a single nanosystem and are considered new-generation NPs [105]. The basic structure is composed of a polymeric core surrounded by a lipid monolayer (Figure 14.1) with the apossibility of further functionalization by PEGylation for prolonged stability and ligand functionalisation for enhanced targetability The main polymers used in LPH-NPs include poly β-amino ester, polycaprolactone, poly(lactic-*co*-glycolic acid), and polylactic acid while commonly used lipids are DOTAP/DOPE, stearic acid, cholesterol, lecithin, lipoid GmbH, DSPE-PEG and PEG2000-Mal [106].

RGD peptide conjugated DOX and mitomycin (MMC) dual loaded LPH NPs (DMPLN) were developed by Zhang et al. and treated in TNBC cells. There was cellular accumulation, reduced lung metastasis, and decreased liver and heart toxicity with prolongation of life with RGD-DMPLN as compared to drugs only when evaluated *in vivo* in lung metastases model of TNBC [107]. Zhou et al. also showed that calcium-phosphate-based-LPH-NPs loaded with inhibitors to miRNA-221/222 and PTX for TNBC treatment decreased cell viability by 80% compared to PTX alone at the same drug concentration ($0.67 \mu g\,ml^{-1}$) [108].

Additionally, fucose-anchored-LPH-NPs loaded with MTX and aceclofenac (ACL) were fabricated by Garg and coauthors and treated on TNBC cells. They observed a quick internalization (within two hours) of drug-NPs conjugate by MDA-MB-231 cells with a 10-fold increased systemic circulation compared to free drug. Also, there was approximately 25% decrease in cellular growth in MDA-MB-231 and five to six times increase in mean residence time caused by ACL-LPH-NPs compared to free drug [109]. To support concept further, Bakar-Ates et al. synthesized cucurbitacin B loaded LPH-NPs (CuB-LPH-NPs) to treat TNBC. CuB-LPH-NPs were found to induce apoptosis in MDA-MB-231 cells and decreased cell viability at 1 and 5 µM drug concentrations compared to controls and drug only treatment [110].

The main NP-based therapeutics for treatment interventions in TNBC have been discussed, however, newer areas in biomedical sciences are being repurposed and explored further as nanotechnology-based treatment for TNBC. They include exosomes (Exo), CRISPR and RNA and DNA based nanotherapeutics. They will be briefly outlined herein and then various ligands that can be used to enhance NP internalization, uptake and targetability will be discussed thereafter.

14.3.4 CRISPR Nanoparticles

CRISPR, also known as clustered regularly interspaced short palindromic repeats (usually associated with cas9) has been employed widely in genome and biomedical engineering to target resistance in diseases including BC [58]. The potential for the CRISPR/Cas9 technology is unlimited and promising as it can easily correct genetic alterations in cells and systems that drives disease progression and/or recurrence. TNBC treatment and resistance genes that might have been disrupted could be repaired by CRISPR. These developments could assist with the existing challenging problems observed in TNBC, thereby furthering the current era of precision medicine. In TNBC, this powerful technology has been used to repair genetic information utilizing different nanoparticle systems for encapsulation to overcome the challenges of delivering the unstable CRISPR/Cas9 across cellular barriers to the target site [111]. Research in CRISPR technology has made it applicable in both clinical and experimental gene therapy in cancers. The expression of *BRCA1* has been associated with a worse prognosis of patients with early breast cancer indicating the potential use of CRISPR/Cas9 targeting PI3K to reverse chemoresistance [112].

14.3.5 Exosomes (Exo)

Exosomes are extracellular vesicles measuring in the nanoscale (30–150 nm), enclosed by a lipid bilayer and are released by all body cells [113]. They are mainly enriched by cholesterol and diacylglycerol as opposed to other extracellular vesicles like microvesicles and apoptotic bodies [114]. They are said to possess numerous surface proteins unique to the endosomal pathway and can encapsulate drugs, cytosolic proteins, receptors, nucleic acids, etc., [113, 115]. Their natural biodistribution and the possibility of surface proteins modification to avoid systemic clearance and cell-specific targeting has made them attractive carrier for cancer therapy [116].

Naseri et al. isolated exosomes from bone marrow-derived mesenchymal stem cells (MSCs-Exo) and loaded them with locked nucleic acid-modified anti-miR-142-3p oligonucleotides for BC treatment. *In vitro* and *in vivo* analysis of these exosomes showed enhanced cellular uptake and expression regulation of target genes [117]. Similarly, Gong and coauthors co-loaded exosomes with DOX hydrochloride and cholesterol-modified miRNA to target BC. The vesicles were surface modified with integrin and metallopro-teinase 15 (Co-A15-Exo) to enhance targetability. Their flow cytometric analysis revealed increased uptake (78.60%) of A15-Exo compared to Exo only (15.23%). Also, they recorded more apoptosis in Co-A15-Exo-treated MDA-MB-231 cells as compared to DOX only [118].

14.3.6 Nucleic Acid (NA)-Based Therapeutics

NAs are the fundamental information-carrying molecules in biological systems that control protein expression. They can be of therapeutic value where their normal and/or modified sequences can be utilized to regulate gene expression, and be applicable in disease treatment and prevention. RNA-based therapeutics have been used to target intrinsic cancer pathways and the current main RNA therapeutic approach in clinical trials are interference (RNAi) for gene silencing. However, the major limiting factor is the ability to specifically deliver these therapeutic modules to affected cells and tissues. Nanotechnology holds great promise in this regard and several nanoplatforms have been pursued and RNA nanotherapy for drug delivery is currently under investigation for TNBC [119].

RNAi as a therapeutic tool can either be siRNA or miRNA (miR). miRNA have been widely applied in TNBC treatment where they regulate gene expression. A study utilized RNA nanotechnology in TNBC treatment in which miR-21, an oncogene involved in tumorigenesis and metastasis in cancers including TNBC was silenced by delivery of anti-miRNA-21, targeting only tumor cells, not their healthy counterpart [120]. In an another study liposome-based hydrogel nanoparticles have been utilized to deliver MiR-1296-5p to decrease ERRB2 gene expression in BC [121]. Similarly, Exo have shown good potential to deliver miRNA and target to BC for improved outcomes [117, 118].

14.4 Ligands Used to Enhance Nanoparticle Therapeutics in TNBC

As mentioned earlier, specific ligand inclusion for NP targeted delivery further functionalises their activities and is classed as active targeting as opposed to passive delivery due EPR effect. Ligands are generally stretches of small molecules, peptides or nucleotides specific to surface and intracellular receptors which interact by ligand-receptor binding [51].

14.4.1 Antibodies

Antibodies are Y-shaped proteins with dual epitopes in their basic structure with high affinity and selectivity for their receptors. They have been applicable in TNBC treatment and diagnostics. Rousseau et al., in a preclinical study on TNBC xeno-grafted murine model radiolabelled B-B4 antibodies with I-131. Their results showed good treatment and diagnostic imaging [122].

14.4.2 Peptides

They are classified as low molecular weight ligands and have boosted diagnosis and treatment of disease by being able to target intracellular components with high specificity [123]. Crisp and coauthors in a study developed an activable cell-penetrating peptide that binds tissue metalloproteinase-2 enzymes when linked to cyclic-RGD peptides covalently. Their TNBC *in vivo* experiment showed good contrast in imaging and responded to chemotherapy [124].

14.4.3 Aptamers

Aptamers are single-stranded RNA/DNA short oligonucleotide stretches and due to their unique 3-D conformation, they bind their target with high affinity and strength [51]. However, they face the difficulty of nuclease degradation both extracellularly and intracellularly. Aptamers have been applied mainly in imaging in TNBC research and diagnosis. Li et al. targeted membrane proteins on TNBC cells in their study with a recently found LXL-1 aptamer by SELEX method [125]. In like manner, Huang et al., targeted platelet-derived growth factors differentially expressed on TNBC with PDGF-aptamer aided by Au-NPs conjugation [126].

14.4.4 Small Molecules

They are ligands weighted <500 Da with great potential in cancer imaging with ^{18}F-FDG, a glucose analog being the most widely applicable clinically [51]. Thapa et al. reported significantly higher cellular uptake of folic acid (FR)-decorated cisplatin, and DTX co-loaded liquid crystalline NPs by FR-overexpressing MDA-MB-231 TNBC cells, which was attributed to folate receptor-mediated endocytosis of the targeted NPs. The increased expression of various apoptotic markers including Bax, p21, and cleaved caspase-3 along with improved antimigration effects in MDA-MB-231 TNBC cells following treatment demonstrated that the folate decorated NPs can be used for successful treatment of metastatic breast cancer. The folate-decorated NPs exhibited significantly higher anticancer efficacy both *in vitro* as well as in the MDA-MB-231 tumor xenograft model compared to non-targeted NPs [127].

In a study by Misra et al., AMD3100 or Plerixafor (CXCR4 ligand) was conjugated to poly(lactide-*co*-glycolide) NPs and shown to have an enhanced uptake by MDA-MB-321 TNBC cells and mediated siRNA gene silencing [128]. Additionally, hyaluronic acid with affinity to CD44 receptor was conjugated with cationic liposome nanocomplexes to target gemcitabine and DTX to CD44-overexpressed TNBC [129].

14.5 Conclusion

BC incidence is still on the rise with poor survival and clinical outcomes. Data for TNBC showed even poorer outcomes due to lack of specific receptors that can be targeted therapeutically like for other BC subtypes. NP-based therapeutics have revolutionized approaches in disease targeting in recent times and previous and current data on their application in TNBC is promising. Future research is encouraged in which bioinformatics and single-cell RNA sequencing data could guide on potential target for TNBC which can then be an approach with appropriate ligand functionalized NPs for treatment.

References

1 Wilkinson, L. and Gathani, T. (2022). Understanding breast cancer as a global health concern. *Br. J. Radiol.* 95 (1130): 20211033.

2 Gucalp, A., Traina, T.A., Eisner, J.R. et al. (2019). Male breast cancer: a disease distinct from female breast cancer. *Breast Cancer Res. Treatment* 173 (1): 37.

3 Sung, H., Ferlay, J., Siegel, R.L. et al. (2021). Global cancer statistics 2020: GLOBOCAN estimates of incidence and mortality worldwide for 36 cancers in 185 countries. *CA Cancer J. Clin.* 71 (3): 209–249.

4 Bray, F., Jemal, A., Grey, N. et al. (2012). Global cancer transitions according to the Human Development Index (2008-2030): a population-based study. *Lancet Oncol.* 13 (8): 790–801.

5 Brand, N.R., Qu, L.G., Chao, A., and Ilbawi, A.M. (2019). Delays and barriers to cancer care in low- and middle-income countries: a systematic review. *Oncologist* 24 (12): e1371–e1380.

6 Denny, L., de Sanjose, S., Mutebi, M. et al. (2017). Interventions to close the divide for women with breast and cervical cancer between low-income and middle-income countries and high-income countries. *Lancet* 389 (10071): 861–870.

7 WHO (2021). *New Global Breast Cancer Initiative Highlights Renewed Commitment to Improve Survival. WHO News* https://www.who.int/news/item/08-03-2021-new-global-breast-cancer-initiative-highlights-renewed-commitment-to-improve-survival.

8 Johnson, K.S., Conant, E.F., and Soo, M.S. (2021). Molecular subtypes of breast cancer: a review for breast radiologists. *J. Breast Imaging* 3 (1): 12–24.

9 Cheung, K.L. (2020). Treatment strategies and survival outcomes in breast cancer. *Cancers* 12 (3): 735.

10 Lairson, D.R., Parikh, R.C., Cormier, J.N. et al. (2015). Cost-effectiveness of chemotherapy for breast cancer and age effect in older women. *Value Health* 18 (8): 1070–1078.

11 David, S., Tan, J., Siva, S. et al. (2022). Combining radiotherapy and immunotherapy in metastatic breast cancer: current status and future directions. *Biomedicines* 10 (4): 821.

12 Afzal, M., Ameeduzzafar, Alharbi, K.S. et al. (2021). Nanomedicine in treatment of breast cancer – a challenge to conventional therapy. *Semin. Cancer Biol.* 69: 279–292.

13 Liu, Y., Xu, J., Choi, H.H. et al. (2018). Targeting 17q23 amplicon to overcome the resistance to anti-HER2 therapy in HER2+ breast cancer. *Nat. Commun.* 9 (1): 1–16.

14 Zheng, G., Zheng, M., Yang, B. et al. (2019). Improving breast cancer therapy using doxorubicin loaded solid lipid nanoparticles: synthesis of a novel arginine-glycine-aspartic tripeptide conjugated, pH sensitive lipid and evaluation of the nanomedicine in vitro and in vivo. *Biomed. Pharmacother.* 116: 109006.

15 Xia, T., He, Q., Shi, K. et al. (2016). Losartan loaded liposomes improve the antitumor efficacy of liposomal paclitaxel modified with pH sensitive peptides by inhibition of collagen in breast cancer. *Pharm. Dev. Technol.* 23 (1): 13–21.

16 Jadon, R.S. and Sharma, M. (2019). Docetaxel-loaded lipid-polymer hybrid nanoparticles for breast cancer therapeutics. *J. Drug Delivery Sci. Technol.* 51: 475–484.

17 Soe, Z.C., Kwon, J.B., Thapa, R.K. et al. (2019). Transferrin-conjugated polymeric nanoparticle for receptor-mediated delivery of doxorubicin in doxorubicin-resistant breast cancer cells. *Pharmaceutics* 11 (2): 63.

18 Yang, J., Lu, W., Xiao, J. et al. (2018). A positron emission tomography image-guidable unimolecular micelle nanoplatform for cancer theranostic applications. *Acta Biomater.* 79: 306–316.

19 Jin, G., He, R., Liu, Q. et al. (2019). Near-infrared light-regulated cancer theranostic nanoplatform based on aggregation-induced emission luminogen encapsulated upconversion nanoparticles. *Theranostics* 9 (1): 246.

20 Siddhartha, V.T., Pindiprolu, S.K.S.S., Chintamaneni, P.K. et al. (2017a). RAGE receptor targeted bioconjuguate lipid nanoparticles of diallyl disulfide for improved apoptotic activity in triple negative breast cancer: in vitro studies. *Artif. Cells Nanomed. Biotechnol.* 46 (2): 387–397.

21 Barenholz, Y. (2012). Doxil® — The first FDA-approved nano-drug: lessons learned. *J. Controlled Release* 160 (2): 117–134.

22 Hawkins, M.J., Soon-Shiong, P., and Desai, N. (2008). Protein nanoparticles as drug carriers in clinical medicine. *Adv. Drug Delivery Rev.* 60 (8): 876–885.

23 Cannon, S. and Yardley, D.A. (2015). A Study of NK012 in patients with advanced, metastatic triple negative breast cancer (Retrieved 23 October 2022). https://clinicaltrials.gov/ct2/show/NCT00951054.

24 Fujiwara, Y., Mukai, H., Saeki, T. et al. (2019a). A multi-national, randomised, open-label, parallel, phase III non-inferiority study comparing NK105 and paclitaxel in metastatic or recurrent breast cancer patients. *Br. J. Cancer* 120 (5): 475–480.

25 Tang, X., Loc, W.S., Dong, C. et al. (2017). The use of nanoparticulates to treat breast cancer. *Nanomedicine* 12 (19): 2367.

26 Baselga, J., Manikhas, A., Cortés, J. et al. (2014). Phase III trial of nonpegylated liposomal doxorubicin in combination with trastuzumab and paclitaxel in HER2-positive metastatic breast cancer. *Ann. Oncol.* 25 (3): 592–598.

27 Samyang Biopharmaceuticals Corporation (2013). Evaluate the efficacy and safety of Genexol®-PM compared to Genexol® in recurrent or metastatic breast cancer (Retrieved 30 October 2022). https://clinicaltrials.gov/ct2/show/NCT00876486.

28 Caraglia, M., De Rosa, G., Salzano, G. et al. (2012). Nanotech revolution for the anti-cancer drug delivery through blood-brain-barrier. *Curr. Cancer Drug Targets 12* (3): 186–196.

29 Shulman, L.N. (2007). Study of Xyotax (CT-2103) in patients with metastatic breast cancer (Retrieved 30 October 2022). https://clinicaltrials.gov/ct2/show/NCT00148707.

30 Zagar, T.M., Vujaskovic, Z., Formenti, S. et al. (2014). Two phase I dose-escalation/pharmacokinetics studies of low temperature liposomal doxorubicin (LTLD) and mild local hyperthermia in heavily pretreated patients with local regionally recurrent breast cancer. *Int. J. Hyperth.* 30 (5): 285–294.

31 Volm, M.D. (2001). Chemotherapy in treating patients with breast cancer (Retrieved 30 October 2022). https://clinicaltrials.gov/ct2/show/NCT00012129.

32 Chawla, S.P. and Bruckner, H.W. (2011). Safety and efficacy study using Rexin-G for breast cancer (Retrieved 30 October 2022). https://clinicaltrials.gov/ct2/show/NCT00505271.

33 Lassen, U. and Nielsen, D. (2021). Phase I/II study to evaluate the safety and tolerability of LiPlaCis in patients with advanced or refractory tumours (Retrieved 30 October 2022). https://clinicaltrials.gov/ct2/show/NCT01861496.

34 Ramanathan, R.K. (2006). Safety study of S-CKD602 in patients with advanced malignancies (Retrieved 30 October 2022). https://clinicaltrials.gov/ct2/show/NCT00177281.

35 Mishra, S.K. and Sharma, A. (2010). A clinical trial to study the effects of nanoparticle based paclitaxel drug, which does not contain the solvent cremophor. Advanced Breast Cancer (Retrieved 30 October 2022). https://clinicaltrials.gov/ct2/show/NCT00915369.

36 BIND Therapeutics (2016). A study of BIND-014 given to patients with advanced or metastatic cancer (Retrieved October 30, 2022). https://clinicaltrials.gov/ct2/show/NCT01300533.

37 Gagnon, J., Lévesque, E., Borduas, F. et al. (2016). Recommendations on breast cancer screening and prevention in the context of implementing risk stratification: impending changes to current policies. *Curr. Oncol.* 23 (6): e615.

38 Han, Y., Wu, Y., Xu, H. et al. (2022). The impact of hormone receptor on the clinical outcomes of HER2-positive breast cancer: a population-based study. *Int. J. Clin. Oncol.* 27 (4): 707–716.

39 Yin, L., Duan, J.J., Bian, X.W., and Yu, S.C. (2020). Triple-negative breast cancer molecular subtyping and treatment progress. *Breast Cancer Res.* 22 (1): 61.

40 Chaudhary, L.N., Wilkinson, K.H., and Kong, A. (2018). Triple-negative breast cancer: who should receive neoadjuvant chemotherapy? *Surg. Oncol. Clin. North Am.* 27 (1): 141–153.

41 Collignon, J., Lousberg, L., Schroeder, H., and Jerusalem, G. (2016). Triple-negative breast cancer: treatment challenges and solutions. *Breast Cancer: Targets Ther.* 8: 93.

42 Morris, G.J., Naidu, S., Topham, A.K. et al. (2007). Differences in breast carcinoma characteristics in newly diagnosed African–American and Caucasian patients. *Cancer* 110 (4): 876–884.

43 Dent, R., Trudeau, M., Pritchard, K.I. et al. (2007). Triple-negative breast cancer: clinical features and patterns of recurrence. *Clin. Cancer Res.* 13 (15): 4429–4434.

44 Lin, N.U., Claus, E., Sohl, J. et al. (2008). Sites of distant recurrence and clinical outcomes in patients with metastatic triple-negative breast cancer. *Cancer* 113 (10): 2638–2645.

45 Zhang, L., Fang, C., Xu, X. et al. (2015). Androgen receptor, EGFR, and BRCA1 as biomarkers in triple-negative breast cancer: a meta-analysis. *BioMed Res. Int.* 2015: 357485.

46 Lehmann, B.D., Bauer, J.A., Chen, X. et al. (2011). Identification of human triple-negative breast cancer subtypes and preclinical models for selection of targeted therapies. *J. Clin. Invest.* 121 (7): 2750–2767.

47 Lehmann, B.D. and Pietenpol, J.A. (2014). Identification and use of biomarkers in treatment strategies for triple-negative breast cancer subtypes. *J. Pathol.* 232 (2): 142–150.

48 Murthy, S.K. (2007). Nanoparticles in modern medicine: state of the art and future challenges. *Int. J. Nanomed.* 2 (2): 129.

49 Hou, K., Ning, Z., Chen, H., and Wu, Y. (2021). Nanomaterial technology and triple negative breast cancer. *Front. Oncol.* 11: 828810.

50 Sharma, A., Goyal, A.K., and Rath, G. (2017). Recent advances in metal nanoparticles in cancer therapy. *J. Drug. Target.* 26 (8): 617–632.

51 Thakur, V. and Kutty, R.V. (2019). Recent advances in nanotheranostics for triple negative breast cancer treatment. *J. Exp. Clin. Cancer Res.* 38 (1): 1–22.

52 Yuan, Y. (2019). Carboplatin and paclitaxel albumin-stabilized nanoparticle formulation before surgery in treating patients with locally advanced or inflammatory triple negative breast cancer (Retrieved 29 October 2022). https://clinicaltrials.gov/ct2/show/NCT01525966.

53 Corcept Therapeutics (2020). Study to evaluate CORT125134 in combination with Nab-paclitaxel in patients with solid tumors (Retrieved 29 October 2022). https://clinicaltrials.gov/ct2/show/NCT02762981.

54 Ouyang, Q. (2016). Weekly and every 3 week administration of paclitaxel liposome injection in metastatic breast cancer (Retrieved 29 October 2022). https://clinicaltrials.gov/ct2/show/NCT02142790.

55 Puhalla, S. (2017). Veliparib in treating patients with malignant solid tumors that do not respond to previous therapy (Retrieved 29 October 2022). https://clinicaltrials.gov/ct2/show/NCT00892736.

56 Fasching, P.A. (2021). Neoadjuvant pembrolizumab (Pbr)/nab-paclitaxel followed by Pbr/Epirubicin/Cyclophosphamide in TNBC (Retrieved 29 October 2022). https://clinicaltrials.gov/ct2/show/NCT03289819.

57 Lee, J.-M. (2019). ZD2281 plus carboplatin to treat breast and ovarian cancer (Retrieved 29 October 2022). https://clinicaltrials.gov/ct2/show/NCT01445418.

58 Nahvi, I., Belkahla, S., Biswas, S., and Chakraborty, S. (2022). A review on nanocarrier mediated treatment and management of triple negative breast cancer: a Saudi Arabian scenario. *Front. Oncol.* 12: 953865.

59 Umar, K., Haque, M.M., Mir, N.A. et al. (2013). Titanium dioxide-mediated photocatalysed mineralization of two selected organic pollutants in aqueous suspensions. *J. Adv. Oxid. Technol.* 16 (2): 252–260.

60 Fahrenholtz, C.D., Swanner, J., Ramirez-Perez, M., and Singh, R.N. (2017). Heterogeneous responses of ovarian cancer cells to silver nanoparticles as a single agent and in combination with cisplatin. *J. Nanomater.* 2017: 5107485.

61 Akter, Z., Khan, F.Z., and Khan, M.A. (2023). Gold nanoparticles in triple-negative breast cancer therapeutics. *Curr. Med. Chem.* 30 (3): 316–334.

62 Vemuri, S.K., Halder, S., Banala, R.R. et al. (2022). Modulatory effects of biosynthesized gold nanoparticles conjugated with curcumin and paclitaxel on tumorigenesis and metastatic pathways—in vitro and in vivo studies. *Int. J. Mol. Sci.* 23 (4): 2150.

63 Kong, T., Zeng, J., Wang, X. et al. (2008). Enhancement of radiation cytotoxicity in breast-cancer cells by localized attachment of gold nanoparticles. *Small* 4 (9): 1537–1543.

64 Guo, D., Zhao, Y., Zhang, Y. et al. (2014). The cellular uptake and cytotoxic effect of silver nanoparticles on chronic myeloid leukemia cells. *J. Biomed. Nanotechnol.* 10 (4): 669–678.

65 Sharma, S., Chockalingam, S., Sanpui, P. et al. (2014). Silver nanoparticles impregnated alginate–chitosan-blended nanocarrier induces apoptosis in human glioblastoma cells. *Adv. Healthcare Mater.* 3 (1): 106–114.

66 Swanner, J., Fahrenholtz, C.D., Tenvooren, I. et al. (2019). Silver nanoparticles selectively treat triple-negative breast cancer cells without affecting non-malignant breast epithelial cells in vitro and in vivo. *FASEB BioAdv.* 1 (10): 639–660.

67 Liu, J., Li, J., Liu, N. et al. (2017). In vitro studies of phospholipid-modified PAMAM-siMDR1 complexes for the reversal of multidrug resistance in human breast cancer cells. *Int. J. Pharm.* 530 (1, 2): 291–299.

68 Chittasupho, C., Anuchapreeda, S., and Sarisuta, N. (2017). CXCR4 targeted dendrimer for anti-cancer drug delivery and breast cancer cell migration inhibition. *Eur. J. Pharm. Biopharm.* 119: 310–321.

69 Kaur, A., Jain, K., Mehra, N.K., and Jain, N.K. (2017). Development and characterization of surface engineered PPI dendrimers for targeted drug delivery. *Artif. Cells Nanomed. Biotechnol.* 45 (3): 414–425.

70 Khurana, R.K., Beg, S., Burrow, A.J. et al. (2017). Enhancing biopharmaceutical performance of an anticancer drug by long chain PUFA based self-nanoemulsifying lipidic nanomicellar system. *Eur. J. Pharm. Biopharm.* 121: 42–60.

71 Khurana, R.K., Gaspar, B.L., Welsby, G. et al. (2018). Improving the biopharmaceutical attributes of mangiferin using vitamin E-TPGS co-loaded self-assembled phosholipidic nano-mixed micellar systems. *Drug Deliv. Transl. Res.* 8 (3): 617–632.

72 Allen, T.M. and Martin, F.J. (2004). Advantages of liposomal delivery systems for anthracyclines. *Semin. Oncol.* 31 (6 Suppl 13): 5–15.

73 Gabizon, A.A., Shmeeda, H., and Zalipsky, S. (2006). Pros and cons of the liposome platform in cancer drug targeting. *J. Liposome Res.* 16 (3): 175–183.

74 Mukherjee, S., Ray, S., and Thakur, R.S. (2009). Solid lipid nanoparticles: a modern formulation approach in drug delivery system. *Indian J. Pharm. Sci.* 71 (4): 349.

75 Iqbal, M.A., Md, S., Sahni, J.K. et al. (2012). Nanostructured lipid carriers system: recent advances in drug delivery. *J. Drug. Target.* 20 (10): 813–830.

76 Müller, R.H., Radtke, M., and Wissing, S.A. (2002). Nanostructured lipid matrices for improved microencapsulation of drugs. *Int. J. Pharm.* 242 (1–2): 121–128.

77 Hadinoto, K., Sundaresan, A., and Cheow, W.S. (2013). Lipid–polymer hybrid nanoparticles as a new generation therapeutic delivery platform: a review. *Eur. J. Pharm. Biopharm.* 85 (3): 427–443.

78 Mandal, B., Bhattacharjee, H., Mittal, N. et al. (2013). Core–shell-type lipid–polymer hybrid nanoparticles as a drug delivery platform. *Nanomed. Nanotechnol. Biol. Med.* 9 (4): 474–491.

79 Arias, J., Clares, B., Morales, E., and M., Gallardo, V., & A. Ruiz, M. (2011). Lipid-based drug delivery systems for cancer treatment. *Curr. Drug Targets* 12 (8): 1151–1165.

80 Olusanya, T.O.B., Ahmad, R.R.H., Ibegbu, D.M. et al. (2018). Liposomal drug delivery systems and anticancer drugs. *Molecules* 23 (4): 907.

81 Guo, P., Yang, J., Liu, D. et al. (2019). Dual complementary liposomes inhibit triple-negative breast tumor progression and metastasis. *Sci. Adv.* 5 (3): eaav5010.

82 Yan, Y., Li, X.Q., Duan, J.L. et al. (2019). Nanosized functional miRNA liposomes and application in the treatment of TNBC by silencing Slug gene. *Int. J. Nanomed.* 14: 3645.

83 Chen, M., Miao, Y., Qian, K. et al. (2021). Detachable liposomes combined immunochemotherapy for enhanced triple-negative breast cancer treatment through reprogramming of tumor-associated macrophages. *Nano Lett.* 21 (14): 6031–6041.

84 Alawak, M., Abu Dayyih, A., Mahmoud, G. et al. (2021). ADAM 8 as a novel target for doxorubicin delivery to TNBC cells using magnetic thermosensitive liposomes. *Eur. J. Pharm. Biopharm.* 158: 390–400.

85 El-Senduny, F.F., Altouhamy, M., Zayed, G. et al. (2021). Azadiradione-loaded liposomes with improved bioavailability and anticancer efficacy against triple negative breast cancer. *J. Drug Delivery Sci. Technol.* 65: 102665.

86 Kim, B., Pena, C.D., and Auguste, D.T. (2019). Targeted lipid nanoemulsions encapsulating epigenetic drugs exhibit selective cytotoxicity on CDH1−/FOXM1+ triple negative breast cancer cells. *Mol. Pharmaceutics* 16 (5): 1813–1826.

87 Xu, H., Hu, M., Liu, M. et al. (2020). Nano-puerarin regulates tumor microenvironment and facilitates chemo- and immunotherapy in murine triple negative breast cancer model. *Biomaterials* 235: 119769.

88 Han, B., Wang, T., Xue, Z. et al. (2021). Elemene nanoemulsion inhibits metastasis of breast cancer by ROS scavenging. *Int. J. Nanomed.* 16: 6035.

89 Saraiva, S.M., Gutiérrez-Lovera, C., Martínez-Val, J. et al. (2021). Edelfosine nanoemulsions inhibit tumor growth of triple negative breast cancer in zebrafish xenograft model. *Sci. Rep.* 11 (1): 1–13.

90 Khurana, R.K., Kumar, R., Gaspar, B.L. et al. (2018). Clathrin-mediated endocytic uptake of PUFA enriched self-nanoemulsifying lipidic systems (SNELS) of an anticancer drug against triple negative cancer and DMBA induced preclinical tumor model. *Mater. Sci. Eng. C Mater. Biol. Appl.* 91: 645–658.

91 Shegokar, R., Singh, K.K., and Mueller, R.H. (2011). Production & stability of stavudine solid lipid nanoparticles-from lab to industrial scale. *Int. J. Pharm.* 416 (2): 461–470.

92 Athawale, R.B., Jain, D.S., Singh, K.K., and Gude, R.P. (2014). Etoposide loaded solid lipid nanoparticles for curtailing B16F10 melanoma colonization in lung. *Biomed. Pharmacother.* 68 (2): 231–240.

93 Singh, K.K. and Pople, P.V. (2011). Safer than safe: lipid nanoparticulate encapsulation of tacrolimus with enhanced targeting and improved safety for atopic dermatitis. *J. Biomed. Nanotechnol.* 7 (1): 40–41.

94 Talluri, S.V., Kuppusamy, G., Karri, V.V.S.R. et al. (2015). Lipid-based nanocarriers for breast cancer treatment – comprehensive review. *Drug. Deliv.* 23 (4): 1291–1305.

95 Guney Eskiler, G., Cecener, G., Egeli, U., and Tunca, B. (2018). Synthetically lethal BMN 673 (Talazoparib) loaded solid lipid nanoparticles for BRCA1 mutant triple negative breast cancer. *Pharm. Res.* 35 (11): 1–20.

96 Kothari, I.R., Mazumdar, S., Sharma, S. et al. (2019). Docetaxel and alpha-lipoic acid co-loaded nanoparticles for cancer therapy. *Ther. Deliv.* 10 (4): 227–240.

97 Pindiprolu, S.K.S.S., Chintamaneni, P.K., Krishnamurthy, P.T. et al. (2018). Formulation-optimization of solid lipid nanocarrier system of STAT3 inhibitor to improve its activity in triple negative breast cancer cells. *Drug Dev. Ind. Pharm.* 45 (2): 304–313.

98 Khurana, R.K., Bansal, A., Beg, S. et al. (2017). Enhancing biopharmaceutical attributes of phospholipid complex-loaded nanostructured lipidic carriers of mangiferin: systematic development, characterization and evaluation. *Int. J. Pharm.* 518 (1-2): 289–306.

99 Houacine, C., Adams, D., and Singh, K.K. (2020). Impact of liquid lipid on development and stability of trimyristin nanostructured lipid carriers for oral delivery of resveratrol. *J. Mol. Liq.* 316: 113734.

100 Pedro, I.D.R., Almeida, O.P., Martins, H.R. et al. (2019). Optimization and in vitro/in vivo performance of paclitaxel-loaded nanostructured lipid carriers for breast cancer treatment. *J. Drug Delivery Sci. Technol.* 54: 101370.

101 Zhang, Q., Zhao, J., Hu, H. et al. (2019). Construction and in vitro and in vivo evaluation of folic acid-modified nanostructured lipid carriers loaded with paclitaxel and chlorin e6. *Int. J. Pharm.* 569: 118595.

102 Lages, E.B., Fernandes, R.S., de Oliveira Silva, J. et al. (2020). Co-delivery of doxorubicin, docosahexaenoic acid, and α-tocopherol succinate by nanostructured lipid carriers has a synergistic effect to enhance antitumor activity and reduce toxicity. *Biomed. Pharmacother.* 132: 110876.

103 Gadag, S., Narayan, R., Nayak, A.S. et al. (2021). Development and preclinical evaluation of microneedle-assisted resveratrol loaded nanostructured lipid carriers for localized delivery to breast cancer therapy. *Int. J. Pharm.* 606: 120877.

104 Gilani, S.J., Bin-Jumah, M., Rizwanullah, M. et al. (2021). Chitosan coated luteolin nanostructured lipid carriers: optimization, in vitro-ex vivo assessments and cytotoxicity study in breast cancer cells. *Coatings* 11 (2): 158.

105 Jain, S., Kumar, S., Agrawal, A.K. et al. (2013). Enhanced transfection efficiency and reduced cytotoxicity of novel lipid-polymer hybrid nanoplexes. *Mol. Pharmaceutics* 10 (6): 2416–2425.

106 Persano, F., Gigli, G., and Leporatti, S. (2021). Lipid-polymer hybrid nanoparticles in cancer therapy: current overview and future directions. *Nano Express* 2 (1): 012006.

107 Zhang, T., Prasad, P., Cai, P. et al. (2017). Dual-targeted hybrid nanoparticles of synergistic drugs for treating lung metastases of triple negative breast cancer in mice. *Acta Pharmacol. Sin.* 38 (6): 835–847.

108 Zhou, Z., Kennell, C., Lee, J.Y. et al. (2017). Calcium phosphate-polymer hybrid nanoparticles for enhanced triple negative breast cancer treatment via co-delivery of paclitaxel and miR-221/222 inhibitors. *Nanomed. Nanotechnol. Biol. Med.* 13 (2): 403–410.

109 Garg, N.K., Tyagi, R.K., Sharma, G. et al. (2017). Functionalized lipid-polymer hybrid nanoparticles mediated codelivery of methotrexate and aceclofenac: a synergistic effect in breast cancer with improved pharmacokinetics attributes. *Mol. Pharmaceutics* 14 (6): 1883–1897.

110 Bakar-Ates, F., Ozkan, E., and Sengel-Turk, C.T. (2020). Encapsulation of cucurbitacin B into lipid polymer hybrid nanocarriers induced apoptosis of MDAMB231 cells through PARP cleavage. *Int. J. Pharm.* 586: 119565.

111 Chen, Y., Zhang, Y., Chen, Y.N., and Zhang, Y.M. (2018). Application of the CRISPR/Cas9 system to drug resistance in breast cancer. *Adv. Sci.* 5 (6): 1700964.

112 Singh, D., Han, I., Choi, E.H., and Yadav, D.K. (2021). CRISPR/Cas9 based genome editing for targeted transcriptional control in triple-negative breast cancer. *Comput. Struct. Biotechnol. J.* 19: 2384–2397.

113 Kumar, D.N., Chaudhuri, A., Aqil, F. et al. (2022). Exosomes as emerging drug delivery and diagnostic modality for breast cancer: recent advances in isolation and application. *Cancers* 14 (6): 1435.

114 Antimisiaris, S.G., Mourtas, S., and Marazioti, A. (2018). Exosomes and exosome-inspired vesicles for targeted drug delivery. *Pharmaceutics* 10 (4): 218.

115 Kumar, D.N., Chaudhuri, A., Singh, S., and Agrawal, A.K. (2022). Extracellular vesicles for nucleic acid delivery: progress and prospects for safe RNA-based gene therapy. In: *Gene Delivery* (ed. Y.V. Pathak), 31–50. Boca Raton, Florida, USA: CRC Press.

116 Aqil, F., Munagala, R., Jeyabalan, J. et al. (2017). Exosomes for the enhanced tissue bioavailability and efficacy of curcumin. *AAPS J.* 19 (6): 1691–1702.

117 Naseri, Z., Oskuee, R.K., Jaafari, M.R., and Moghadam, M.F. (2018). Exosome-mediated delivery of functionally active miRNA-142-3p inhibitor reduces tumorigenicity of breast cancer in vitro and in vivo. *Int. J. Nanomed.* 13: 7727.

118 Gong, C., Tian, J., Wang, Z. et al. (2019). Functional exosome-mediated co-delivery of doxorubicin and hydrophobically modified microRNA 159 for triple-negative breast cancer therapy. *J. Nanobiotechnol.* 17 (1): 1–18.

119 Haque, S., Cook, K., Sahay, G., and Sun, C. (2021). RNA-based therapeutics: current developments in targeted molecular therapy of triple-negative breast cancer. *Pharmaceutics* 13 (10): 1694.

120 Shu, D., Li, H., Shu, Y. et al. (2015). Systemic delivery of anti-miRNA for suppression of triple negative breast cancer utilizing RNA nanotechnology. *ACS Nano* 9 (10): 9731–9740.

121 Chen, G., He, M., Yin, Y. et al. (2017). MiR-1296-5p decreases ERBB2 expression to inhibit the cell proliferation in ERBB2-positive breast cancer. *Cancer Cell Int.* 17 (1): 1–8.

122 Rousseau, C., Ruellan, A.L., Bernardeau, K. et al. (2011). Syndecan-1 antigen, a promising new target for triple-negative breast cancer immuno-PET and radioimmunotherapy. A preclinical study on MDA-MB-468 xenograft tumors. *EJNMMI Res.* 1 (1): 1–11.

123 Reubi, J.C. and Maecke, H.R. (2008). Peptide-based probes for cancer imaging. *J. Nucl. Med.* 49 (11): 1735–1738.

124 Crisp, J.L., Savariar, E.N., Glasgow, H.L. et al. (2014). Dual targeting of integrin αvβ3 and matrix metalloproteinase-2 for optical imaging of tumors and chemotherapeutic delivery. *Mol. Cancer Therap.* 13 (6): 1514–1525.

125 Li, X., Zhang, W., Liu, L. et al. (2014). In vitro selection of DNA aptamers for metastatic breast cancer cell recognition and tissue imaging. *Anal. Chem.* 86 (13): 6596–6603.

126 Huang, Y.F., Lin, Y.W., Lin, Z.H., and Chang, H.T. (2008). Aptamer-modified gold nanoparticles for targeting breast cancer cells through light scattering. *J. Nanopart. Res.* 11 (4): 775–783.

127 Thakur, V. and Kutty, R.V. (2019). Recent advances in nanotheranostics for triple negative breast cancer treatment. *J. Exp. Clin. Cancer Res.* 38 (1): 430.

128 Misra, A.C., Luker, K.E., Durmaz, H. et al. (2015). CXCR4-targeted nanocarriers for triple negative breast cancers. *Biomacromolecules* 16 (8): 2412–2417.

129 Fan, Y., Wang, Q., Lin, G. et al. (2017). Combination of using prodrug-modified cationic liposome nanocomplexes and a potentiating strategy via targeted co-delivery of gemcitabine and docetaxel for CD44-overexpressed triple negative breast cancer therapy. *Acta Biomater.* 62: 257–272.

15

Current Updates in Breast Cancer Drugs

Nitu L. Wankhede[1], Mayur B. Kale[1], Pranali A. Chandurkar[1], Manish M. Aglawe[1], Ashwini K. Bawankule[2], Brijesh G. Taksande[1], Milind J. Umekar[1], Rupesh K. Gautam[3], and Aman B. Upaganlawar[4]

[1] *RTM Nagpur University, Nagpur, Division of Neuroscience, Smt. Kishoritai Bhoyar College of Pharmacy, Kamptee, Nagpur, Maharashtra, 441002, India*
[2] *Babasaheb Ambedkar Technical University, S. K. B. College of Pharmacy, Gada, Kamptee, Nagpur, Maharashtra, 441001, India*
[3] *Indore Institute of Pharmacy, Department of Pharmacology, IIST Campus, Rau, Indore, Madhya Pradesh, 453331, India*
[4] *Savitribai Phule Pune University, SNJB's Shriman Sureshdada Jain College of Pharmacy, Neminagar, Chandwad, Nashik, Maharashtra, 423101, India*

15.1 Introduction

For many years, cancer has continuously been one of the leading global causes of death. Breast cancer (BC) is the second-leading cause of death among women [1]. The prevalent rate of BC increasing consistently in high-income countries, attributes to the advancement in the diagnostic tools including mammographic screening, along with failure of strategies for prevention of BC [2]. Numerous studies have been conducted to understand the causes and progression of this lethal illness. Despite this, early diagnosis has been found to be the best approach for prevention of BC. Further, the heterogeneous nature of the disease put forth the major challenges for determining the treatment approaches [3]. With the advancement of cancer, the metastasis of BC to the vital organs further complicates the available treatment approach [4].

Histological and molecular research are used to establish the clinical subgroups of BC, categorized into three different group: BC expressing progesterone receptor (PR) or estrogen receptor (ER), triple-negative breast cancer (TNBC), and human epidermal receptor 2 (HER2) (Table 15.1) [5, 6]. The precise pathway for the initiation and progression BC is not clear [5]. Though the variety of risk factor contribute to the disease development includes reproductive, genetics, factors related to dietary and lifestyle, and environmental factors [7]. While involvement with these risk

Drug and Therapy Development for Triple Negative Breast Cancer, First Edition.
Edited by Pravin Kendrekar, Vinayak Adimule, and Tara Hurst.
© 2023 WILEY-VCH GmbH. Published 2023 by WILEY-VCH GmbH.

Table 15.1 Subtypes of breast cancer.

Types	Molecular expression	Subcategories	Prevalence
Hormone receptor	ER or PR	Luminal A – low levels of Ki-67 Luminal B – high levels of Ki-67	40%<20%
Human epidermal receptor 2 (HER2)	HER2		10–20%
Triple-negative breast cancer (TNBC)	—	Basal like-1 (BL-1) Basal like-2 (BL-2) Mesenchymal (M) Immunomodulatory (IM) Mesenchymal stem cell-like (MSL) Luminal androgen receptor (LAR)	<20%

factors associated with BC differs with the type cancers, genetic mutation in BRCA1/2 in predisposed women has been an important factor BC associated malignancy [6]. Normal development of mammary stem cells (MaSC) is controlled via complex signaling mechanisms, such as HER2 pathway, hormonal receptors, and canonical Wnt signaling pathway, which tightly regulates cell proliferation, differentiation, and cell death. Moreover, newly available evidence suggests that the impairment in these signaling pathways via cancerous cells and cancerous stem cells (CSC) along with the genetic mutation may play essential function in heterogenicity as well as in metastasis of BC.

Surgery, as well as chemotherapy, are the principal approach for BC therapy [8, 9]. However, the traditional chemotherapeutic agents possess highly toxic over efficacy with lack of tumor cell targeting activity and also the conventional preparations have low rate of drug utilization. Therefore, appropriate selection of drug has become critical for the treatment with better outcomes for BC. Furthermore, the development of resistance and metastasis of BC to vital organs makes the treatment complicated and ineffective [10–12].

In this chapter, we summarize the current studies about drug therapy and discuss the associations of mechanisms. More importantly, implications of potential targeted strategies with the current therapies. The goal of these data is to highlight potential of targeted treatment over conventional therapy to improve treatment response for BC therapy.

15.2 Therapeutic Approaches

The precise pathway for the pathogenesis of BC is not clear due to heterogenicity of the disease. Thus, therapeutic approach majorly depends on the type of cancer, for instance, hormonotherapy is most suited approach for the therapy of PR and ER-positive BC [13]. While the cases with HER2 expression can be treated with the use of specific antibodies targeting different site at HER2 along with chemotherapy [14]. The treatment approach for the TNBC is somewhat complex in nature owning to dynamics and metastasis of disease. Though the conventional therapy for the TNBC found to be effective in early cases, the effectiveness remains doubtable in advance cases. The current regimen for TNBC approved by FDA includes chemotherapy with taxanes, anthracyclines, and antimetabolites [15]. Although the extensive research proposed that there is poor chemosensitivity for regimen than former BC, owing to the drug resistance and higher rate of metastasis in TNBC [16]. The therapeutic efficiency of the traditional and targeted drugs could be further improved via controlled release formulation by site-specific delivery using a nanocarrier vehicle. The technology also helps to understand the microenvironment around the malignant cell [7, 17]. Additionally, a receptor-based targeting approach has become milestone in treatment of cancer and has been undergoing clinical trials. The use of nanoparticle (NP) delivery has been potential approach owing to its high drug loading capacity and stability with lower incidence of toxicity of chemotherapeutic drug compared to conventional drugs. Numerous types of nanocarriers are available including nanocapsules, nanospheres, micelles, liposomes, microspores nanoemulsions, hydrogel, quantum dots, and gold nanoparticles [18]. Thus, the novel treatment therapeutic approaches with targeted delivery would be more useful for managing the progression of TNBC.

15.2.1 Hormonotherapy

For ER-positive BC patients, endocrine therapy is the gold standard of care, and it typically refers to treatments that block the connection between estrogen production and the dependent pathway [19]. These strategies involve direct ERα inhibition or modify its action via selective modulation or degradations of ER (SERM and SERD) and aromatase inhibitors (AIs).

The most common drug used in BC treatment includes tamoxifen, raloxifene, and toremifene. Tamoxifen has been the clinically approved principal option for early and advanced cases of BC. Tamoxifen, a SERM that fights with estrogen for active site on ERα which suppresses coactivator recruitment [20]. In addition, it also promotes weak transcriptional activation invto during poor or incomplete block in E2-stimulated conditions. Despite the success of antiestrogen therapy with tamoxifen, there are higher chances of recurrence of disease in one-third of patients due to retention of ERα expression. Thus, the drug is used as adjuvant therapy in the ERBC patients [21]. On contrary, the SERDs like Fulvestrant have been effective against the retained ERα expression, it disturbs the dimerization of receptor and thereby nuclear localization resulting in more complete blockage of transcription activity of ERα subunit [22]. Although the drug has been effective in luminal A BC but has marked side effects such as osteoporosis and poor physicochemical characteristics limiting its clinical application [23]. Numerous studies have demonstrated the positive effect with raloxifene treatment in positive ER cases but does not have any influence on ER-negative BC. However, the drug also shares the similar adverse effect profile to the tamoxifen [24, 25]. Few studies have demonstrated that raloxifene has comparable efficacy to the tamoxifen and has related to minor degree of thromboembolism and cataract more than the tamoxifen. Thus, raloxifene approved by FDA for the therapy of ERBC in postmenopausal females [26].

Furthermore, the randomized clinical trials with AI have proven effective in treatment of BC. Currently, only third-generation AIs such as exemestane, letrozole, and anastrozole are being used to treat advanced cases of ERBC due to their higher therapeutic efficacy and specificity, as the agents directly suppresses the conversion of androgens into estrogen thereby depleting the estrogenic stores [27]. The first FDA-approved drug anastrozole used for the adjuvant therapy of ER-positive BC [20]. Letrozole in 2004, was considered a neoadjuvant in second-line therapy following tamoxifen.

15.2.2 Chemotherapy

The chemotherapy has been used for early and advanced cases of BC and made substantial progress in the last 20 years. In spite of the treatment choice with the chemotherapeutic agents depends on the efficacy to toxicity profile of the cytotoxic agent [28]. Nevertheless, chemotherapy has been used alongside with the newer targets to manage advanced and metastatic cases [21]. Clinically, Taxol has been used for treatment of BC, found to be critically important in managing metastatic BC in clinical trials [29]. At present Docetaxel and Paclitaxel are used as adjuvant treatments of BC. Numerous clinical trials have been conducted with these drugs in different regimen cycle for better efficacy in advanced cases. But the choice for taxane-based regimen is based on pharmacokinetics, dosing schedule, and clinical activity for BC patients. Currently, docetaxel has consistent positive results in clinical trials with short-infusion schedule with BC patients [30]. Anthracycline is second most common drug implied in the BC treatment. Anthracyclines, including doxorubicin and epirubicin, act via inhibiting cell division in cancerous cells via interacting with enzyme desired for cell separation [31]. However, they also possess some toxicity profile that indeed limits their use in influencing the quality of life [32, 33]. Based on the clinical evidences, the agents are still imparted in adjuvant regimens in high-risk early BC patients.

15.3 Targeted Therapy

At present, there are numerous novel treatment approaches that are critically designed to target BC during the development stages. These drugs act via different mechanism to irradiate cancerous cell and prevent its progression, which includes molecular components of signaling pathways, and cell cycle, monoclonal antibodies, immunotherapy, and antibody–drug conjugates (ADCs). They also work to induce genomic changes to prevent drug resistance and metastasis associated with conventional BC treatment.

15.3.1 Drug Repurposing

The concept of drug repositioning has become attractive novel approach in cancer treatment. It mainly involves use of drug showing high off-target efficacy in one

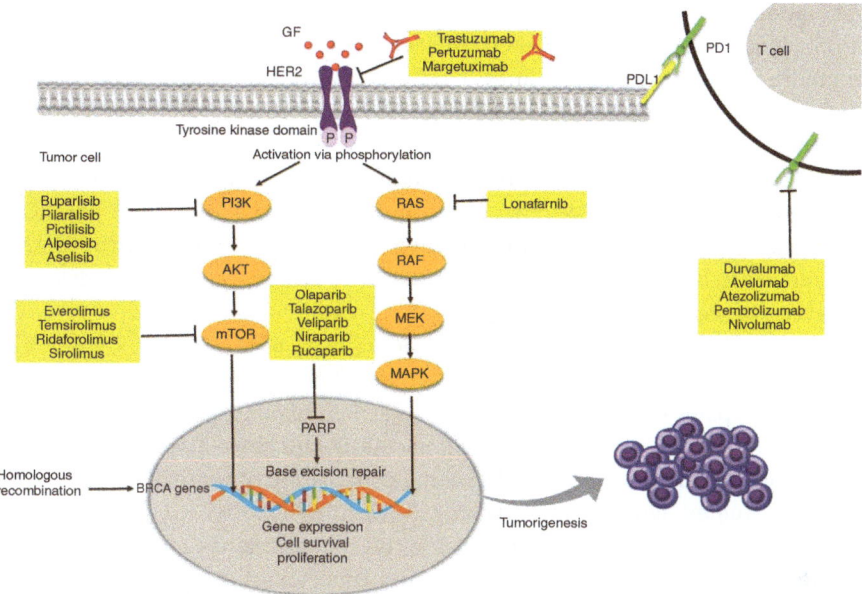

Figure 15.1 Novel therapeutic targets for treatment of breast cancer. GF, growth factor; HER2, human epidermal receptor 2; MAPK, mitogen-activated protein kinase; PDL1, programmed cell death ligand 1; BRCA, breast cancer gene 1; PARP, poly ADP-ribose polymerase; PI3K, phosphatidylinositol-3-kinase; mTOR, mammalian target of rapamycin.

disease and thus can be explored further and used as an innovative indication for other diseases. The method adds advantage over the new discovery including reduced time and cost for the research and has well established pharmacokinetic profile [34–36]. Numerous drugs are repurposed in the therapy of BC includes CDK4/6 inhibitor, alkylating agents, PI3K/AKT/mTOR inhibitor, antimetabolites, and mitotic inhibitors (Figure 15.1) [37].

These drugs disrupt the cancer cell growth via inhibition of DNA replication process, they damage the DNA by alkylating guanine and thereby prevent its binding to complementary base pair. The cyclophosphamide and thiotepa that was previously indicated as immunosuppressant in solid tumors have been used for BC treatment [37–40]. The drug has been administered intravenously in combination with taxols and anthracyclines. The recommended regimen of cyclophosphamide for BC patients includes six to four cycles of cyclophosphamides with antimetabolites or Adriamycin [41, 42]. While the thiotepa is used in combination with Vinblastine, Halotestin, and Adriamycin in relapsing and metastatic BC patients [43].

Another popular category of drug that has been repurposed for the treatment of BC are Antimetabolites, which have analogous structure to that of cellular metabolites thereby hindering the metabolic pathways. An uracil analog, 5-Flourouracil (5-FU) that interferes with the thymidine conversion in DNA and also inhibits RNA synthesis via forming toxic metabolite in cancer [44]. The drug has been subsequently used for BC treatment alone or in combination, which hinders the metabolic pathways [45–47]. The drug has been also indicated in metastatic BC in combination with cyclophosphamide, doxorubicin, cisplatin, and epirubicin [48]. The other combinations of 5-FU are paclitaxel and doxorubicin with paclitaxel [47]. The prodrug of 5-FU, capecitabine has been used due to the selective conversion in active form in the vicinity of tumor cells, thus preventing systemic exposure of 5-FU [49]. Generally, capecitabine is implied alone or in combination with cabazitaxel, ixabepilone, or vinorelbine in the treatment of taxol-resistant BC and metastatic cancer [14, 50–52]. Second most common antimetabolite used for BC is Methotrexate, which inhibits DNA and RNA synthesis by binding with dihydrofolate reductase. The effective regimen for the advanced cases of BC includes six to eight cycles of methotrexate, cyclophosphamide, and 5-FU [53].

Currently Palbociclib, an CDK4/6 inhibitor has been used against ER and HER2 positive BC. The drug demonstrated overall improvement in postmenopausal and metastatic BC patients. The enzyme, CDK4/6 kinase regulates retinoblastoma protein phosphorylation via interacting with cyclin D, thereby activating the expression of gene for progression of cell division, i.e. transition of G1 to S phase [54, 55]. Numerous studies demonstrated the involvement of CDK4/6 in tumorigenesis, thus can be a good therapeutic target for the management of cancer [56, 57]. Additionally, FDA-approved Abemaciclib and Ribociclib for the management of ERBC and HER2 metastatic BC, arresting cell cycle in G1 phase, thereby enhancing apoptosis [58]. All these drugs are used in combination with AI or hormonotherapy to delay the development of resistance thereby improving the survival in BS patients [50, 59–62].

The immunosuppressant, everolimus initially used for organ transplantation, was reported to have beneficial effect in pancreatic cancer (Figure 15.1). The drug was additionally found to impair the PI3K pathway through mTOR inhibition. The clinical studies also revealed its therapeutic potential in combination with anastrozole or letrozole for ER or PR-positive and HER2 advanced cases and metastatic BC [63, 64].

Mitosis is critical process in cell division. Mitotic inhibitors, such as paclitaxel, docetaxel, and vinblastine, disrupt the microtubule dynamics in cell division thereby arresting the cell in G2-M phase. Taxol is used as an adjuvant treatment option in combination with the therapy of BC in postmenopausal as well as in advanced and metastatic cases [30].

Additionally, vinblastine, a vinca alkaloid inhibits spindle formation. The alkaloid was discovered for diabetic treatment but was effective in sarcoma and adenocarcinoma and sarcoma [65, 66]. The drug has been approved in advanced and metastatic cases of BC in combination with mitomycin and cisplatin [67, 68].

15.3.2 HER2 Inhibitors

The HER2 positive BC has higher degree recurrence and poor prognosis [50, 69, 70]. In HER2-BC, overexpressed HER2 leads to dimerization of receptor resulting in autophosphorylation of intracellular subunits that further activates downstream signaling pathways, the major includes inositol phosphate 3-kinase (PI3K)/AKT kinase-B and Ras-Raf-Mek-MAPK that mediates proliferation, metastasis, and invasion of tumor, as well as, responsible for poor prognosis (Figure 15.1) [71, 72].

In addition, activated signaling pathways can further shorten cell cycle and resist the process of apoptosis as well as trigger expression of tumor-promoting genes. Thus, making HER2 a potential target for HER2-BC. The molecular mechanism for targeting HER2 could be divided into different subcategories.

The first category involves targeting HER2 directly with antibodies (monoclonal), includes pertuzumab and trastuzumab. The antibodies attach to the specific region at extracellular cite thereby inhibiting the ligand binding over HER2 and inhibiting its subsequent activation. Furthermore, the autophosphorylation of the intracellular subunits can be inhibited with tyrosinase inhibitors including lapatinib, which inhibits the intracellular tyrosine kinase domain phosphorylation, and blocks downstream signaling pathway [46]. In addition, Margetuximab has demonstrated promising effect in advanced cases of HER2-BC. Furthermore, the ongoing clinical trials find the efficacy of Margetuximab along with pembrolizumab [73].

Nowadays, targeted therapy has been achieved with hybrid molecules formed via conjugating tumor-specific antibody with cytotoxic drugs, thereby directing cytotoxic drugs onto the tumorous cell. This newer approach has shown great potential and has been termed ADCs [74]. The clinical efficacy of the use of ADCs relies on the pharmacokinetic parameters, stability of ADCs, and target antigen [7]. The approved ADCs for HER2 positive BC therapy include, trastuzumab–emtansine conjugate, formed by conjugating monoclonal antibody and anti-tubulin agent with disulfide bond. The ADCs bind to the overexpressed HER2 on tumor surface and release emtansine, resulting in tumor necrosis [70].

The subsequent category includes inhibitors of downstream HER2 signaling pathways such as PI3K/AKT/mTOR inhibitors. These downstream PI3K/AKT pathway activated via HER2 is critical for the cell cycle progression, differentiation, and cell proliferation. Thus, activation of this downstream pathway could be an important mechanism for tumorigenesis and therapeutic resistance, posing a potential therapeutic target for resistant HER2-BC [71, 75]. Clinically the hyperactivation of mTOR through AKT that are related to the resistance development toward hormonotherapy, thus reducing their clinical benefits. Therefore combination of mTOR inhibitors with antiestrogen could restore their therapeutic efficacy [71, 76, 77].

The PI3K Inhibitors has been used in combination with AI, which is the second-line therapy option for the treatment of resistant and advanced cases of HER2. However, the clinical efficacy of PI3K inhibitors has been outweighed by the higher incidence of toxicity associated with buparlisib, pilaralisib, and pictilisib [78]. Thus the more selective agents with lower incidence of toxicity are being tested currently, which include alpelisib and taselisib [79]. In addition, antitumor activity has been demonstrated with the help of pictilisib and taselisib in combination with letrozole or anastrozole in early and resistant HER2 patients [80]. An mTOR inhibitor, everolimus with exemestane has been approved by FDA for advanced and resistant BC [81]. Moreover, temsirolimus has not shown any clinical benefits alone or in combination with any adjuvant therapy advanced and resistance HER2 [82]. Contrarily, the most recent mTOR inhibitors, like sirolimus and ridaforolimus, have demonstrated encouraging refractory action in HER2-BC patients [83, 84]. The research study has also demonstrated the positive effect with inhibition of

Ras-Raf-Mek-MAPK downstream pathway in HER2 advanced patients. Lonafarnib inhibits Ras functioning by blocking the action of farnesyl transferase thereby blocking the downstream signaling [81, 85–87].

15.3.3 PARP Inhibitors

Among the cases of BC, about 10–20% of females has genetic history of BC with mutation in BRCA1 or BRCA2 susceptibility genes. The significance of this gene in BC can be demonstrated from the tumor-suppressing action of BRCA proteins via repairing damage in DNA through homologous recombination repair, thereby inhibiting tumorigenesis [88]. Thus, mutations could result in loss of DNA repair function associated with BRCA genes and can lead to expansion of tumorous cells [42]. The enzyme, PARP works in DNA repairs via resuming DNA replication which also helps in BRCA1 or BRCA2 genes-mediated homologous recombination DNA repair [14, 42, 49, 89]. Therefore, PARP inhibitors could be novel therapeutic target for BC related to BRCA gene mutation. The inhibitors block the DNA repair, resting cell cycle that results in subsequent death of the tumor cells through apoptosis [90, 91]. Thus, PARP inhibitors can be clinically efficient in BC associated with BRCA mutation (Figure 15.1). These agents are critical for the treatment of TNBC, involving high recurrence and poor prognosis with lower treatment outcomes. At same instance, TNBC patients have high risk of BRCA mutations making PARP inhibitors a wise strategy for the treatment.

The PARP inhibitors in combination with the other traditional and novel treatment are in clinical development. Olaparib, has shown promising antitumor activity against BRCA mutation and has been alone and in combination has been under the clinical trials for the treatment of BRCA mutated and advanced cases of TNBC [92, 93]. Recently, Talazoparib has been found to be clinically effective in metastatic and advanced cases of TNBC [94]. Combination of drug veliparib along with carboplatin, paclitaxel, and temozolomide has demonstrated positive clinical outcomes in BC with BRCA mutation and metastatic cases [95]. The therapeutic efficacy of niraparib and rucaparib has been recently studied as monotherapy in advanced BRCA-BC [96–98].

15.3.4 Immunotherapy

This is an emerging concept that works to enhance the immune responses in host against the cancerous cells for its recognition and destruction [99]. The concept of immunotherapy has been tuned to be significant in management of BC. The approach works to enhance and restore the tumor suppressive responses in the vicinity of the tumor cell. As the numerous factors of immune system influence the antitumor responses, targeting it alone or together could provide a novel approach in context of BC. Thus, the immunotherapeutic protocol has been applied based on recruiting regulatory T cells onto the tumor, subsequently increasing the infiltration of lymphocytes and dendritic cells, which are required for the antitumor activity [100]. The concept has been further evolved to design the combinatorial treatment of immunotherapeutic agents with traditional cytotoxic agents [101] to target advanced and metastatic states. The concept majorly involves intrinsic checkpoints which are necessary for the maintenance of normal equilibrium immunological response.

The inhibitors of checkpoint have been already implicated clinically approved by FDA in treatment of numerous tumors such as lung cancer, melanoma, and colon cancer [102, 103]. The BC demonstrated progressive loss of activated cytotoxic T cells (CD8+), suggesting poor immune response during BC evolution into invasive cancer. The immunotherapy of BC included molecules that are immune checkpoint inhibitors (ICIs) majorly targeting programmed cell death receptor 1 (PD-1) with its ligand-1 (PD-L1) and Cytotoxic T-Lymphocyte associated Antigen (CTLA-4) (Figure 15.1; Table 15.2). With immunotherapy, breakthrough research has been reported with subsequent approval by regulating authorities for use of atezolizumab along with nab-paclitaxel in therapy of metastatic TNBC. However, the efficacy of PD-L1 inhibitor, i.e. atezolizumab lies in the cases expressing PD-1 cells [108, 110]. Other PD-L1 inhibitors that are under clinical trials for treatment of advanced and metastatic TNBC includes pembrolizumab, durvalumab, avelumab, and nivolumab [105–108, 110, 113–115]. The clinical data demonstrated that pembrolizumab or nivolumab in combination with chemotherapy has been proven effective against advanced cases of TNBC as well as HER2-negative BC [113, 116]. With positive results in advanced and metastatic settings, trials with the curative strategies targeting checkpoint inhibitors including CTLA-4 or in addition along with chemotherapy are intended to improve outcomes in preoperative radiotherapy. These include combining tremelimumab, an anti-CTLA-4, with trastuzumab in BC [117]. Another study evaluating the potential of anti-CTLA-4, ipilimumab at early-stage BC has been found to be clinically effective in inducing immunologic responses systemically in BC [118].

15.3.5 Others Novel Targets

15.3.5.1 Histone Deacetylase (HDAC) Inhibitors
Among the mechanism elaborated in resistance therapy involves loss of ER expression mediated via histone deacetylation (HDAC) ER-positive patients [119]. With the HDAC inhibitors expression of ERα could be upregulated along with aromatase and additionally inhibits growth factor signaling pathways [120, 121]. The HDAC inhibitors mostly available as second therapy option for the advanced HR-positive BC, includes entinostat and vorinostat, in combination with tamoxifen and exemestane [122, 123].

15.3.5.2 Angiogenesis Inhibitors
Angiogenesis is crucial for the maintenance of tumor biology and mainly triggered via tumor hypoxia, which involves upregulation of vascular endothelial growth factor (VEGF) as well as its receptor (VEGDR). Expression of VEGF is responsible for viability of blood vessels into tumor vicinity and thus improves survival of tumor cells [124]. Therefore, targeting VEGD with the drugs could be an efficient approach

Table 15.2 Current immunotherapy for breast cancer.

Antibody	Combination	Mechanism	Treatment	Status	References
Tremelimumab	Aromatase inhibitor	Inhibition of CTL-associated protein-4 (CTLA-4)	Advanced ER-positive breast cancer	—	[104]
	—	—		Completed	NCT02536794NCT02536794
Durvalumab	—	Inhibitor of PD-1	ER+/HER2− breast cancer	Completed	NCT02997995
	OlaparibPaclitaxel	PD-1 inhibitor	High-risk HER2-negative BCTNBCBRCA-mutated metastatic BC	—	[105][106][107]
Durvalumab and Tremelimumab	—	PD-1 inhibitorAnti-CTLA-4	HR-positiveHypermutated metastatic breast cancer	Completed Completed Active	NCT03608865 NCT02649686 NCT03132467
Avelumab		PD-1 inhibitor	Advanced BC or metastatic BC		[108, 109]
Atezolizumab	Nab-paclitaxel	PD-1 inhibitor	Advanced or metastatic TNBC		[110]
Pembrolizumab	Eribulin mesylate	PD-1 inhibitor	HR-positive, HER2-negative metastatic BC		[111]
Pembrolizumab		PD-1 inhibitor	TNBC		[112]
Atezolizumab + pembrolizumab		PD-1 inhibitor	TNBC		[113]

Drug	Combination	Mechanism	Condition	Status	NCT Number
Ipilimumab	—	Anti-CTLA-4	Early stage/resectable breast cancer	Completed	NCT01502592
Ipilimumab + nivolumab	—	Inhibit PD-1Anti-CTLA-4	Metastatic hypermutated HER2-negative BCTNBC	ActivePhase 2	NCT03789110NCT03546686
Nivolumab + ipilimumab	Capecitabine	PD-1 inhibitorAnti-CTLA-4	TNBC	TNB	NCT03818685
	DoxorubicinCyclophosphamide	Inhibit PD-1Anti-CTLA-4	Hormonal receptor +ve tumorMetastatic BC	Active	NCT03409198
	Bicalutamide	Inhibit PD-1Anti-CTLA-4	HER2-Negative BC	Recruiting	NCT03650894
	Nab-paclitaxel	Inhibit PD-1Anti-CTLA-4	Positive ER, HER2 negative-advanced BC	Completed	NCT04132817
	INCAGN01876	PD-1 inhibitorAnti-CTLA-4	Triple-negative breast cancer (TNBC)	Completed	NCT03126110

Table 15.3 Angiogenesis inhibitors used in breast cancer.

Drug	Disease	Mechanism	References
Bevacizumab	Metastatic breast cancer	Angiogenesis inhibitors Angiogenesis modulating agents	NCT00423917 NCT00217672
Ramucirumab or icrucumab	Metastatic breast cancer	Inhibits VEGFR1 and VEGFR2	NCT01234402
Capecitabine + pazopanib	Resistance breast cancer	Inhibits tyrosine kinases including VEGFR1, VEGFR2, and VEGFR3	NCT01498458
Dasatinib	Triple-negative breast cancer	Decreased phosphorylation of VEGFR	NCT00371254
Sunitinib	Neoadjuvant therapy for breast cancer	Inhibits platelet-derived growth factor receptor-β and VEGFR2	NCT00656669

to inhibit angiogenesis and survival of tumor. Use of antiangiogenetic antibodies has been approved treatment for variety of cancer [125]. The VEGD gene was mostly found to be more unregulated in TNBC patients than non-TNBC patients.

An antiangiogenetic antibody, Bevacizumab suppresses neo-vasculature growth in tumor and thereby inhibits metastasis of tumor cells. Bevacizumab along with chemotherapy greatly enhances the response ratio clinically. Moreover the combination was reported to be significantly safe compared to previous regimen, and being recommended as fine-line therapy for HER2-resistant cases [126] (Table 15.3).

15.4 Conclusion

With the advancements in targeted approaches in BC, the rate of mortality has been decreased drastically. The targeted approach has been linked with the improved treatment modalities against all types of BC. Moreover, the success of the targeted delivery with the use of biological drugs receptor-specific monoclonal antibody have highlighted the molecular targeting approach in BC therapy. However, the treatment for metastatic TNBC has limited feasibility in advanced cases. But with the recent advancement in identification of associated signaling pathways and underlying molecular mechanisms, treatment responses in resistant HER2 and metastatic TNBC are better elucidated. This has led to the development of novel approaches, including inhibitors of HER2, PARP, PI3K/AKT/mTOR, and ICIs, for the specificity in the different subcategories of BC.

Acknowledgment

Not Declared

Conflict of Interest

None

Authors Contribution

Nitu Wankhede, Mayur Kale, Ashwini Bawankule, and Pranali Chandurkar performed extensive literature survey and prepared primary draft of chapter. Manish Aglawe, Brijesh Taksande, Aman Upaganlawar, and Milind Umekar checked the primary draft and corrected it. Nitu Wankhede prepared figures and tables.

References

1 Siegel, R.L., Miller, K.D., and Jemal, A. (2017). Cancer statistics. *Cancer J. Clin.* 67: 7–30. https://doi.org/10.3322/CAAC.21387.

2 Chen, T., Sepanlou, S.G., Zeeb, H. et al. (2020). The burden and trends of breast cancer from 1990 to 2017 at the global, regional, and national levels: results from the Global Burden of Disease Study 2017. *Front. Oncol.* www.frontiersin.org 1: 650. https://doi.org/10.3389/fonc.2020.00650.

3 Siegel, R.L., Miller, K.D., and Jemal, A. (2019). Cancer statistics. *Cancer J. Clin.* 69: 7–34. https://doi.org/10.3322/CAAC.21551.

4 Wang, L., Zhang, S., and Wang, X. (2021). The metabolic mechanisms of breast cancer metastasis. *Front. Oncol.* 10: 1–21. https://doi.org/10.3389/fonc.2020.602416.

5 Lehmann, B.D., Bauer, J.A., Chen, X. et al. (2011). Identification of human triple-negative breast cancer subtypes and preclinical models for selection of targeted therapies. *J. Clin. Invest.* 121: 2750–2767. https://doi.org/10.1172/JCI45014DS1.

6 Feng, Y., Spezia, M., Huang, S. et al. (2018). Breast cancer development and progression: risk factors, cancer stem cells, signaling pathways, genomics, and molecular pathogenesis. *Genes Dis.* 5: 77–106. https://doi.org/10.1016/J.GENDIS.2018.05.001.

7 Barzaman, K., Karami, J., Zarei, Z. et al. (2020). Breast cancer: biology, biomarkers, and treatments. *Int. Immunopharmacol.* 84: https://doi.org/10.1016/j.intimp.2020.106535.

8 Buchholz, T.A. (2009). Radiation therapy for early-stage breast cancer after breast-conserving surgery. *N. Engl. J. Med.* 360: 63–70. https://doi.org/10.1056/NEJMCT0803525.

9 Florescu, A., Amir, E., Bouganim, N., and Clemons, M. (2011). Immune therapy for breast cancer in 2010 – hype or hope? *Curr. Oncol.* 18: 9–18.

10 Carty, N.J., Foggitt, A., Hamilton, C.R. et al. (1995). Patterns of clinical metastasis in breast cancer: an analysis of 100 patients. *Eur. J. Surg. Oncol.* 21: 607–608. https://doi.org/10.1016/S0748-7983(95)95176-8.

11 Mego, M., Mani, S.A., and Cristofanilli, M. (2010). Molecular mechanisms of metastasis in breast cancer – clinical applications. *Nat. Rev. Clin. Oncol.* 7: 693–701. https://doi.org/10.1038/nrclinonc.2010.171.

12 Tang, Y., Wang, Y., Kiani, M.F., and Wang, B. (2016). Classification, treatment strategy, and associated drug resistance in breast cancer. *Clin. Breast Cancer* 16: 335–343. Elsevier Ltd. https://doi.org/10.1016/j.clbc.2016.05.012.

13 Reinert, T. and Barrios, C.H. (2015). Optimal management of hormone receptor positive metastatic breast cancer in 2016. *Ther. Adv. Med. Oncol.* 7: 304–320. https://doi.org/10.1177/1758834015608993.

14 Nandi, S.S., Suryavanshi, A., Adimule, V., and Yallur, B.C. (2020). Super capacitor characteristics of novel rare earth perovskite nanomaterials of $Sr_{0.5}$, $Cu_{0.4}$, $Y_{0.1}$. *AIP Conf. Proc.* 2274: 020007. https://doi.org/10.1063/5.0022454.

15 Berrada, N., Delaloge, S., and André, F. (2010). Treatment of triple-negative metastatic breast cancer: toward individualized targeted treatments or chemosensitization? *Ann. Oncol.* 21: 30–35. https://doi.org/10.1093/annonc/mdq279.

16 Pascual, J. and Turner, N.C. (2019). Targeting the PI3-kinase pathway in triple-negative breast cancer. *Ann. Oncol.* 30: 1051–1060. https://doi.org/10.1093/annonc/mdz133.

17 Adimule, V., Suryavanshi, A., Yallur, B.C., and Nandi, S.S. (2020). A facile synthesis of poly(3-octyl thiophene):$Ni_{0.4}Sr_{0.6}TiO_3$ hybrid nanocomposites for solar cell applications. *Macromol. Symp.* 392: 2000001.

18 Adimule, V., Yallur, B.C., Pai, M.M. et al. (2022). Biogenic synthesis of magnetic palladium nanoparticles decorated over reduced graphene oxide using *Piper betle* Petiole extract ($Pd-rGO@Fe_3O_4$ NPs) as heterogeneous hybrid nanocatalyst for applications in Suzuki–Miyaura coupling reactions of biphenyl compounds. *Top. Catal.* https://doi.org/10.1007/s11244-022-01672-9.

19 Garcia-Martinez, L., Zhang, Y., Nakata, Y. et al. (2021). Epigenetic mechanisms in breast cancer therapy and resistance. *Nat. Commun.* 12: 1–14. https://doi.org/10.1038/s41467-021-22024-3.

20 Nandi, S.S., Adimule, V., and Yallur, B.C. (2022). Synthesis, structural and optical properties of Co doped Sm_2O_3 nanostructures. *Adv. Mater. Res.* 1173: 59–69. Trans Tech Publications, Ltd.

21 Adimule, V., Yallur, B.C., Batakurki, S., and Nandi, S.S. (2022). Synthesis, morphology and enhanced optical properties of novel $Gd_xCo_3O_4$ nanostructures. *Adv. Mater. Res.* 1173. Trans Tech Publications, Ltd.: 71–82.

22 Journé, F., Body, J.J., Leclercq, G., and Laurent, G. (2008). Hormone therapy for breast cancer, with an emphasis on the pure antiestrogen fulvestrant: mode of action, antitumor efficacy and effects on bone health. *Expert Opin. Drug Saf.* 7: 241–258. https://doi.org/10.1517/14740338.7.3.241.

23 Guan, J., Zhou, W., Hafner, M. et al. (2019). Therapeutic ligands antagonize estrogen receptor function by impairing its mobility. *Cell* 178: 949–963.e18. https://doi.org/10.1016/J.CELL.2019.06.026.

24 Vogel, V.G. (2007). Chemoprevention strategies 2006. *Curr. Treat. Options Oncol.* 8: 74–88. https://doi.org/10.1007/S11864-007-0019-Z.

25 Mortimer, J. and Urban, J.H. (2003). Long-term toxicities of selective estrogen-receptor modulators and antiaromatase agents. *Oncology (Williston Park, NY)*. Europepmc.Org. 17 (5): 652–659. https://europepmc.org/article/med/12800793.

26 Adimule, V., Batakurki, S., Yallur, B.C. et al. (2022). Enhanced photoluminescence, optical, structural properties of ZrO_2-incorporated Sm_2O_3:Co_3O_4 nanocomposite and their applications in photocatalytic degradation of methylene blue. *J. Mater. Res.* 37: 2396–2405.

27 Boughey, J.C. and Nguyen, T. (2016). Axillary staging after neoadjuvant chemotherapy for breast cancer: a pilot study combining sentinel lymph node biopsy with radioactive seed localization of pre-treatment positive axillary lymph nodes. *Breast Diseases: A Year Book Quarterly* 27: 282–284. https://doi.org/10.1016/j.breastdis.2016.09.005.

28 Shapiro, C.L. and Recht, A. (2001). Side effects of adjuvant treatment of breast cancer. *N. Engl. J. Med.* 344: 1997–2008. https://doi.org/10.1056/NEJM200106283442607.

29 Miller, K., Wang, M., Gralow, J. et al. (2007). Paclitaxel plus bevacizumab versus paclitaxel alone for metastatic breast cancer. *N. Engl. J. Med.* 357: 2666–2676. https://doi.org/10.1056/NEJMOA072113.

30 Crown, J., O'Leary, M., and Ooi, W.-S. (2004). Docetaxel and paclitaxel in the treatment of breast cancer: a review of clinical experience. *Oncologist* 9: 24–32. https://doi.org/10.1634/theoncologist.9-suppl_2-24.

31 Tan, D.S.P., Marchió, C., Jones, R.L. et al. (2007). Triple negative breast cancer: molecular profiling and prognostic impact in adjuvant anthracycline-treated patients. *Breast Cancer Res. Treat.* 111: 27–44. https://doi.org/10.1007/S10549-007-9756-8.

32 Monsuez, J.J., Charniot, J.C., Vignat, N., and Artigou, J.Y. (2010). Cardiac side-effects of cancer chemotherapy. *Int. J. Cardiol.* 144: 3–15. https://www.sciencedirect.com/science/article/pii/S0167527310001786.

33 Grevelman, E. and Breed, W.P.M. (2005). Prevention of chemotherapy-induced hair loss by scalp cooling. *Ann. Oncol.* 16: 352–358. https://www.sciencedirect.com/science/article/pii/S0923753419478769.

34 Nosengo, N. (2016). Can you teach old drugs new tricks? *Nature* 534: 314–316. https://doi.org/10.1038/534314A.

35 Pushpakom, S., Iorio, F., Eyers, P.A. et al. (2018). Drug repurposing: progress, challenges and recommendations. nature.com. *Nat. Rev. Drug Discovery* 18: 41–58. https://doi.org/10.1038/nrd.2018.168.

36 Aggarwal, S., Verma, S.S., Aggarwal, S., and Gupta, S.C. (2021). Drug repurposing for breast cancer therapy: old weapon for new battle. *Semin. Cancer Biol.* 68: 8–20. https://doi.org/10.1016/J.SEMCANCER.2019.09.012.

37 Shim, J.S. and Liu, J.O. (2014). Recent advances in drug repositioning for the discovery of new anticancer drugs. *Int. J. Biol. Sci.* 10: 654–663. https://doi.org/10.7150/ijbs.9224.

38 De Jonge, M.E., Huitema, A.D.R., Rodenhuis, S., and Beijnen, J.H. (2012). Clinical pharmacokinetics of cyclophosphamide. *Clin. Pharmacokinet.* 44: 1135–1164. https://doi.org/10.2165/00003088-200544110-00003.

39 Moya, E., Arnold Lyons, B., Rcsi, F., and George Edelstyn, D. (1962). Thiotepa in treatment of advanced breast cancer. *BMJ* 2: 1280–1283. https://doi.org/10.1136/BMJ.2.5315.1280.

40 Keri, R.S., Adimule, V., Kendrekar, P. et al. (2022). The nano-based catalyst for the synthesis of benzimidazoles. *Top. Catal.*

41 Adimule, V., Nandi, S.S., Yallur, B.C., and Shaikh, N. (2021). CNT/graphene-assisted flexible thin-film preparation for stretchable electronics and superconductors. In: *Sensors for Stretchable Electronics in Nanotechnology*, 89–103. CRC Press.

42 Adimule, V., Medapa, S., Rao, P.K., and Kumar, L.S. (2014). Synthesis of Schiff bases of 5-[5-(4-fluorophenyl) thiophen-2-yl]-1,3,4-thiadiazol-2-amine and its anticancer activity. *Int. J. Adv. Pharm. Sci.* 5 (1): 1761–1768.

43 Batakurki, S.R., Adimule, V., Pai, M.M. et al. (2022). Synthesis of Cs–Ag/Fe_2O_3 nanoparticles using *Vitis labrusca* Rachis extract as green hybrid nanocatalyst for the reduction of arylnitro compounds. *Top. Catal.* https://doi.org/10.1007/s11244-022-01593-7.

44 Yen Moore, A. (2009). Clinical applications for topical 5-fluorouracil in the treatment of dermatological disorders. *J. Dermatol. Treat.* 20: 328–335. https://doi.org/10.3109/09546630902789326.

45 Adimule, V., Bhat, V.S., Yallur, B.C. et al. (2022). Facile synthesis of novel $SrO_{0.5}$:$MnO_{0.5}$ bimetallic oxide nanostructure as a high-performance electrode material for supercapacitors. *Nanomater. Nanotechnol.* 12: 1–14. https://doi.org/10.1177/18479804211064028.

46 Shaikh, N.M., Bagihalli, G.B., Adimule, V. et al. (2022). A novel silica immobilized acidic ionic liquid [BMIM][$AlCl_4$] as an effective catalyst for biscoumarine synthesis. *Top. Catal.* https://doi.org/10.1007/s11244-022-01591-9.

47 Zoli, W., Ulivi, P., Tesei, A. et al. (2005). Addition of 5-fluorouracil to doxorubicin-paclitaxel sequence increases caspase-dependent apoptosis in breast cancer cell lines. *Breast Cancer Res.* 7: 1–9. https://doi.org/10.1186/BCR1274.

48 Cameron, D.A., Gabra, H., and Leonard, R.C.F. (1994). Continuous 5-fluorouracil in the treatment of breast cancer. *Br. J. Cancer* 70: 120–124. https://doi.org/10.1038/bjc.1994.259.

49 Bryant, H.E., Petermann, E., Schultz, N. et al. (2009). PARP is activated at stalled forks to mediate Mre11-dependent replication restart and recombination. *EMBO J.* 28: 2601–2615. https://doi.org/10.1038/emboj.2009.206.

50 Adimule, V., Nandi, S.S., and Yallur, B.C. (2022). Devices and sensors based on additively manufactured shape-memory of hybrid nanocomposites. In: *Shape Memory Composites Based on Polymers and Metals for 4D Printing* (ed. M.R. Maurya, K.K. Sadasivuni, J.J. Cabibihan, et al.). Cham: Springer https://doi.org/10.1007/978-3-030-94114-7_15.

51 Adimule, V., Yallur, B.C., and Gowda, A.H.J. (2022). Advanced sensors based on carbon nanomaterials. In: *Carbon Nanomaterials-Based Sensors* (Chapter 14) (ed. J.G. Manjunatha and C.M. Hussain), 259–268. Elsevier https://doi.org/10.1016/B978-0-323-91174-0.00004-4.

52 Adimule, V., Yallur, B., and Gowda, A. (2022). Crystal structure, morphology, optical and super-capacitor properties of Srx: α-Sb_2O_4 nanostructures. *Anal. Bioanal. Electrochem.* 14 (1): 1–17.

53 Fracchia, A.A., Farrow, J.H., Adam, Y.G. et al. (1970). Systemic chemotherapy for advanced breast cancer. *Cancer* 26: 642–649. https://doi.org/10.1002/1097-0142(197009)26:3<642::AID-CNCR2820260323>3.0.CO;2-4.

54 Xiong, Y., Li, T., Assani, G. et al. (2019). Ribociclib, a selective cyclin D kinase 4/6 inhibitor, inhibits proliferation and induces apoptosis of human cervical cancer in vitro and in vivo. *Biomed. Pharmacother.* 112: https://doi.org/10.1016/J .BIOPHA.2019.108602.

55 Chou, A., Froio, D., Nagrial, A.M. et al. (2018). Tailored first-line and second-line CDK4-targeting treatment combinations in mouse models of pancreatic cancer. *Gut* 67: 2142–2155. https://doi.org/10.1136/GUTJNL-2017-315144.

56 Roskoski, R. Jr. (2016). Cyclin-dependent protein kinase inhibitors including palbociclib as anticancer drugs. *Pharmacol. Res.* 107: 249–275. https://doi .org/10.1016/j.phrs.2016.03.012.

57 Adimule, V., Medapa, S., Adarsha, H.J. et al. (2014). Synthesis, characterization and in vitro anticancer properties of 1-{5-aryl-2-[5-(4-fluoro-phenyl)-thiophen-2-yl]- [1,3,4] oxadiazol-3-yl}-ethanone. *Int. J. Pharm. Res. Rev.* 3 (12): 20–25.

58 Kwapisz, D. (2017). Cyclin-dependent kinase 4/6 inhibitors in breast cancer: palbociclib, ribociclib, and abemaciclib. *Breast Cancer Res. Treat.* 166: 41–54. https://doi.org/10.1007/S10549-017-4385-3.

59 Finn, R.S., Martin, M., Rugo, H.S. et al. (2016). Palbociclib and letrozole in advanced breast cancer. *N. Engl. J. Med.* 375: 1925–1936. https://doi.org/10.1056/ NEJMOA1607303/SUPPL_FILE/NEJMOA1607303_DISCLOSURES.PDF.

60 Adimule, V., Suryavanshi, A., and Nandi, S. (2020). Synthesis, characterization and impedance studies of novel nanocomposites of gadolinium titanate. *IOP Conf. Ser.: Mater. Sci. Eng.* 872: 012099.

61 Turner, N.C., Slamon, D.J., Ro, J. et al. (2018). Overall survival with palbociclib and fulvestrant in advanced breast cancer. *N. Engl. J. Med.* 379: 1926–1936. https://doi .org/10.1056/NEJMOA1810527/SUPPL_FILE/NEJMOA1810527_DATA-SHARING.PDF.

62 Masuda, N., Inoue, K., Nakamura, R. et al. (2019). Palbociclib in combination with fulvestrant in patients with hormone receptor-positive, human epidermal growth factor receptor 2-negative advanced breast cancer: PALOMA-3 subgroup analysis of Japanese patients. *Int. J. Clin. Oncol.* 24: 262–273. https://doi.org/10.1007/ S10147-018-1359-3/FIGURES/5.

63 Baselga, J., Campone, M., Piccart, M. et al. (2012). Everolimus in postmenopausal hormone-receptor–positive advanced breast cancer. *N. Engl. J. Med.* 366: 520. https://doi.org/10.1056/NEJMOA1109653.

64 Royce, M.E. and Osman, D. (2015). Everolimus in the treatment of metastatic breast cancer. *Breast Cancer (Auckl)* 9: 73. https://doi.org/10.4137/BCBCR.S29268.

65 Radford, J.A., Knight, R.K., and Rubens, R.D. (1985). Mitomycin C and vinblastine in the treatment of advanced breast cancer. *Eur. J. Cancer Clin. Oncol.* 21: 1475–1477. https://doi.org/10.1016/0277-5379(85)90241-X.

66 Noble, R.L., Beer, C.T., and Cutts, J.H. (1958). Role of chance observations in chemotherapy: *Vinca rosea. Ann. N.Y. Acad. Sci.* 76: 882–894. https://doi .org/10.1111/J.1749-6632.1958.TB54906.X.

67 Adimule, V., Vageesha, P., Bagihalli, G. et al. (2019). Synthesis, characterization of hybrid nanomaterials of strontium, yttrium, copper doped with indole Schiff base derivatives possessing dielectric and semiconductor properties. In: *Emerging Research in Electronics, Computer Science and Technology*, Lecture Notes in

Electrical Engineering, vol. 545 (ed. V. Sridhar, M. Padma, and K. Rao). Singapore: Springer. https://doi.org/10.1007/978-981-13-5802-9_97.

68 Sedlacek, S.M. (1993). First-line and salvage therapy of metastatic breast cancer with mitomycin/vinblastine. *Oncology* 50: 16–23. https://doi.org/10.1159/000227243.

69 Slamon, D.J., Godolphin, W., Jones, L.A. et al. (1989). Studies of the HER-2/neu proto-oncogene in human breast and ovarian cancer. *Science (80-.)* 244: 707–712. https://doi.org/10.1126/SCIENCE.2470152.

70 Guo, J., Li, Q., Zhang, P. et al. (2019). Trastuzumab plus adjuvant chemotherapy for human epidermal growth factor receptor 2 (HER2)-positive early-stage breast cancer: a real-world retrospective study in Chinese patients. *Chin. J. Cancer Res.* 31: 759–770. https://doi.org/10.21147/j.issn.1000-9604.2019.05.06.

71 Adimule, V., Kerur, S.S., Chinnam, S. et al. (2022). Guar gum and its nanocomposites as prospective materials for miscellaneous applications: a short review. *Top. Catal.* https://doi.org/10.1007/s11244-022-01587-5.

72 Elster, N., Collins, D.M., Toomey, S. et al. (2015). HER2-family signalling mechanisms, clinical implications and targeting in breast cancer. *Breast Cancer Res. Treat.* 149: 5–15. https://doi.org/10.1007/s10549-014-3250-x.

73 Burris, H.A., Giaccone, G., Im, S.-A. et al. (2013). Phase I study of margetuximab (MGAH22), an FC-modified chimeric monoclonal antibody (MAb), in patients (pts) with advanced solid tumors expressing the HER2 oncoprotein. *J. Clin. Oncol.* 31: 3004. https://doi.org/10.1200/JCO.2013.31.15_SUPPL.3004.

74 Adimule, V., Revaigh, M.G., and Adarsha, H.J. (2020). Synthesis and fabrication of Y-doped ZnO nanoparticles and their application as a gas sensor for the detection of ammonia. *J. Mater. Eng. Perform.* 29: 4586–4596. https://doi.org/10.1007/s11665-020-04979-4.

75 Koren, S., Reavie, L., Couto, J.P. et al. (2015). PIK3CAH1047R induces multipotency and multi-lineage mammary tumours. *Nature* 525: 114–118. https://doi.org/10.1038/nature14669.

76 Kim, E.K., Kim, H.A., Koh, J.S. et al. (2011). Phosphorylated S6K1 is a possible marker for endocrine therapy resistance in hormone receptor-positive breast cancer. *Breast Cancer Res. Treat.* 126: 93–99. https://doi.org/10.1007/S10549-010-1315-Z.

77 Pérez-Tenorio, G. and Stål, O. (2002). Activation of AKT/PKB in breast cancer predicts a worse outcome among endocrine treated patients. *Br. J. Cancer* 86: 540–545. https://doi.org/10.1038/sj/bjc/6600126.

78 Nandi, S.S., Suryavanshi, A., Adimule, V., and Maradur, S.R. (2020). Semiconductor current–voltage characteristics of some novel perovskite ionic nanocomposites of $Sr_{0.5}$, $Cu_{0.4}$, $Y_{0.1}$ and $Sr_{0.5}$, $Mn_{0.5}$ and their electronic sensor applications. *AIP Conf. Proc.* 2274: 020006: https://doi.org/10.1063/5.0022453.

79 Dickler, M.N., Saura, C., Richards, D.A. et al. (2018). Phase II study of Taselisib (GDC-0032) in combination with fulvestrant in patients with HER2-negative, hormone receptor–positive advanced breast cancer. *Clin. Cancer Res.* 24: 4380–4387. https://doi.org/10.1158/1078-0432.CCR-18-0613.

80 Adimule, V., Batakurki, S., Yallur, B.C. et al. (2022). Samarium-decorated ZrO_2@SnO_2 nanostructures, their electrical, optical and enhanced photoluminescence properties. *J. Mater. Sci. – Mater. Electron.* 33: 18699–18715. https://doi.org/10.1007/s10854-022-08718-4.

81 Baselga, J., Campone, M., Piccart, M. et al. (2012). Everolimus in postmenopausal hormone-receptor–positive advanced breast cancer. *N. Engl. J. Med.* 366: 520–529.

82 Fleming, G.F., Ma, C.X., Huo, D. et al. (2012). Phase II trial of temsirolimus in patients with metastatic breast cancer. *Breast Cancer Res. Treat.* 136: 355–363. https://doi.org/10.1007/s10549-011-1910-7.

83 Acevedo-Gadea, C., Hatzis, C., Chung, G. et al. (2015). Sirolimus and trastuzumab combination therapy for HER2-positive metastatic breast cancer after progression on prior trastuzumab therapy. *Breast Cancer Res. Treat.* 150: 157–167. https://doi.org/10.1007/s10549-015-3292-8.

84 Vinayak, A., Sudha, M., Jaadeesha, A.H. et al. (2014). Synthesis, characterization of some novel 1,3,4-oxadiazole compounds containing 8-hydroxy quinolone moiety as potential antibacterial and anticancer agents. *Int. J. Pharm. Res. 4* (4): 180–185.

85 Bagchi, S., Rathee, P., Jayaprakash, V., and Banerjee, S. (2018). Farnesyl transferase inhibitors as potential anticancer agents, mini-reviews. *Med. Chem.* 18: 1611–1623. https://doi.org/10.2174/1389557518666180801110342.

86 Zhumakayeva, A.M., Rakhimov, K.D., Omarova, I.M. et al. (2019). Experimental, clinical and morphological analysis of H-Ras oncoproteins for locally advanced breast cancer. *Maced. J. Med. Sci.* 7: 3153–3157. https://doi.org/10.3889/oamjms.2019.708.

87 Klochkov, S.G., Neganova, M.E., Yarla, N.S. et al. (2019). Implications of farnesyltransferase and its inhibitors as a promising strategy for cancer therapy. *Semin. Cancer Biol.* 56: 128–134. https://doi.org/10.1016/j.semcancer.2017.10.010.

88 Trainer, A.H., Lewis, C.R., Tucker, K. et al. (2010). The role of BRCA mutation testing in determining breast cancer therapy. *Nat. Rev. Clin. Oncol.* 7: 708–715. https://doi.org/10.1038/nrclinonc.2010.175.

89 Evers, B., Helleday, T., and Jonkers, J. (2010). Targeting homologous recombination repair defects in cancer. *Trends Pharmacol. Sci.* 31: 372–380. https://www.sciencedirect.com/science/article/pii/S0165614710000969.

90 Turk, A.A. and Wisinski, K.B. (2018). PARP inhibitors in breast cancer: bringing synthetic lethality to the bedside. *Cancer* 124: 2498–2506. https://doi.org/10.1002/cncr.31307.

91 Farmer, H., McCabe, H., Lord, C.J. et al. (2005). Targeting the DNA repair defect in BRCA mutant cells as a therapeutic strategy. *Nature* 434: 917–921. https://doi.org/10.1038/nature03445.

92 ClinicalTrials.gov, (n.d.). Olaparib. https://clinicaltrials.gov/ct2/results?cond=&term=Olaparib+&cntry=&state=&city=&dist= (accessed 9 November 2022).

93 Robson, M., Im, S.-A., Senkus, E. et al. (2017). Olaparib for metastatic breast cancer in patients with a germline BRCA mutation. *N. Engl. J. Med.* 377: 523–533. https://doi.org/10.1056/nejmoa1706450.

94 Brown, J.S., Kaye, S.B., and Yap, T.A. (2016). PARP inhibitors: the race is on. *Br. J. Cancer* 114: 713–715. https://doi.org/10.1038/bjc.2016.67.

95 Han, H.S., Diéras, V., Robson, M. et al. (2018). Veliparib with temozolomide or carboplatin/paclitaxel versus placebo with carboplatin/paclitaxel in patients with

BRCA1/2 locally recurrent/metastatic breast cancer: randomized phase II study. *Ann. Oncol.* 29: 154–161. https://doi.org/10.1093/annonc/mdx505.

96 Spring, L.M., Han, H., Liu, M.C. et al. (2022). Neoadjuvant study of niraparib in patients with HER2-negative, BRCA-mutated, resectable breast cancer. *Nat. Cancer* 3: 927–931. https://doi.org/10.1038/S43018-022-00400-2.

97 ClinicalTrials.gov (n.d.). Evaluation of the safety and tolerability of niraparib with everolimus in advanced gynecologic malignancies and breast. https://clinicaltrials.gov/ct2/show/NCT03154281?term=Niraparib&cond=Breast+Cancer&draw=2&rank=11 (accessed 9 November 2022).

98 Patsouris, A., Diop, K., Tredan, O. et al. (2021). Rucaparib in patients presenting a metastatic breast cancer with homologous recombination deficiency, without germline BRCA1/2 mutation. *Eur. J. Cancer* 159: 283–295. https://doi.org/10.1016/J.EJCA.2021.09.028.

99 Hadden, J.W. (1999). The immunology and immunotherapy of breast cancer: an update. *Int. J. Immunopharmacol* 21: 79–101. https://www.sciencedirect.com/science/article/pii/S0192056198000770.

100 Luen, S., Savas, P., Fox, S. et al. (2017). Tumour-infiltrating lymphocytes and the emerging role of immunotherapy in breast cancer. *Pathology* 49: 141–155. https://www.sciencedirect.com/science/article/pii/S0031302516404137.

101 Baxevanis, C.N., Fortis, S.P., and Perez, S.A. (2021). The balance between breast cancer and the immune system: challenges for prognosis and clinical benefit from immunotherapies. *Semin. Cancer Biol.* 72: 76–89. https://doi.org/10.1016/j.semcancer.2019.12.018.

102 Adams, S., Gatti-Mays, M.E., Kalinsky, K. et al. (2019). Current landscape of immunotherapy in breast cancer: a review. *JAMA Oncol.* 5: 1205–1214. https://doi.org/10.1001/jamaoncol.2018.7147.

103 Ribas, A. and Wolchok, J.D. (2018). Cancer immunotherapy using checkpoint blockade. *Science (80-.)* 359: 1350–1355. https://doi.org/10.1126/SCIENCE.AAR4060.

104 Vonderheide, R.H., Lorusso, P.M., Khalil, M. et al. (2010). Tremelimumab in combination with exemestane in patients with advanced breast cancer and treatment-associated modulation of inducible costimulator expression on patient T cells. *Clin. Cancer Res.* 16: 3485–3494. https://doi.org/10.1158/1078-0432.CCR-10-0505/83456/AM/TREMELIMUMAB-IN-COMBINATION-WITH-EXEMESTANE-IN.

105 Pusztai, L., Yau, C., Wolf, D.M. et al. (2021). Durvalumab with olaparib and paclitaxel for high-risk HER2-negative stage II/III breast cancer: results from the adaptively randomized I-SPY2 trial. *Cancer Cell* 39: 989–998.e5. https://doi.org/10.1016/J.CCELL.2021.05.009.

106 Ghebeh, H., Al-Sayed, A., Eiada, R. et al. (2021). Weekly paclitaxel given concurrently with durvalumab has a favorable safety profile in triple-negative metastatic breast cancer. *Sci. Rep.* 11: https://doi.org/10.1038/S41598-021-98113-6.

107 Domchek, S.M., Postel-Vinay, S., Im, S.A. et al. (2020). Olaparib and durvalumab in patients with germline BRCA-mutated metastatic breast cancer (MEDIOLA): an open-label, multicentre, phase 1/2, basket study. *Lancet Oncol.* 21: 1155–1164. https://doi.org/10.1016/S1470-2045(20)30324-7.

108 Juliá, E.P., Amante, A., Pampena, M.B. et al. (2018). Avelumab, an IgG1 anti-PD-L1 immune checkpoint inhibitor, triggers NK cell-mediated cytotoxicity and cytokine production against triple negative breast cancer cells. *Front. Immunol.* 9: 2140. https://doi.org/10.3389/FIMMU.2018.02140/BIBTEX.

109 Dirix, L.Y., Takacs, I., Jerusalem, G. et al. (2018). Avelumab, an anti-PD-L1 antibody, in patients with locally advanced or metastatic breast cancer: a phase 1b JAVELIN solid tumor study. *Breast Cancer Res. Treat.* 167: 671–686. https://doi.org/10.1007/s10549-017-4537-5.

110 Schmid, P., Adams, S., Rugo, H.S. et al. (2018). Atezolizumab and nab-paclitaxel in advanced triple-negative breast cancer. *N. Engl. J. Med.* 379: 2108–2121. https://doi.org/10.1056/nejmoa1809615.

111 Tolaney, S.M., Barroso-Sousa, R., Keenan, T. et al. (2020). Effect of eribulin with or without pembrolizumab on progression-free survival for patients with hormone receptor-positive, ERBB2-negative metastatic breast cancer: a randomized clinical trial. *JAMA Oncol.* 6: 1598–1605. https://doi.org/10.1001/jamaoncol.2020.3524.

112 Schmid, P., Cortes, J., Pusztai, L. et al. (2020). Pembrolizumab for early triple-negative breast cancer. *N. Engl. J. Med.* 382: 810–821. https://doi.org/10.1056/nejmoa1910549.

113 Kwapisz, D. (2021). Pembrolizumab and atezolizumab in triple-negative breast cancer. *Cancer Immunol., Immunother.* 70: 607–617. https://doi.org/10.1007/s00262-020-02736-z.

114 Santa-Maria, C.A., Kato, T., Park, J.H. et al. (2018). A pilot study of durvalumab and tremelimumab and immunogenomic dynamics in metastatic breast cancer. *Oncotarget* 9: 18985–18996. https://doi.org/10.18632/ONCOTARGET.24867.

115 ClinicalTrials.gov (n.d.). Pre-operative, single-dose ipilimumab and/or cryoablation in early stage/resectable breast cancer. https://clinicaltrials.gov/ct2/show/NCT01502592?term=ipilimumab&cond=breast+cancer&draw=2&rank=3 (accessed 10 November 2022).

116 ClinicalTrials.gov (n.d.). NIMBUS: nivolumab plus ipilimumab in metastatic hypermutated HER2-negative breast cancer. https://clinicaltrials.gov/ct2/show/results/NCT03789110?term=ipilimumab&cond=breast+cancer&draw=2&rank=1 (accessed 10 November 2022).

117 McArthur, H., Beal, K., Halpenny, D. et al. (2017). Abstract 4705: CTLA4 blockade with HER2-directed therapy (H) yields clinical benefit in women undergoing radiation therapy (RT) for HER2-positive (HER2+) breast cancer brain metastases (BCBM). *Cancer Res.* 77: 4705–4705. https://doi.org/10.1158/1538-7445.AM2017-4705.

118 McArthur, H.L., Diab, A., Page, D.B. et al. (2016). A pilot study of preoperative single-dose ipilimumab and/or cryoablation in women with early-stage breast cancer with comprehensive immune profiling. *Clin. Cancer Res.* 22: 5729–5737. https://doi.org/10.1158/1078-0432.CCR-16-0190.

119 Zhuang, J., Huo, Q., Yang, F., and Xie, N. (2020). Perspectives on the role of histone modification in breast cancer progression and the advanced technological tools to study epigenetic determinants of metastasis. *Front. Genet.* 11: 1353. https://doi.org/10.3389/FGENE.2020.603552/BIBTEX.

120 Zucchetti, B., Shimada, A.K., Katz, A., and Curigliano, G. (2019). The role of histone deacetylase inhibitors in metastatic breast cancer. *Breast* 43: 130–134. https://doi.org/10.1016/j.breast.2018.12.001.

121 Guo, P., Chen, W., Li, H. et al. (2018). The histone acetylation modifications of breast cancer and their therapeutic implications. *Pathol. Oncol. Res.* 24: 807–813. https://doi.org/10.1007/S12253-018-0433-5.

122 Munster, P.N., Thurn, K.T., Thomas, S. et al. (2011). A phase II study of the histone deacetylase inhibitor vorinostat combined with tamoxifen for the treatment of patients with hormone therapy-resistant breast cancer. *Br. J. Cancer* 104: 1828–1835. https://doi.org/10.1038/bjc.2011.156.

123 Yardley, D.A., Ismail-Khan, R.R., Melichar, B. et al. (2013). Randomized phase II, double-blind, placebo-controlled study of exemestane with or without entinostat in postmenopausal women with locally recurrent or metastatic estrogen receptor-positive breast cancer progressing on treatment with a nonsteroidal aromatase inhibitor. *J. Clin. Oncol.* 31: 2128. https://doi.org/10.1200/JCO.2012.43.7251.

124 Linderholm, B.K., Hellborg, H., Johansson, U. et al. (2009). Significantly higher levels of vascular endothelial growth factor (VEGF) and shorter survival times for patients with primary operable triple-negative breast cancer. *Ann. Oncol.* 20: 1639–1646. https://doi.org/10.1093/ANNONC/MDP062.

125 Oguntade, A.S., Al-Amodi, F., Alrumayh, A. et al. (2021). Anti-angiogenesis in cancer therapeutics: the magic bullet. *J. Egypt. Natl. Cancer Inst.* 33: https://doi.org/10.1186/s43046-021-00072-6.

126 Miles, D.W., De Haas, S.L., Dirix, L.Y. et al. (2013). Biomarker results from the AVADO phase 3 trial of first-line bevacizumab plus docetaxel for HER2-negative metastatic breast cancer. *Br. J. Cancer* 108: 1052–1060. https://doi.org/10.1038/bjc.2013.69.

Index

Drug and Therapy Development for Triple Negative Breast Cancer, First Edition.
Edited by Pravin Kendrekar, Vinayak Adimule, and Tara Hurst.
© 2023 WILEY-VCH GmbH. Published 2023 by WILEY-VCH GmbH.